Solar Power Generation: Concepts and Technology

Solar Power Generation: Concepts and Technology

Edited by Catherine Waltz

SYRAWOOD
PUBLISHING HOUSE

New York

Published by Syrawood Publishing House,
750 Third Avenue, 9th Floor,
New York, NY 10017, USA
www.syrawoodpublishinghouse.com

Solar Power Generation: Concepts and Technology
Edited by Catherine Waltz

International Standard Book Number: 978-1-68286-486-9 (Hardback)

.

The publisher's policy is to use permanent paper from mills that operate a sustainable forestry policy. Furthermore, the publisher ensures that the text paper and cover boards used have met acceptable environmental accreditation standards.

Trademark Notice: Registered trademark of products or corporate names are used only for explanation and identification without intent to infringe.

Cataloging-in-Publication Data

Solar power generation : concepts and technology / edited by Catherine Waltz.
    p. cm.
Includes bibliographical references and index.
ISBN 978-1-68286-486-9
1. Solar energy. 2. Photovoltaic power generation. I. Waltz, Catherine.
TJ810 .S65 2017
621.47--dc23

Printed in the United States of America.

TABLE OF CONTENTS

1

A New Fast Peak Current Controller for Transient Voltage Faults for Power Converters

Jesús Muñoz-Cruzado-Alba [1,*], Javier Villegas-Núñez [1,‡], José Alberto Vite-Frías [1,‡] and Juan Manuel Carrasco Solís [2,‡]

Academic Editor: Rodolfo Araneo

[1]   R & D Department, GPTech, Av. Camas N26, Bollullos de la Mitacion 41703, Spain; jvillegas@greenpower.es (J.V.-N.); javite@greenpower.es (J.A.V.-F.); jmcarrasco@us.es (J.M.C.S.)
[2]   Electronics Engineering Department, Seville University, Av. de los Descubrimientos S/N, Seville 41092, Spain
*   Correspondence: jmunoz@greenpower.es
†   The original paper was presented in A New Fast Peak Current Controller for Transient Voltage Faults for Power Converters. In Proceedings of the EEEIC 2015, Rome, Italy, 10–13 June 2015.
‡   These authors contributed equally to this work.

Abstract: Power converters are the basic unit for the transient voltage fault ride through capability for most renewable distributed generators (DGs). When a transient fault happens, the grid voltage will drop suddenly and probably will also suffer a phase-jump event as well. State-of-the-art voltage fault control techniques regulate the current injected during the grid fault. However, the beginning of the fault could be too fast for the inner current control loops of the inverter, and transient over-current would be expected. In order to avoid the excessive peak current of the methods presented in the literature, a new fast peak current control (FPCC) technique is proposed. Controlling the peak current magnitude avoids undesirable disconnection of the distributed generator in a fault state and improves the life expectancy of the converter. Experimental and simulation tests with high power converters provide the detailed behaviour of the method with excellent results.

Keywords: distributed generators (DGs); voltage ride through (VRT); fast peak current control (FPCC); phase-jump ride through (PJRT); photo-voltaic (PV) systems; dip voltage

1. Introduction

Power is typically produced at a wide range of generation plants. Some years ago, for renewable power sources, it was allowed to switch off the source when a voltage fault occurred. Back then, disconnection of that power sources had little, if any, impact on the recovery capability of the electric power grid after a fault. Nowadays, a high penetration of renewable distributed generators (DGs) [1–4] has toughened the grid connection minimum technical requirements (MTRs) worldwide [5–13]. MTRs include voltage sags and phase-jump capability, frequency active power regulation and anti-islanding techniques, among others.

The voltage ride through (VRT) capability requirement has been widely described in recent grid codes [5–13]. Tables 1 and 2 point out some of the most popular MTRs for photo-voltaic (PV) plants about VRT.

First, a maximum allowed voltage profile is defined for voltage excursions. If the fault reaches the error profile, the inverter is allowed to disconnect. German legislation [5] usually is taken as the reference for other legislation. A zero-voltage transient fault is required for 0.15 s. Recent legislation imposes a zero-voltage fault, too, like the Puerto Rican or Jordanian legislation [7,13]. However, other legislation is imposing smaller voltage sag magnitude requirements, but longer in time, for example the Chilean and Romanian grid codes [11,12].

Table 1. Minimum technical requirements (MTRs) review for voltage ride through (VRT) capabilities of photo-voltaic (PV) plants (A). German association of energy and water industries: BDEW; Puerto Rico Electric Power Authority: PREPA; National energy regulator of South Africa: NERSA; Spanish electric grid: REE.

Country	Germany	Puerto Rico
Standard	BDEW	PREPA
VRT profile	V_{PCC} (p.u.): 1.00, 0.90 ... 0.30; T(s): 0.0 0.15 0.60 1.50 3.00	V_{PCC} (p.u.): 1.40 1.30 1.25 1.15 1.00 0.90 ... 0.00; T(s): 0.0 0.15 0.6 1.0 3.0
I_q VRT	ΔI_Q (p.u.): 1.00 ... -1.00; V(p.u.): 0.0 0.5 0.9 1.1 1.2	ΔI_Q (p.u.): 1.00 ... -1.00; V(p.u.): 0.0 V_1 0.85
I_d VRT	0	0
Recovery time	5 s	5 s
Phase jump	N/D	N/D

Country	South Africa	Spain
Standard	NERSA	REE
VRT profile	V_{PCC} (p.u.): 1.20 1.10 1.00 0.90 0.80 ... 0.00; T(s): 0.0 0.15 2.0 20.0	V_{PCC} (p.u.): 1.00 0.95 0.80 ... 0.20; T(s): 0.0 0.5 1.0 15.0
I_q VRT	ΔI_Q (p.u.): 1.00 ... -1.00; V(p.u.): 0.0 0.5 0.9 1.1 1.2	I_Q (p.u.): 1.00 0.90 ... -1.00; V(p.u.): 0.00 0.5 0.85
I_d VRT	Previous fault I_d	0
Recovery time	5 s	N/D
Phase jump	Up to 40°	N/D

Table 2. MTRs review for VRT capabilities of PV plants (B). National Commission of Energy: CNE; Electrotechnical Italian Committee: CEI; Electricity Regulatory Commission: ERC; Romanian Electricity Authority: Transelectrica.

Country	Chile	Italy
Standard	CNE	CEI
VRT profile		
I_q VRT		
I_d VRT	Previous fault I_d	0
Recovery time	N/D	N/D
Phase jump	N/D	N/D

Country	Jordan	Romania
Standard	ERC	Transelectrica
VRT profile		
I_q VRT		
I_d VRT	0	0
Recovery time	60 s	350 ms
Phase jump	N/D	N/D

Then, some requirements are imposed over the power generation during the voltage excursions in order to help the system stability. An injection of reactive current (I_q) is always required. Older grid codes, like the Spanish or Italian legislation [8,9], usually require generating the maximum possible capacitive current. However, most recent grid codes usually require a droop relationship between capacity current and the depth of the voltage sag, in order to provide a softener recovery [6,7,12,13].

There are two choices for the active current (I_d) requirement: to follow the previous value to the fault state, for example the South African and Chilean cases [6,12]; or to drop the reference to zero, but consumption is not allowed, like the German and Puerto Rican cases [5,7].

A recovery time after the fault requirements could be needed, as well. It could vary from milliseconds [11] to minutes [13].

Finally, most recent grid codes are also including phase-jump fault requirements, for example the South African grid code [6].

The worst scenarios cover the necessity to remain connected against $40°$ phase-jumps and 0.0 p.u. low voltage excursions, for three-phase and mono-phase faults. A sudden occurrence of this type of fault could cause a peak in the converter output current. Therefore, these current peaks cause unit errors and disconnections, being a hazard to the unit safety.

Together with the operation mode and the imposed limits, response time is crucial in these kinds of events, whose durations are in the order of milliseconds. At the beginning of the fault, any delay could be critical, because the grid voltage could change very fast. Figure 1 shows an uncontrolled peak current due to a severe low voltage excursion in a three-phase power converter.

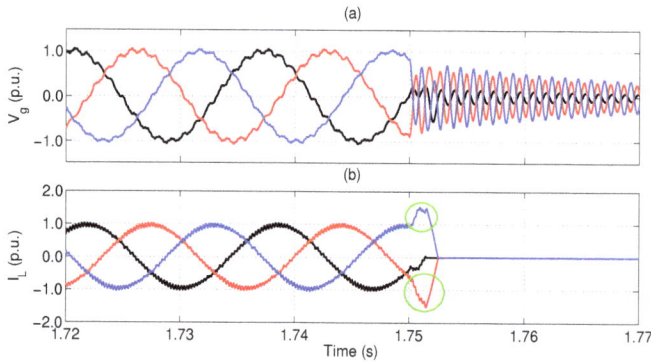

Figure 1. Peak current fault under a severe low voltage excursion. (**a**) Transient low voltage profile; (**b**) stack converter output currents; uncontrolled peak currents marked with green circles.

There is much research on power converter controllers during grid fault conditions [14–19], but unfortunately, there is not enough analysis about the uncontrolled beginning of the fault. A new fast peak current control (FPCC) is proposed to help the converter control limiting the over-current peak of the converter at the beginning of the fault. Consequently, hardware and software current protection could be avoided, improving the MTRs compliance. Further, lower peak current reduces insulated gate bipolar transistors (IGBTs)' degradation and unexpected disconnections from the grid of the power converters, so mean time between failures (MTBF) increases.

This manuscript is organized as follows: Section 2 analyzes in detail the proposed method theory; Sections 3 and 4 show the test results; discussions are given in Section 5; Finally, the materials and methods used are pointed out in Section 6.

2. Fast Peak Current Control Method

2.1. Power Converter Control Strategy

A two-level three-phase topology has been selected for the study. Industrial high-power grid-tie converters usually use a single-stage inverter topology, with an LC output filter [20,21].

Figure 2 shows a classical DG converter control block scheme. The controller is divided into four layers. The highest level controller generates the appropriate references for the middle controller. The middle level controller reacts by modifying the response as a function of the environment agents, which could limit the inverter capability. Typically, special voltage sag control will be placed at this layer. Then, the low level controller includes the inner current control loop that sets the inverter control actions following the references. Finally, the hardware level controller translates the control signals to the physical pulses of the converter.

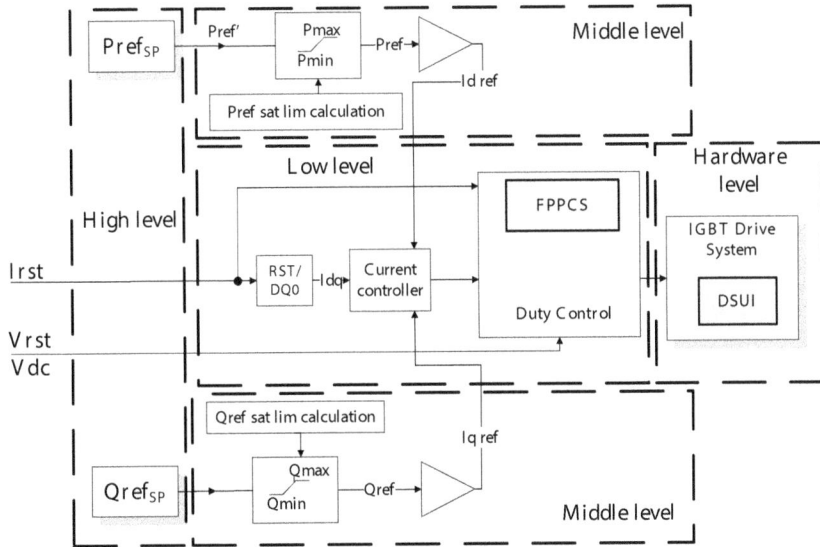

Figure 2. Simplified control block scheme of a distributed generator (DG) power converter.

The proposed FPCC will be improved with two individual actions. Gray boxes in Figure 2 show where these actions take place. On the one hand, in the lower level, the duty cycle control signal is saturated with a theoretical current limit, called the fast predictive peak current saturation (FPPCS) method. On the other hand, at the hardware level, the delay of duty control signal updating is reduced without modifying pulse width modulation (PWM) switching frequency with a technique denoted as duty signal updating improvement (DSUI).

2.2. Fast Predictive Peak Current Saturation Method

Figure 3 shows a simplified single-line model of the converter as an ideal controlled voltage source. This model has been widely presented in the literature [22]. The converter output line-to-line voltage (V_{gRS}, V_{gST}) is defined by an impedance (Z_{sc}) and a grid voltage source. The measured voltage could be used to build up an equivalent voltage source model (V_{nR}, V_{nS}, V_{nT}) connected to the virtual neutral point of the converter model (noted by the dashed lines in Figure 3).

Equation (1) shows the relationship of the inductor voltage (V_L) with the voltage source model of Figure 3 and with the differential equation of an inductor:

$$V_L = \begin{cases} D\frac{V_{dc}}{2} - V_n \\ L\frac{dI_L}{dt} \end{cases} \tag{1}$$

where V_{dc} is the DC-link voltage, V_n is the grid voltage, D is the DG duty control signal in the range of $[-1, 1]$, L is the inductive value of the filter value and I_L is the current across the inductance. Since the controller is executed periodically at a fixed frequency F_s, Equation (1) could be discretized, and D would be given by Equation (2):

$$D_k = 2\frac{L(I_{L_{k+1}} - I_{L_k})F_s + V_{n_k}}{V_{dc_k}} \tag{2}$$

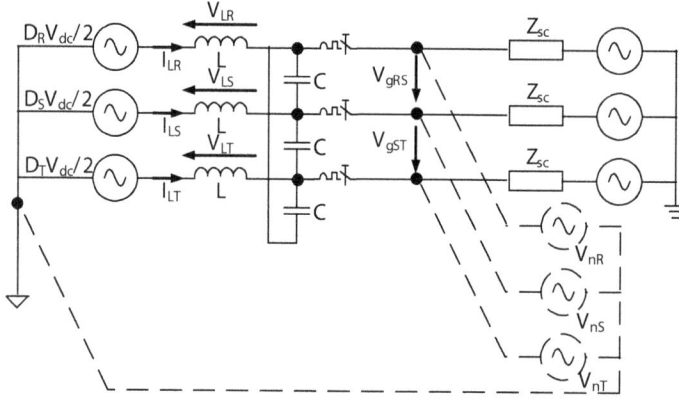

Figure 3. Simplified inverter voltage source model.

Equation (2) gives the relationship between I_L time evolution and the control signal value. Consequently, the current measurement on the next step control could be predicted. Imposing a control law restriction with a maximum current threshold I_{FPPCS}, Equation (3) sets a theoretical maximum control signal:

$$
\begin{aligned}
D_{\max R_k} &= 2\frac{L(I_{FPPCS}-I_{R_k})F_s+V_{g_k}}{V_{dc_k}} \\
D_{\max S_k} &= 2\frac{L(I_{FPPCS}-I_{S_k})F_s+V_{g_k}}{V_{dc_k}} \\
D_{\max T_k} &= 2\frac{L(I_{FPPCS}-I_{T_k})F_s+V_{g_k}}{V_{dc_k}}
\end{aligned}
\tag{3}
$$

where $D_{\max R_k}$, $D_{\max S_k}$ and $D_{\max T_k}$ are the maximum duty allowed control signals for the defined I_{FPPCS} in each phase and I_{R_k}, I_{S_k} and I_{T_k} are the measured currents of the three phases at the k instant.

2.3. Duty Signal Updating Improvement Method

Typically, PWM techniques update only their control signals in the valleys and peaks of the triangular carrier, T_0 and T_2 respectively (see Figure 4), guaranteeing non-desirable firing, the switching frequency remaining constant and avoiding extra power losses [23].

Figure 4 shows a typical delay added in a power converter controller. If the control processor needs the computational time (T_c) since the last sampling time (T_0), then an additional delay of T_m will be inserted before the action will be executed, because the control signal can only be updated in the peaks and the valleys. The proposed technique updates the control signal at T_1 with some restrictions. Then, only T_c delay happens, and the peak current under faulty conditions will drop.

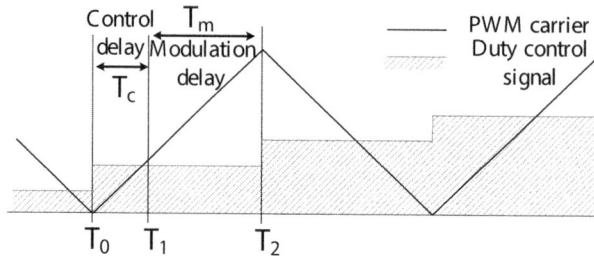

Figure 4. Typical delay added in a power converter controller. The measures are taken at T_0, but T_c is needed to calculate the next control signal. Finally, the control signal is updated and applied at T_2. Pulse width modulation: PWM.

As an example, Figure 5 shows all four possible cases in the up-slope PWM carrier semi-cycle, but similar cases could be exposed in the down-slope semi-cycle. On the one hand, during the up-slope, if the previous control signal (D_0) is greater than the triangular carrier value at T_1, no extra transition is guaranteed, and the new control signal (D_1) could be updated without any additional switching in the semi-cycle (Cases c and d in Figure 5). On the other hand, if D_0 is lower than the triangular carrier at T_1, at least three transitions may occur if D_1 is updated at T_1: the first one belongs to the D_0 level; a second transition happens at T_1; and a third transition will happen at the D_1 level. Consequently, the control signal will be updated in the next valley or peak to avoid extra switching (Case a in Figure 5). Finally, Case b does not produce any extra-switching, but neither modifies the control output.

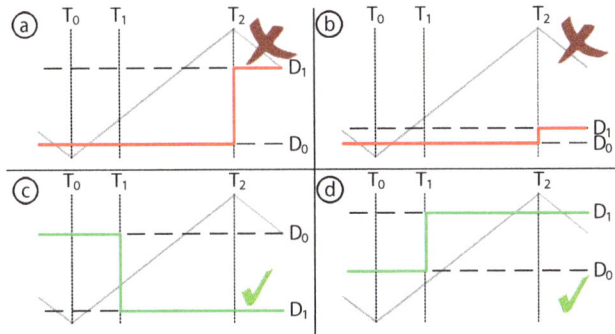

Figure 5. All four possible duty updating cases in the rising PWM semi-cycle. The control signal could be updated at T_1 in Cases c and d, but must be updated at T_2 in Cases a and b.

Graphically, Figure 6 shows an example of PWM signals generated in Cases a and c. In Case c, the duty signal is updated (blue line) without producing extra switching, reducing the on state of the semiconductor. However, in Case a, the control signal must be updated in T_2 (solid blue line); otherwise, two switching events will happen (dotted red line).

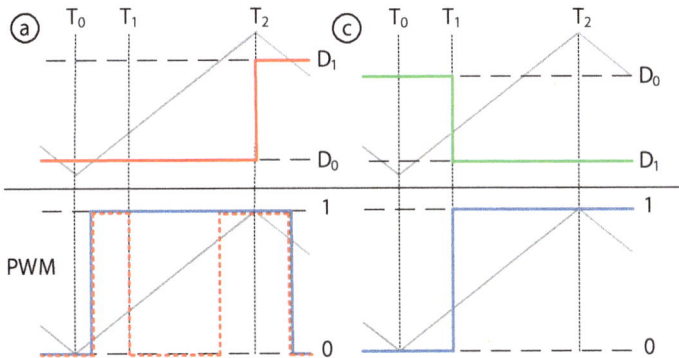

Figure 6. PWM signals generated with Cases a and c of the rising PWM semi-cycle. The red dotted line points out possible malfunctioning with two switchings in the same semi-cycle if the proposed rule is not applied. The blue line points out PWM signals generated with the duty signal updating improvement (DSUI) method.

Following the same steps, along the modulation down-slope cycle, if D_0 is lower than the modulation value at T_1, the control signal could be updated without any change in the switching frequency. Figure 7 shows all four possible cases on the down-slope semi-cycle.

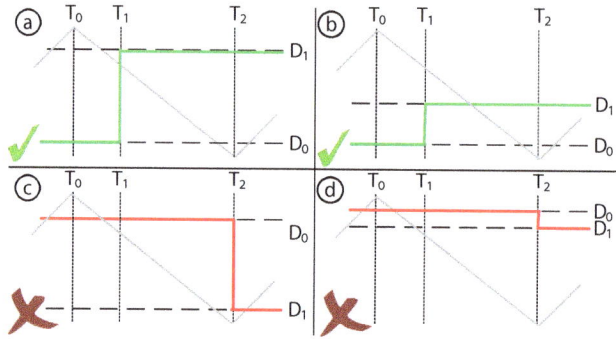

Figure 7. All four possible duty updating cases in the falling PWM semi-cycle. The control signal could be updated at T_1 in Cases a and b, but must be updated at T_2 in Cases c and d.

Fortunately, not all cases are relevant with respect to current faults. Therefore, a study could be made to determine the effectiveness of the improvement in these special cases. There are two fault conditions: a positive and a negative over-current peak. From here, the up-slope case will be analyzed, but a similar reasoning could be done for the down-slope semi-cycle.

According to Equation (2), at instant $k = 1$, the worst over-current peak (I_{L_0}) would happen if the current peak fault was add up over the maximum current value modulated. Consequently, D_0 is expected to be positive and big enough. In addition, if a dangerous peak current happened, the controller would have to drop I_L to a safety region. According to Equation (2), to take down I_{L_1}, D_1 will be very low. Therefore, if a positive over-current peak happens, Case c of Figure 5 is expected. As previously mentioned, this is one of the allowed cases to refresh the control signal, so the over-current peak will be reduced.

A similar reasoning could be made with a negative over-current fault. On the one hand, now, D_0 is expected to be negative and big enough. On the other hand, from Equation (2), the D_1 expected value will be very high. Consequently, if a negative over-current peak happens, Case a or d of Figure 5 is expected. If Case d happens, the control signal will be updated, and the over-current peak will be reduced. Unfortunately, if Case a happens, the method will not act in this semi-cycle.

Figure 5 points out that T_c influences the effectiveness of the method. If T_c is forced to zero, all control steps will be in the c and d cases, so it is important to have a small delay T_c to short the measured peak current in most situations.

Finally, one more action could be performed to reduce the over-current peak. A fault happening in semi-cycle k will be measured at the beginning of the next semi-cycle $k + 1$, and the control action will be placed at T_1 in the best case or at the beginning of $k + 2$ in the worst case. Therefore, the maximum delay could be $2T_s$ or $T_s + T_c$.

If T_c is relatively short, a new control step could be done at the middle of the semi-cycle. This control signal will be applied with the same rules as the others, so in a general way, it will be placed at the final part of the semi-cycle. In this case, if a fault happens at the middle of the semi-cycle k', it will be measured at the beginning of the next control step $k' + 1$, and the control action will be placed at T_1 in the best case or at the beginning of $k' + 2$ in the worst case. Therefore, the maximum delay could be T_s or $0.5 \cdot T_s + T_c$. Assuming V_L constant in a short period of time during the fault condition:

$$I_L = \frac{1}{L} \int V_L dt \longrightarrow \Delta I_L = \frac{V_L}{L} \cdot T_{delay} \tag{4}$$

where ΔI_L is the expected increment on the peak current induced by the fault and T_{delay} is the time necessary to control the fault. Therefore, according to Equation (4), the peak current will be reduced in:

$$
\begin{aligned}
P_{I_{max}} &= \frac{\frac{V_L}{L} \cdot 1.5 T_s}{\frac{V_L}{L} \cdot 2 T_s} = 0.75 \text{ p.u.} \\
P_{I_{min}} &= \lim_{T_c \to 0} \frac{\frac{V_L}{L} \cdot (0.5 T_s + T_c)}{\frac{V_L}{L} \cdot T_s + T_c} = 0.5 \text{ p.u.}
\end{aligned}
\tag{5}
$$

where P_I is the proportion of the peak current reduced with the improvement (between 50% and 75%).

3. Simulations

A high power industrial PV solar inverter has been modeled to test FPCC. However, similar results could be obtained with other applications. The following devices have been modeled in the simulation: a solar panel field; a detailed commercial model of a two-level three-phase power inverter; a medium voltage transformer; the point of interconnection (POI) with the utility grid; a RL divider to generate voltage sags and phase-jumps.

Two types of faults have been analyzed at full power. The worst cases described in the international legislation [5–13] have been selected. The system will be tested against symmetric and asymmetric voltage sags and phase-jump faults. Figure 8 shows line-to-line voltages of the deepest faults of each type used to test the system. Case a shows a three-phase voltage fault with zero remaining voltage. Case b shows an asymmetric voltage fault with two phases overlapped. Case c shows a three-phase 45° voltage phase-jump. Finally, Case d shows the same phase-jump for only one phase.

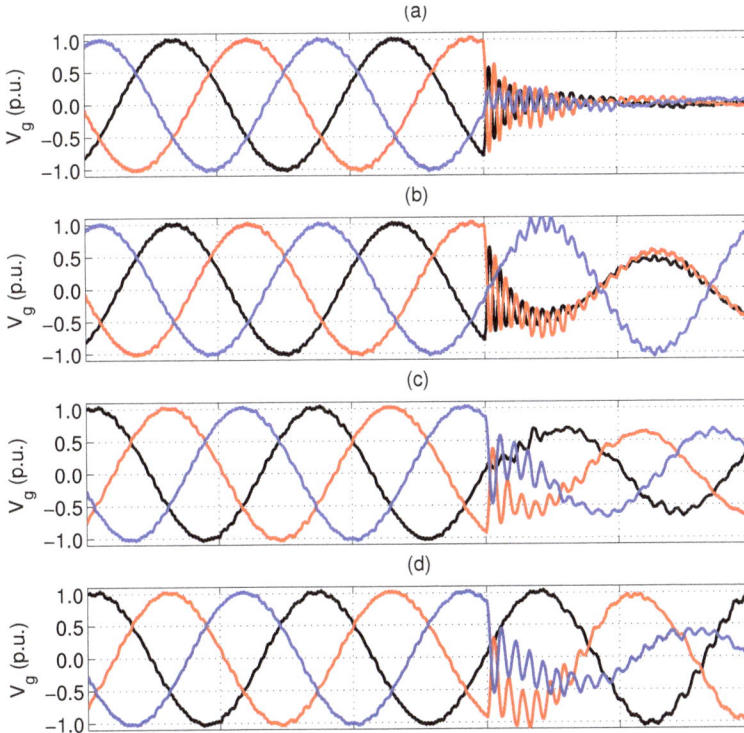

Figure 8. Line-to-line three-phase grid voltage for all types of faults tested. (**a**) Three-phase zero-voltage fault; (**b**) asymmetric zero-voltage fault; (**c**) three-phase 45° voltage phase-jump and (**d**) mono-phase 45° voltage phase-jump.

Figures 9 and 10 show the transient power converter response against a 45° three-phase jump disturbance and a 0.0 p.u. three-phase dip voltage, respectively, fired at 0.00 s. For both figures, (a) shows the transient behavior of one phase voltage (V_g). Figure 9b shows the phase (θ) transient to 45° and Figure 10b the voltage module (m) transient to 0.0 p.u., during the fault. Cases c to e show two curves fir each one; the solid line curve represents the evolution of the system with a classical approach (CA), and the dashed line curve represents the evolution of the system with FPCC. Cases c to e show the duty signal control (D), the stack output current (I_L) and the converter output current (I_{out}), respectively. Note that software protection (SP) and hardware protection (HP) thresholds are pointed out over Case d and how, with the FPCC active, the threshold is not reached. As a result, the unit remains connected. The I_{out} peak is reduced, too, but additional remaining peaks appear in the converter output due to the line capacitor filter.

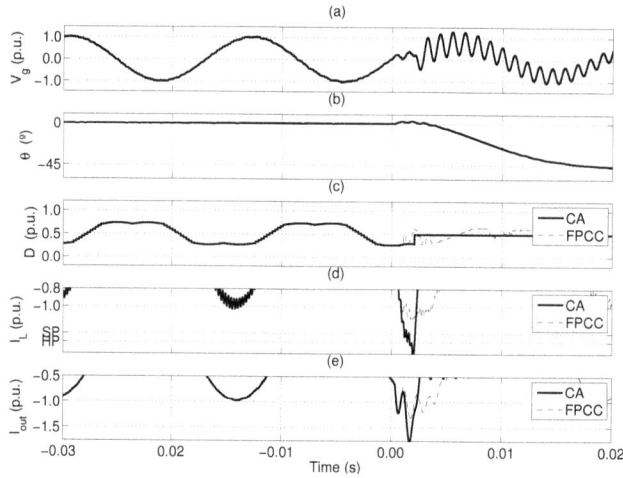

Figure 9. DG response against a 45° phase-jump fault with the classical approach (CA) and FPCC techniques. (**a**) Grid voltage; (**b**) phase; (**c**) duty control signal; (**d**) stack current and (**e**) converter current output.

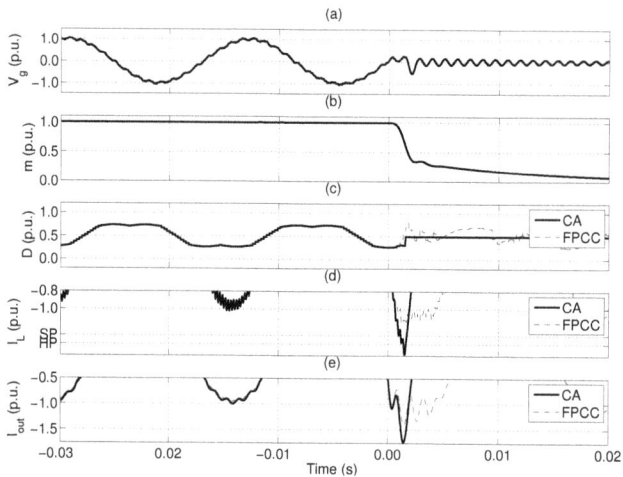

Figure 10. DG response against a 0.00 p.u. three-phase dip voltage with the classical approach (CA) and fast peak current (FPCC) techniques. (**a**) Grid voltage; (**b**) module; (**c**) duty control signal; (**d**) stack current and (**e**) converter current output.

However, results may differ depending the triggering time of the fault. To cope with this effect, all cases have been repeated ten times, firing the fault at different instants to look for the worst case. Table 3 shows a comparison of several phase-jumps and voltage sags. Several values have been tested for both the mono-phase and the three-phase cases. In every cell, values at the left are for I_L, and values at the right are for I_{out}. In order to quantize the weight of each improvement in the overall result, three cases are considered in the study: the classical approach (CA) method, only the FPPCS algorithm and the complete FPCC technique.

Table 3. Simulations results against dip voltage and phase-jump faults. FPPCS, fast predictive peak current saturation.

Fault	Phases	Value p.u./°	Max. I_L-I_{out} p.u. CA	FPPCS	FPCC
Phase-jump	3	10	1.34–1.67	1.29–1.54	1.18–1.51
Phase-jump	3	20	1.42–1.83	1.36–1.60	1.21–1.60
Phase-jump	3	30	1.49–1.93	1.39–1.65	1.23–1.64
Phase-jump	3	40	1.56–2.02	1.41–1.69	1.25–1.66
Phase-jump	3	45	1.59–2.04	1.42–1.73	1.26–1.67
Phase-jump	1	10	1.32–1.65	1.27–1.47	1.18–1.46
Phase-jump	1	20	1.44–1.81	1.35–1.55	1.20–1.55
Phase-jump	1	30	1.50–1.92	1.38–1.60	1.22–1.59
Phase-jump	1	40	1.60–1.99	1.40–1.65	1.24–1.61
Phase-jump	1	45	1.62–2.01	1.40–1.68	1.24–1.62
Dip voltage	3	0.2	1.51–1.79	1.32–1.56	1.18–1.53
Dip voltage	3	0.1	1.65–2.00	1.38–1.65	1.22–1.62
Dip voltage	3	0.0	1.83–2.26	1.45–1.83	1.28–1.79
Dip voltage	1	0.2	1.47–1.75	1.30–1.54	1.18–1.48
Dip voltage	1	0.1	1.61–1.95	1.36–1.64	1.21–1.61
Dip voltage	1	0.0	1.77–2.20	1.42–1.81	1.26–1.78

Table 3 shows FPCC over-currents about 0.4 p.u. lower than the CA method and helps to maintain the range of operation of the converter (SP at 1.3 p.u.). The table also shows results for the influence of every part of the method.

4. Experimental Validation

FPCC results had been verified experimentally. A real high power test bench has been used with a three-phase two-level grid-tied inverter DG for PV applications.

The same voltage sags and phase-jumps of simulation tests have been selected, in accordance with the fault descriptions of the main international legislation [5–13]. All tests have been repeated five times to ensure results with different triggering conditions, and the tests were performed with CA, only FPPCS and FPCC complete improvement to compare the results.

Repeating simulation tests, Figures 11 and 12 show the transient power converter response against a 45° three-phase jump disturbance and a 0.00 p.u. three-phase dip voltage, respectively, fired at 0.00 s. For both figures, Curve a shows the transient behavior of V_g. Cases b and c show I_L with CA and FPCC, respectively. Finally, Cases d and e show I_{out} with CA and FPCC, respectively. The SP threshold is not reached in the case of FPCC active, verifying the simulation results.

Finally, all tests have been repeated ten times, and the worst results are shown in Table 4. Three cases are considered in the study: CA, only FPPCS and FPCC complete methods. Again, the simulation results are validated, and the experimental peak current working under faulty conditions is greatly reduced with FPCC.

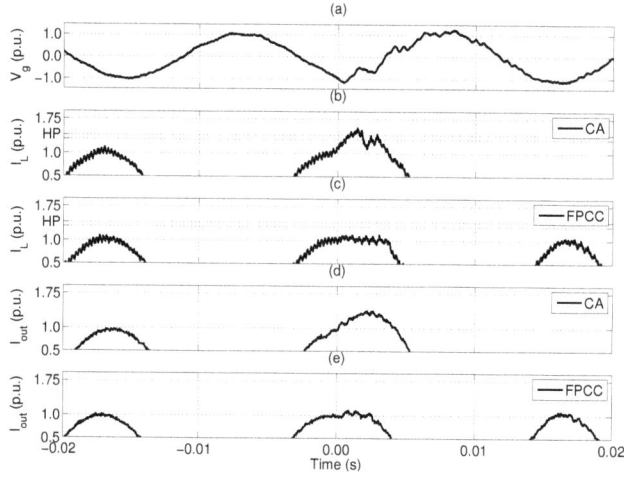

Figure 11. DG response against a 45° phase-jump fault with the CA and FPCC techniques. (**a**) Grid voltage; (**b,c**) stack current and (**d,e**) converter current output.

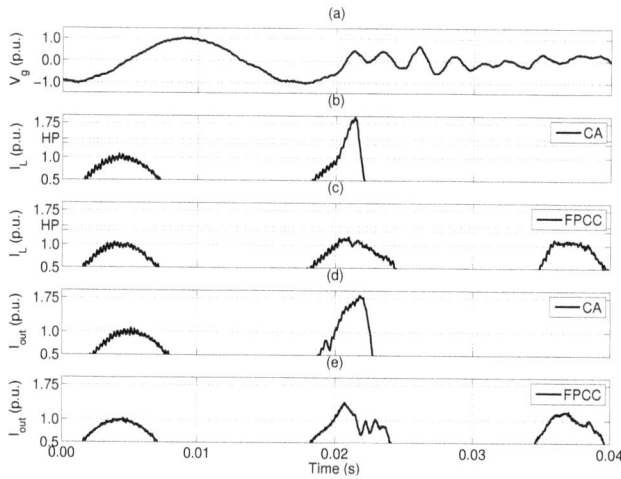

Figure 12. DG response against a 0.00 p.u. three-phase dip voltage with CA and FPCC techniques. (**a**) Grid voltage; (**b,c**) stack current and (**d,e**) converter current output.

Table 4. Experimental results against dip voltage and phase-jump faults.

Fault	Phases	Value	Max. I_L p.u.		
		$p.u./°$	CA	FPPCS	FPCC
Phase-jump	3	10	1.15	1.15	1.12
Phase-jump	3	20	1.21	1.18	1.12
Phase-jump	3	30	1.59	1.17	1.15
Phase-jump	3	40	1.69	1.25	1.17
Phase-jump	3	45	1.73	1.24	1.24
Phase-jump	1	45	1.26	1.16	1.16
Dip voltage	3	0.2	1.65	1.27	1.14
Dip voltage	3	0.1	1.74	1.34	1.16
Dip voltage	3	0.0	1.87	1.52	1.21
Dip voltage	1	0.0	1.46	1.26	1.17

4.1. Model Validation

Finally, a transient comparison between simulations and experimental results could be very meaningful in order to validate the model. Two cases are considered, both for a three-phase zero-voltage remaining fault; one case under a classic control approach and the second with the proposed FPCC method.

Figure 13 shows the voltage and stack current outputs for a 0.00 p.u. three-phase dip voltage with the CA technique. Cases a and b show line-to-line voltage and how the voltage is generated at the same instant and with the same behavior for simulations and experiments, respectively. Additionally, Cases c and d show the stack output currents for the three phases. Results show approximately the same magnitudes in simulations and experiments.

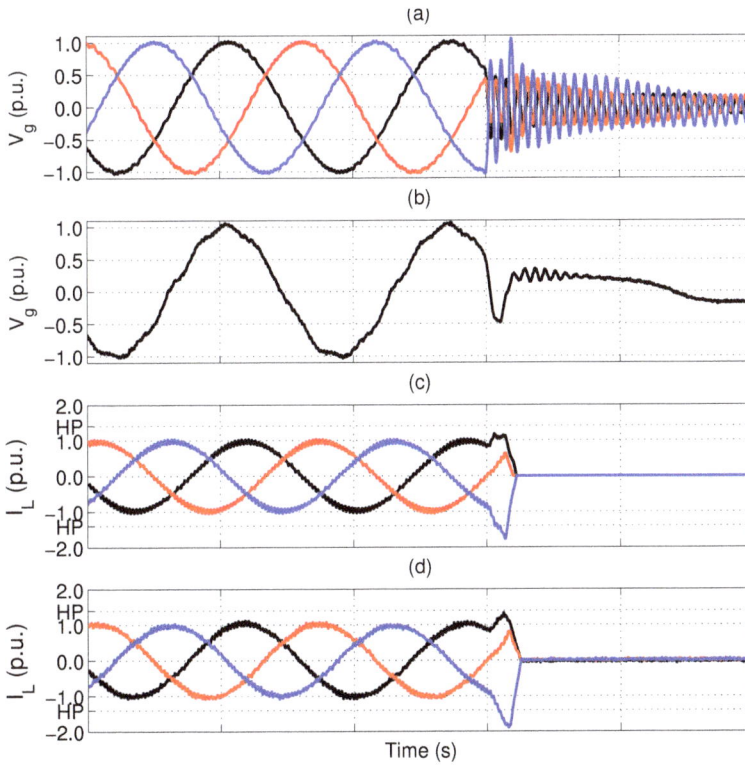

Figure 13. Simulation and experiment response against a 0.00 p.u. three-phase dip voltage with the CA technique. (**a**) Simulation line-to-line three-phase grid voltage; (**b**) experiment line-to-line grid voltage; (**c**) simulation stack current outputs and (**d**) experiment stack current outputs.

Moreover, Figure 14 shows the voltage and stack current outputs for a 0.00 p.u. three-phase dip voltage with the FPCC technique. The same four cases are presented (grid voltage and output stack currents in simulation and experimental tests). Again, the same behavior is proven for simulations and real experiments, observing the same slopes and peak current magnitudes.

In conclusion, Figures 13 and 14 prove that the simulation test model has a very accurate response compared to real experiments. Therefore, the simulation test results presented are validated.

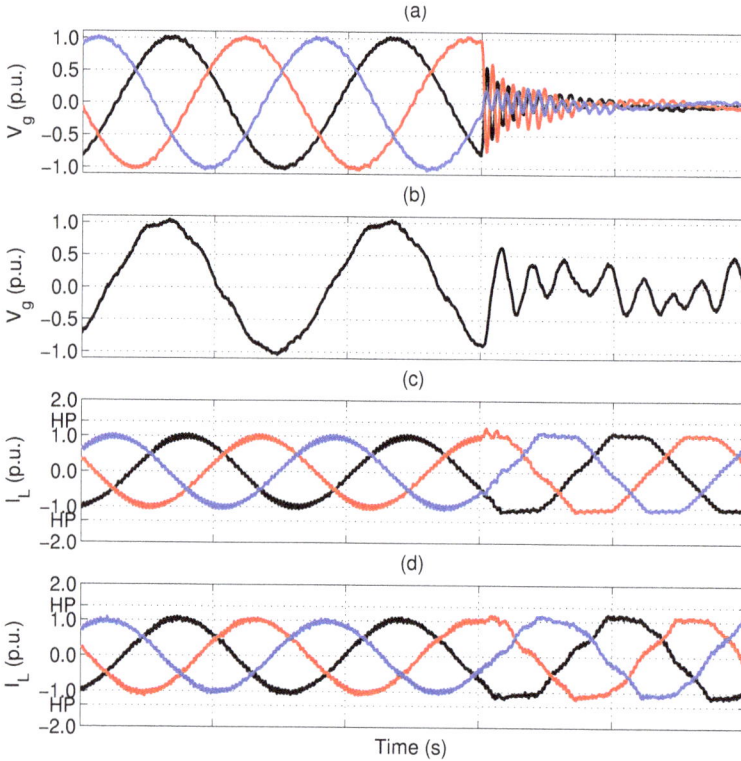

Figure 14. Simulation and experiment response against a 0.00 p.u. three-phase dip voltage with the FPCC technique. (**a**) Simulation line-to-line three-phase grid voltage; (**b**) experiment line-to-line RS grid voltage; (**c**) simulation stack current outputs and (**d**) experiment stack current outputs.

5. Discussion

The proposed FPCC highly reduces the peak currents detected by the converters of DGs. The method theory has been analyzed in detail, and results verify the effectiveness of the method via simulation. Furthermore, an experimental validation with an industrial high power solar power converter has been performed with excellent results. The relationship between simulation and experimental results is consistent, so the models used have proven their efficacy.

Peak currents for all types of faults (voltage sags and phase jumps) have been reduced, avoiding critical thresholds. Severe faults cause larger peaks currents than small faults under a classical control technique. However, peak currents do not increase in the same way with the FPCC technique. The increment of peak current magnitudes is flatter with FPCC than with a classical approach. Consequently, the range of tolerance against faults is increased.

Note that the FPCC threshold is set to 1.05 p.u. for all tests. Results show peak currents between 1.12 p.u. and 1.24 p.u. Therefore, the control threshold has been passed only 0.07–0.17 p.u. Therefore, the efficiency of the method has been measured.

The output current of the converter is greater than the stack output current significantly. This is due to the energy saved in the AC capacitors of the output filter of the converter. The peak current component due to AC capacitors is not reduced with FPCC, because it is not controlled by the converter. Therefor, the output current of the converter is greater than the stack current output. However, the most critical components against peak currents are IGBTs, so the extra current added by the capacitors does not increase the risk of failure.

Finally, the main advantages of FPCC are summarized in the next sentences:

- The peak current has been reduced between 0.4 p.u. and 0.7 p.u. for the worst cases.
- The method helps to comply with international MTRs.
- The method prevents the unit from tripping by over-current, reducing production losses and helping the grid recover from the fault.
- Reducing peak currents prevents unit damage. Consequently, the MTBF of the units is longer.
- The method does not need any additional hardware, so it is very inexpensive and easy to implement in existing units.

6. Experimental Section

6.1. Simulations

A high power industrial PV solar inverter has been modeled to test FPCC. Simulations had been performed with the electric transient power tool EMTDC/PSCAD V4.2.

Figure 15 shows a detailed description of the simulation. The simulation includes:

- Solar panel field model.
- Detailed commercial two-level three-phase grid-tie inverter; the inverter acts as the device under test (DUT) of the simulation.
- Medium voltage transformer.
- POI with the utility grid.
- Variable parallel impedance load to perform voltage sags and phase-jumps.

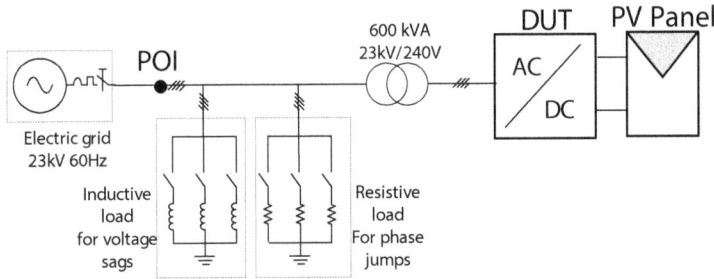

Figure 15. Detailed simulation scheme to test FPCC.

The PV panels have been modeled according to the diode model [24]. The main parameters used to configure the panel are summarized in Table 5. Test conditions have been set to reach maximum power.

Table 5. PV panel model setup.

Panel Nominal Values	Value	Test Conditions	Value
Active power ($P_{MPP;STC}$)	500 kW	Temperature (T)	25 °C
Voltage ($U_{MPP;STC}$)	825 V	Irradiance (G)	1000 Wm^2
Temperature (T_{STC})	25 °C	**Panel technology**	**Value**
Irradiance (G_{STC})	1000 Wm^2	Coefficient of voltage change F_{fu}	0.8 p.u.
Temperature model	**Value**	Coefficient of current change F_{fi}	0.9 p.u.
Temperature correction T_0	0 °C	Technology coefficient C_g	2.514×10^{-3} W/m^2
Irradiance gain k	0.03 m^2/W	Technology coefficient C_v	0.08593 p.u.
Time constant τ	300 s	Technology coefficient C_r	1.088×10^{-4} m^2/W
		Irradiance change $Vl2h$	0.95 p.u.
		Current temperature α	0.0004 p.u.
		Voltage temperature β	−0.004

The main parameters of the commercial two-level three-phase grid-tie inverter are summarized in Table 6. Nominal values, inner current control PIDs, switching frequency, software protections and the FPCC setup are detailed.

Table 6. Grid-tied inverter setup.

Parameter	Value	Parameter	Value
Nominal active power P_{DG}	500 kW	Over-current software protection SP	1.3p.u./0.1 ms
Nominal output current I_n	1202 A	Over-current hardware protection HP	1.4 p.u.
Nominal grid voltage V_n	240 V	Active current control	$kp_D = 0.05$
			$ki_D = 5$
FPPCS limit I_{FPPCS}	1.05 p.u.	Reactive current control	$kp_Q = 0.05$
			$ki_Q = 5$
Delay control time T_c	0.6 p.u.	PWM frequency f_{PWM}	1980 Hz

The POI is simulated with an ideal three-phase voltage source at $V_L = 23$ kV and $f = 60$ Hz, with a short circuit power (S_{sc}) of 500 MVA. Additionally, the medium voltage transformer is detailed in Table 7. Finally, faults are modeled with an RL three-phase impedance, regulating values to desired faults.

Table 7. Medium voltage transformer setup.

Parameter	Value	Parameter	Value
Primary winding voltage	23 kV	Nominal power	0.6 MWe
Secondary winding voltage	0.24 kV	Nominal grid frequency	60 Hz
Winding type	YΔ	Leakage reactance	0.12 p.u.
		Copper losses	0.01 p.u.

6.2. Testbench

FPCC results had been verified experimentally with a real high power test bench. Figure 16 shows a diagram (a) and a photo (b) of the test bench used for the experiments.

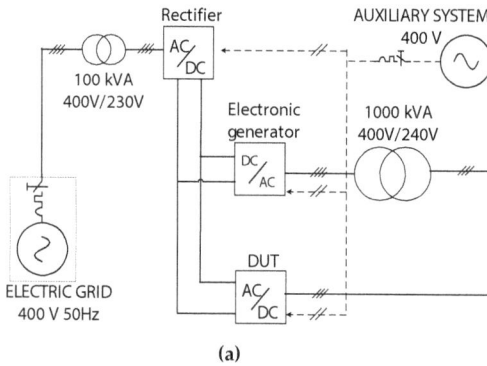

(a)   (b)

Figure 16. Test-bench scheme (**a**) and panoramic view (**b**); the device under test (DUT) is on the right side of the photo, and the rectifier and generator are in the background. The power transformer is inside the metallic jail.

The utility grid was emulated with an electronic power converter generator in order to perform controlled faults. A three-phase two-level converter was used. A grid of 240 V and 60 Hz was generated to perform all tests.

The DUT selected was a commercial high-power three-phase two-level grid-tied inverter DG for PV applications. The converter was set according to Table 6.

PV panels were emulated with a controlled rectifier that provided a suitable DC voltage input for both the DG and the electronic generator. A three-phase two-level converter was used. The rectifier could provide a DC-link input voltage from 425 V–825 V.

The transformer has a YY configuration, with winding voltages of 400 V:400 V and a short-circuit impedance of 0.09 p.u. Finally, all wiring in the test bench was enough to fulfil the power demands, and no significant inductance was added (three wires of 240 mm in diameter per phase).

Acknowledgments: This work was supported by GPTech Spain (http://www.greenpower.es) and the Electronic Engineering Department of the University of Seville.

Author Contributions: Jesús Muñoz-Cruzado-Alba conceived of and designed the proposed control strategy and significantly contributed to the implementation of the simulation and test bench. Javier Villegas-Núñez and José Alberto Vite-Frías helped in the laboratory tests and the writing of the paper. Juan Manuel Carrasco Solís responsible for guidance and a number of key suggestions.

Conflicts of Interest: The authors declare no conflict of interest.

References

1. Hashemi, S.; Yang, G.; Ostergaard, J.; You, S.; Cha, S. Storage application in smart grid with high PV and EV penetration. In Proceedings of the 2013 4th IEEE/PES Innovative Smart Grid Technologies Europe (ISGT EUROPE), Lyngby, Denmark, 6–9 October 2013; pp. 1–5.
2. Ma, T.; Lashway, C.R.; Song, Y.; Mohammed, O. Optimal renewable energy farm and energy storage sizing method for future hybrid power system. In Proceedings of the 2014 17th International Conference on Electrical Machines and Systems (ICEMS), Hangzhou, China, 22–25 October 2014; pp. 2827 – 2832
3. Ziadi, Z. Optimal Power Scheduling for Smart Grids Considering Controllable Loads and High Penetration of Photovoltaic Generation. *IEEE Trans. Smart Grid* **2014**, *5*, 2350–2359.
4. Hejazi, H.A.; Mohsenian-Rad, H. Optimal Operation of Independent Storage Systems in Energy and Reserve Markets With High Wind Penetration. *IEEE Trans. Smart Grid* **2014**, *5*, 1949–3053.
5. *Guideline for Generating Plants, Connection to and Parallel Operation with the Medium-Voltage Network*; German Association of Energy and Water Industries (BDEW): Berlin, Germany, 2008.
6. *Grid Connection Code for Renewable Power Plants (RPPs) Connected to the Electricity Transmission System (TS) or the Distribution System (DS) in South Africa*; rev 2.6; National energy regulator of South Africa (NERSA): Pretoria, South Africa, 2012.
7. *Minimum Technical Requirements for Photovoltaic Generation (PV) Projects*; Puerto Rico Electric Power Authority (PREPA): San Juan, Puerto Rico, 2012.
8. *Requisitos de Respuesta Frente a Huecos de Tensión de las Instalaciones de Producción de Régimen Especial.* Spanish electric grid (REE): Madrid, Spain, 2006. (In Spanish)
9. *Allegato A.70: Regolazione Tecnica dei Requisiti di Sistema Della Generazione Distribuita*; Rev. 01; TERNA: Rome, Italy, 2012. (In Italian)
10. *Inverters, Converters, Controllers and Interconnection System Equipment for Use With Distributed Energy Resources*, UL Standard 1741, rev02. Underwriters Laboratories Inc (UL): Northbrook, IL, USA, 2010.
11. *Technical Requirements for Connecting Photovoltaic Power Plants to the National Energy System*; Transelectrica: Bucarest, Romania, 2013.
12. *Norma Técnica de Seguridad y Calidad de Servicio.* Electric national commission (CNE): Santiago de Chile, Chile, 2015. (In Spanish)
13. *Technical Requirements and Evaluation of Grid Code Compliance for Photovoltaic Power Plants Connected to the Medium Voltage Level in Jordan*; Electricity Regulatory Commission (ERC): Amman, Jordan, 2014.
14. Rodriguez, P.; Timbus, A.V.; Teodorescu, R.; Liserre, M.; Blaabjerg, F. Flexible Active Power Control of Distributed Power Generation Systems During Grid Faults. *IEEE Trans. Ind. Electron.* **2007**, *54*, 2583–2592.
15. Castilla, M.; Miret, J.; Camacho, A.; de Vicuna, L.G.; Matas, J. Modeling and Design of Voltage Support Control Schemes for Three-Phase Inverters Operating Under Unbalanced Grid Conditions. *IEEE Trans. Power Electron.* **2014**, *29*, 6139–6150.

16. Miret, J.; Castilla, M.; Camacho, A.; de VicuÃśa, L.G.; Matas, J. Control Scheme for Photovoltaic Three-Phase Inverters to Minimize Peak Currents During Unbalanced Grid-Voltage Sags. *IEEE Trans. Power Electron.* **2012**, *27*, 4262–4271.

17. Yaramasu, V.; Wu, B.; Alepuz, S.; Kouro, S. Predictive Control for Low-Voltage Ride-Through Enhancement of Three-Level-Boost and NPC-Converter-Based PMSG Wind Turbine. *IEEE Trans. Ind. Electron.* **2014**, *61*, 6832–6843.

18. Vite, J.A.; Múñoz-Cruzado Alba, J.; Villegas Núñez, J.; Hernandez, E.R. PV inverter requirement updated for LVRT capabilities under recent grid codes legislation. In Proceedings of the International Exhibition and Conference for Power Electronics, Intelligent Motion, Renewable Energy and Energy Management (PCIM Europe 2014), Nuremberg, Germany, 20–22 May 2014; pp. 1–6.

19. Vrionis, T.D.; Koutiva, X.I.; Vovos, N.A. A Genetic Algorithm-Based Low Voltage Ride-Through Control Strategy for Grid Connected Doubly Fed Induction Wind Generators. *IEEE Trans. Power Syst.* **2014**, *29*, 1325–1334.

20. Tamai, S. High power converter technologies for saving and sustaining energy. In Proceedings of the Power Semiconductor Devices & IC's (ISPSD), Waikoloa, HI, USA, 15–19 June 2014; pp. 12–18.

21. Burkart, R.; Kolar, J.W.; Griepentrog, G. Comprehensive comparative evaluation of single- and multi-stage three-phase power converters for photovoltaic applications. In Proceedings of the the the 2012 IEEE 34th International Telecommunications Energy Conference (INTELEC), Scottsdale, AZ, USA, 30 September–4 October 2012; pp. 1–8.

22. Yazdani, A.; Iravani, R. *Voltage-Sourced Converters in Power Systems; Modeling, Control and Applications*; John Wiley & Sons: Hoboken, NJ, USA, 2010.

23. Monmasson, E. *Power Electronic Converters: PWM Strategies and Current Control Techniques*; John Wiley & Sons: Hoboken, NJ, USA, 2011.

24. *Overall Efficiency of Grid Connected Photovoltaic Inverters*; European Committee for Electrotechnical Standardization (CENELEC): Brussels, Belgium, 2010.

2

Empirical Validation of a Thermal Model of a Complex Roof Including Phase Change Materials

Stéphane Guichard [1,*]**, Frédéric Miranville** [2]**, Dimitri Bigot** [2]**, Bruno Malet-Damour** [2]**, Teddy Libelle** [2] **and Harry Boyer** [2]

Academic Editor: Chi-Ming Lai

[1]   Research Institute in Innovation and Business Sciences (IRISE) Laboratory/Superior Industrial Center Study (CESI)-Reunion Chamber of Commerce and Industry (CCIR)/Regional Centre for the Innovation and Transfer of Technologies (CRITT), The CESI engineering school, Campus Pro—CCIR 65 rue du Père Lafosse-Boîte n°4, Saint-Pierre 97410, France

[2]   Physics and Mathematical Engineering Laboratory for Energy, Environment and Building (PIMENT), University of Reunion, 117, rue du Général Ailleret Le Tampon 97430, France; frederic.miranville@univ-reunion.fr (F.M.); dimitri.bigot@univ-reunion.fr (D.B.); bruno.malet-damour@univ-reunion.fr (B.M.-D.); teddy.libelle@me.com (T.L.); harry.boyer@univ-reunion.fr (H.B.)

*   Correspondence: sguichard@cesi.fr

Abstract: This paper deals with the empirical validation of a building thermal model of a complex roof including a phase change material (PCM). A mathematical model dedicated to PCMs based on the heat apparent capacity method was implemented in a multi-zone building simulation code, the aim being to increase the understanding of the thermal behavior of the whole building with PCM technologies. In order to empirically validate the model, the methodology is based both on numerical and experimental studies. A parametric sensitivity analysis was performed and a set of parameters of the thermal model has been identified for optimization. The use of the generic optimization program called GenOpt® coupled to the building simulation code enabled to determine the set of adequate parameters. We first present the empirical validation methodology and main results of previous work. We then give an overview of GenOpt® and its coupling with the building simulation code. Finally, once the optimization results are obtained, comparisons of the thermal predictions with measurements are found to be acceptable and are presented.

Keywords: phase change materials (PCMs); building thermal simulation; model optimization; model validation

1. Introduction

Buildings are indisputably considered as one of the largest energy consuming sectors. According to the International Energy Agency (IEA), the average energy consumed by buildings represents 32% of worldwide energy consumption, with approximately 40% of the primary energy used in most countries. In France, the energetic consumption in the building sector is approximately about 43%, representing a quarter of the nation's carbon dioxide emissions. The use of energy-hungry appliances to improve the thermal comfort is responsible for both the electrical energy consumption and the increase of greenhouse gas emissions [1].

For this reason, some actions are performed to curb energy consumption and to protect the environment, for example, the use of renewable energies, passive energy buildings and the use of appropriate building codes for new or retrofitted buildings. Moreover, several studies and applications have shown that the building thermal inertia is among the possible solutions and should be improved, in order to achieve high performance and low-energy buildings [2–4]. As a result, the question of the increase of the thermal energy capacity storage of materials used in the building sector arises. With this target in mind, a technology such as the use of phase change materials (PCMs) may be integrated into building envelopes, both to enhance the thermal energy storage [5] and to improve the thermal comfort. Indeed, because of their higher thermal energy storage densities than other heat storage materials, these materials are able to store and release thermal energy as latent heat, when the phase change occurs. It is important to highlight the fact that the latent heat storage is much larger than the sensible heat storage [6]. Usually, organic and inorganic PCMs are often used and solid-liquid phases are chosen [7]. Despite the fact its thermal conductivity should be improved, paraffin is often used for latent heat thermal energy storage [8]. Indeed, it has useful thermal properties such as absence of supercooling, chemical stability and a low vapor pressure [3,9].

The use of PCMs in the building envelope may reduce the peak loads due to heating, ventilating and air conditioning (HVAC) energy consumptions by increasing the thermal inertia of each wall of the building. The peak load may thus be shifted to the off-peak energy use load periods [10]. In addition, the results of PCM-oriented research on buildings have shown that the thermal comfort was improved and energy savings could be achieved.

Among all the PCM applications to achieve high energy efficiency buildings, for example PCM-integrated walls, PCM-assisted ceiling heating and cooling, photovoltaic systems coupled with a PCM-based heat storage system [11], and so on, this paper deals with the inclusion of PCMs in the roof system. Generally, the roof is considered as a thermal buffer between the indoor and outdoor environment. This is the part of the building most exposed to solar radiation in hot climates and considered as the weakest part of a building's thermal performance [12,13].

One possible solution to reduce heat transfer from outside to inside a building may be found in the increased use of mass insulation. Nevertheless, this type of thermal insulation allows one to reduce heat transfer due to conduction, but does not reduce heat transfer by infrared radiation through the roofing. To overcome this drawback, a solution based on the use of radiant barrier systems, which requires the presence of air layers in order to minimize radiation heat transfer, is proposed [12]. Based on the same approach, PCM under a flexible sheet form laminated on both sides with an aluminum sheet is used in the lower part of the roof between the air layer and the drywall, both to enhance the thermal energy storage and to reduce the solar radiation through the roof system [14]. Furthermore, before integrating PCM into new or retrofitted buildings, it is interesting to predict the thermal effects of these materials on the whole building performance. To contribute to the energy efficiency policy recommendations for buildings' energy consumption, a one-dimensional simplified numerical model for PCMs has been developed and implemented in a prototype building simulation code named ISOLAB. ISOLAB is able to take into account the building envelopes, including PCM or not, and their actual impacts on energy consumption. Nevertheless, before using the PCMs model integrated in the ISOLAB code and to ensure the reliability of the results, this article focuses on the empirically validation of the latter. We first present the tools used, building simulation code and the generic optimization program GenOpt®. We then briefly describe the studied system as well as the main results of previous work. Finally, results of the experimental validation of the thermal model are presented and discussed.

2. Presentation of Tools

2.1. A Building Simulation Code: ISOLAB

Developed by Miranville [15], ISOLAB is a prototype building simulation software developed with the Matlab environment. Nodal descriptions of buildings and a finite difference scheme of the time variable in one-dimension are used to simulate the dynamic thermal behavior of a monozone or multi-zone building according to its environment (weather data and location). In order to determine temperatures of the whole building, the following matrix formalism Equation (1) is solved by using an implicit finite-difference method [16]:

$$C_w \dot{T}_w = A_w T_w + B_w \qquad (1)$$

where the index w is used for walls and windows. A_w is the state matrix including the terms linked to heat conduction and the interior linearized convective exchanges. B_w is the vector containing outside or internal solicitations of the system. C_w is the capacities matrix. T_w is the state vector containing every temperature of each wall and $\dot{T}_w$ is the temperature derivative of T_w. The ISOLAB code has already been validated with the International Energy Agency, Building Energy Simulation Test (IEA BESTEST) procedure, concerned with buildings without PCMs [15].

2.2. A Generic Optimization Program: GenOpt®

The optimization process consists in running several model simulations with different parameters sets. The chosen parameters are identified by using a mathematical tool like parametric sensitivity analysis. Thereafter, each simulation is launched with a different set of parameters (continuous parameters, discrete parameters, or both), and the associated cost function value, with or without constraints, is recorded. Finally, to ensure convergence to the best set of parameters, it is necessary to run many simulations varying each parameter within the right range.

A generic optimization program called GenOpt® from the Lawrence Berkeley National Laboratory of the University of California, developed with the Java environment by Wetter, is used to optimize the unknown parameters [17]. The choice of GenOpt® has been made according to its simulation program interface that allows the coupling with any simulation program, without requiring code modifications. Indeed, the coupling is done by creating files required by GenOpt® to run an optimization and also an auto-executable version of the model [18]. In addition, GenOpt® is dedicated for the thermal building simulation. For more details about the generic optimization program, the interested reader may refer to [17]. Before performing the optimization sequences with GenOpt®, the studied system with the main results of previous work are presented in the following section.

3. Experimental Set up of a Complex Standard Roof Including Phase Change Materials

3.1. Introduction

The objective of this part is to show the main results of experimental investigations on a complex roof incorporating PCM panels. The use of such a database allows us to empirically validate the numerical code developed and implemented in a prototype building simulation software to determinate the thermal performance of any building envelopes with PCM. First, a description of the roofing complex and the associated instrumentation is presented below.

3.2. Localization and Structure of the Test Cell

With dimensions of 3 m (height) × 3 m (width) × 3 m (length) and an internal volume about 30 m^3, the test cell, called industrial engineering laboratory (LGI), can be considered a typical room of existing buildings on Reunion Island. It has been designed with a flexible structure in order to study several configurations and phenomena. Thanks to its modular structure, the movable walls allow testing PCM panels included in the standard roof inclined 20° with respect to the horizontal. The whole building envelope is constituted of vertical opaque walls, a jalousie type window, a glass door and a complex roof with PCM panels. The roofing complex is an assembly of homogeneous or inhomogeneous materials, separated by one or several air layers [12,14,18]. The building components of LGI are given in Table 1, and an overview is depicted in Figure 1a. In steady state conditions and applying the French thermal rules [19], the U-value of each wall is calculated using the following formula:

$$\frac{1}{U} = \frac{1}{h_{ext}} + \sum_{i=1}^{n} \frac{e_i}{\lambda_i} + \frac{1}{h_{in}},\tag{2}$$

with $h_{ext} = h_{re} + h_{ce}$ and $h_{in} = h_{ri} + h_{ci}$. The U-values of the construction are also given in Table 1.

Table 1. Arrangement of the LGI test cell [13]. Phase change material: PCM.

Element	Composition	Remark (s)
Opaque vertical walls	Sandwich board 80 mm thick cement-fiber/polyurethane/cement-fiber	$U_{walls} = 0.4\ \mathrm{W \cdot m^{-2} \cdot K^{-1}}$
Window	Aluminium frame, 8 mm clear glass	Blind-type 0.8 m × 0.8 m
Glass door	Aluminium frame, 8 mm clear glass	Glass in upper and lower parts, 0.7 m × 2.2 m
Roofing complex	Corrugated galvanised steel 1 mm/air layer of 280 mm thick/PCM 5.26 mm thick/Plasterboard 12.5 mm	PCM is laminated to aluminium protective foils. Roof inclined at 20°. $U_{roof} = 2\ \mathrm{W \cdot m^{-2} \cdot K^{-1}}$
Floor	Concrete slabs 80 mm thick on 60 mm thick polystyrene	$U_{floor} = 0.7\ \mathrm{W \cdot m^{-2} \cdot K^{-1}}$

(a)   (b)

Figure 1. (a) Industrial engineering laboratory (LGI) cell; and (b) experimental platform.

Located at a low altitude (68 m) above sea level, the experimental set-up was erected on the experimental platform of the University Institute of Technology of Saint-Pierre (Reunion Island). This choice of location implies the study of a tropical climate with strong solar radiation and humidity. Besides, the test cell is oriented north in order to receive symmetrical solar radiation during the day [14]. To obtain extreme solicitations inputs from the roof, a dark blue color has been chosen for the corrugated iron. During the experimental validation procedure, the blind windows and the window panes were masked, and the test cell was kept closed, without using mechanical ventilation or an air-conditioning system [13].

Furthermore, near the experimental devices, two meteorological stations (the weather station is depicted in a red circle in Figure 1b) are installed in order to collect data from the environment, such as solar radiation (global, direct and diffuse, on a horizontal plane), ambient air temperature, exterior relativity humidity, wind speed and direction. As depicted in Figures 2 and 3 and because of the significant duration of the chosen experimental sequence (approximately 92 days), only five days of data from August to October 2012 are presented in this paper.

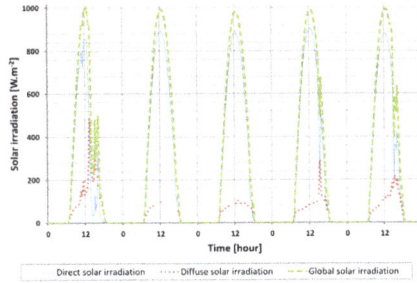

Figure 2. Solar irradiation from 20 to 24 September 2012 [13].

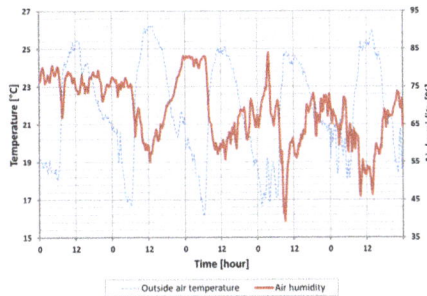

Figure 3. Outside air temperature and air humidity from 20 to 24 September 2012 [13].

3.2.1. Instrumentation of the Enclosure

Sixty sensors located both in the enclosure and on the roof of the LGI test cell have been used. The walls (north, south, east and west, inside and outside) are equipped with thermal sensors on the surfaces, such as T-type thermocouples and flux meters in order to measure the inside or outside surface temperature of each wall. For air temperature, the thermocouples are inserted into aluminum cylinders. For radiant temperature, the thermocouples are contained in a black globe.

To assess the stratification of the air, the interior volume was measured at three different heights from floor to ceiling (Figure 4). Many heat sensors have been sealed in the concrete slabs supporting the cell, in order to assess the information on the boundary conditions from the ground. Indeed, modeling errors concerning this part of the building, the ground measurements database is used during the code validation step. For more details on sensors' locations and set-up and the on the associated errors, the interested reader may refer to [13].

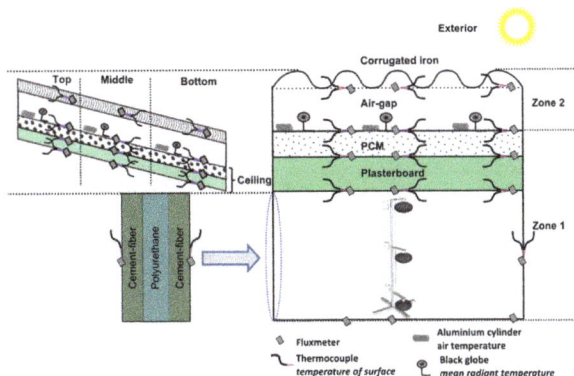

Figure 4. Simplified schematic of the LGI cell and instrumentation.

3.2.2. Instrumentation of the Roof

To ensure reliable measurements of the complex roof, sensors have been located as depicted in Figure 4. To assess temperatures and heat flux measurements from the corrugated iron to the ceiling, each component of the complex roof was instrumented. The heat sensors are spread on both sides of the surfaces of the corrugated iron, on the PCM panels and on the drywall. Between the PCM panels and the corrugated iron, the air layer is not ventilated, and the dry-air and black globe temperatures were also measured.

Before using the heat sensors, the thermocouples were calibrated and verified on site according to a convenient procedure. Heat flux sensors were calibrated by the manufacturer. The accuracy of the thermocouples is estimated to ± 0.5 °C, and according to manufacturer, the relative error of flux meters is approximately about 5%. In order to avoid air bubbles for both thermocouples and flux meters, thus causing thermal resistances of contact, a conductive heat paste was applied on the surfaces. A data logger was installed in the LGI cell in order to collect data from all the sensors every 15 min. All data were saved on a computer [13].

4. Previous Investigation

4.1. Introduction

Before May 2013, both the thermal behavior of building envelopes including PCM and time-varying thermal properties of materials were not taken into account by ISOLAB. For this, a simplified numerical model of the thermal behavior of PCMs was developed and implemented. However, many constraints on the PCM model have been imposed, such as the respect of the state system formalism and the use of an implicit one-dimensional scheme according to the finite difference approach. Moreover, the apparent heat capacity method has been used.

4.2. Description of the Roofing Complex

The model is intended to predict temperature evolutions of each components of the whole building, including the inside air of enclosures. The studied LGI cell is considered as well-isolated and divided into two thermal zones. The ceiling is a specific wall separating the two vertical zones. It plays the role of ceiling for the first zone and of floor for the second zone. We note that the air gap simulated in zone 2 is considered as a thermal zone in our multi-zone building model (Figure 4).

The presence of an air layer between the corrugated iron and PCM panels allows one to benefit from the principle of action of reflective insulation, which is closely linked to the radiative properties of the surfaces of PCM panels. Taking into account the air layer with the combination of homogenous and inhomogeneous materials, the roof system can be qualified as complex with coupled heat transfers involved (conduction, convection and radiative transfers) [20]. Therefore, the chosen configuration may complicate the determination of thermal performances due to the multiple configurations of the air layer: opened or closed, naturally ventilated or forced-ventilated [14]. This is the reason that explains why the enclosed air space is considered as a thermal zone in the approach proposed in this paper.

4.3. Description of Phase Change Material Test

Tested PCMs are commercial products from Dupont™ Energain®. They consist of 5 mm thick flexible sheets, made of 60% microencapsulated paraffin wax within a copolymer laminated on both sides with an aluminum sheet [13,14]. The panels' characteristics are summarized in Table 2.

Values of the heat capacity in each phase and melting point have been determined by differential scanning calorimetry (DSC) measurements. These parameters are exposed in details in [21,22].

Table 2. Characteristics of PCMs used [13].

	Parameters	Value	Unit
Thermal properties	Thermal conductivity: λ_s/λ_l	0.22/0.18	$W \cdot m^{-1} \cdot K^{-1}$
	Heat capacity: C_{ps}/C_{pl}	3134/2833	$J \cdot kg^{-1} \cdot K^{-1}$
	Latent heat: $L_{melting}$	71	$KJ \cdot kg^{-1}$
	Melting temperature	23.4	°C
Descriptive properties	Thickness	5.26	mm
	Width	1000	mm
	Length	1198	mm

4.4. Mathematical Model for Phase Change Material

The solidification and melting process are the most studied in building applications. Usually, numerical modelling of these phenomena is either based on the first or second law of thermodynamics. For more details, the interested reader may refer to [23].

The thermal model for phase change is based on the apparent heat capacity method from the enthalpy method. This method allows one to obtain the general form of a heat conduction equation with a nonlinear specific heat, without needing to know ahead of time the location of the phase interface. To simplify the mathematical model, some assumptions were made [13,14,24]. Through the solid (or liquid) fraction term, called f_s, the final expression of transient heat conduction can be written in 1-D along the $\overrightarrow{x}$ direction as follows [13,24]:

$$C_{app}(T) \frac{\partial T(x,t)}{\partial t} = \lambda_{PCM}(T) \frac{\partial^2 T(x,t)}{\partial x^2}$$

With:
$$\begin{cases} C_{app}(T) = \rho_s c_s + \Delta(\rho c) f_s + \dfrac{df_s(T)}{dT}(\rho_l L_m + \Delta(\rho c) \cdot (T(x,t) - T_m)) \\ \Delta(\rho c) = \rho_l c_l - \rho_s c_s \\ \lambda_{PCM}(T) = (1 - f_s(T))\lambda_s + \lambda_l f_s(T) \\ f_s(T) = \dfrac{1}{2} - \dfrac{1}{2}\tanh\left(\gamma \dfrac{T_m - T(x,t)}{4\delta T}\right) \\ \dfrac{df_s(T)}{dT} = \dfrac{\gamma}{8\Delta T\left[\cosh\left(\gamma\dfrac{T_m - T(x,t)}{4\delta T}\right)\right]^2} \end{cases}$$

(3)

The governing equation in terms of the apparent heat capacity can be solved using a standard heat transfer code, and a wide range of discretization approaches can be used. As a result, the one-dimensional finite difference method can be chosen [20]. To definitely be in accordance with the formalism of the ISOLAB equation, the use of a backward Euler scheme is possible thanks to the solid (or liquid) fraction term. Generally, the expression of the solid fraction is given by using an approximation of the Heaviside function. In the PCM model, a specific parameter called γ appears in the final expression. The latter is usually equal to 1, for instance in [25]. In order to not underestimate or overestimate the latent heat value during the phase change process, a proposed method was given by [13]. For a given phase change interval δT, and at $T = T_m$, the expression of the apparent heat capacity (C_{app}) leads to the first determination of γ, according to the following process [13]:

If $T(x,t) \geqslant T_m - \delta T$ and $T(x,t) \leqslant T_m + \delta T$ then :

$$\left| \max(\gamma) = \left(\max(C_{DSC}) - \left(\frac{c_s + c_l}{2}\right) \right) \cdot \frac{8\delta T}{L_m} \right.$$

else
$|\gamma$ must be determined
end

where max(C_{DSC}) is given by DSC, and δT is chosen very small. In the numerical simulation, $\delta T = 0.01C$. In the first approach, γ is evaluated as follows:

$$\gamma = \frac{L_m}{C_{ps} \cdot (T_m - \delta T)} \tag{4}$$

It is important to highlight that the model of the test cell with PCMs takes into account all heat transfer modes (radiative heat transfer, convection heat transfer and conduction heat transfer), the related values being presented in detail in [13]. Besides, the radiative model is based on specific developments for low emissivity walls and thus includes the effects of the reflective surfaces of the PCM panels [26].

4.5. Main Results

The thermal model is considered validated only if the validation criteria are met. These criteria are given by [13,18]:

(1)   To reach 10% validation error between numerical solutions and experimental data from actual building;
(2)   To reach acceptable absolute errors between some numerical solutions and measurements:

- $\pm 2\,°C$ for temperatures on each side of each suspended ceiling
- $\pm 1\,°C$ for indoor air temperature of real building

To definitely validate the numerical thermal model, different steps are required. These steps are illustrated in Figure 5: the two first steps have already been presented in details in [13]. The results showed that the numerical thermal model was able to predict the dynamic thermal behavior of PCMs. Nevertheless, the validity criteria were not respected. According to the validation methodology, it was necessary to highlight the origins of errors. A parametric sensitivity analysis, according to a method derived from the fast Fourier amplitude transform (FAST) method has been performed and this made it possible to put in evidence the parameters with most influence on model outputs [27–29]. Following the sensitivity analysis, the main causes of any differences between the model and experimental data can be explained, but also we can focus on the search of the set of unknown parameters of the model in a restricted range. The following important results were shown [13]:

- The thermal behavior of the complex roof is governed by convective heat transfers, both in the air layer and the faces of the suspended ceiling.
- The absorptivity coefficient of corrugated iron was not the same as new sample.
- The specific parameter γ can be considered as a non-dimensional velocity of phase change because the phase change is assumed to occur slowly. This parameter can also be used to overcome numerical instabilities in the zone near the interface between the two phases of the PCM. Indeed, during the phase change, a mushy zone between the two phases is created. From a numerical point of view, sharp discontinuities at the phase interface are observed, implying some very difficult to overcome numerical instabilities. Moreover, it influences the derivative gradient of the solid fraction. However, its physical interpretation is still in investigation.

The parameters linked to the convective exchange coefficients are depicted in Figure 6 and all parameters are given by the Table 3. However, the unknown parameters have to be determined. For this, optimization sequences were performed.

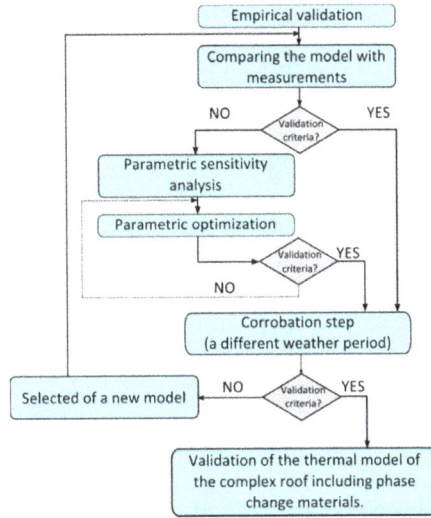

Figure 5. Methodology used for the empirical validation.

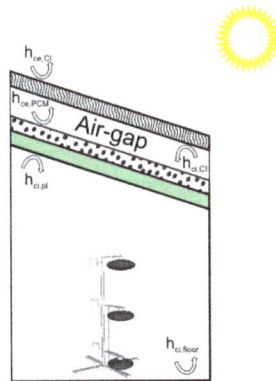

Figure 6. Most influential convective exchange coefficients.

Table 3. Most influential parameters on model outputs at the end of sensitivity analysis step.

Frequency	Parameters	Unit
103 ($h_{ci,floor}$)	Indoor convective exchange coefficient of the floor (zone 1)	$W \cdot m^{-2} \cdot K^{-1}$
107 ($h_{ci,pl}$)	Indoor convective exchange coefficient of the plasterboard	$W \cdot m^{-2} \cdot K^{-1}$
108 ($h$ci,PCM)	Outdoor convective exchange coefficient of the PCM panel	$W \cdot m^{-2} \cdot K^{-1}$
154	γ coefficient	-
199 ($h_{ci,CI}$)	Indoor convective exchange coefficient of the corrugated iron	$W \cdot m^{-2} \cdot K^{-1}$
200 ($h_{ce,CI}$)	Outdoor convective exchange coefficient of the corrugated iron	$W \cdot m^{-2} \cdot K^{-1}$
218 (α_{CI})	Solar absorptivity coefficient of the corrugated iron	-

5. Empirical Validation of the Phase Change Material Model

5.1. Phase Change Material Model Optimization Using GenOpt®

The optimization sequence consists in achieving the best set of parameters by maximizing or minimizing a chosen cost function, subject to a set of constraints (or a defined domain), until optimization criteria are met. Indeed, many simulations are run and when a stopping criterion is reached, like the required error on the studied output for instance, the optimization is stopped.

To optimize the thermal model of PCM implemented in ISOLAB by using GenOpt®, a methodology for the coupling of the building simulation code with the optimization program is required. The study has already been published and the interested reader may see [18] for details.

Among the various methods of fixing sets of parameters implemented in GenOpt®, the generalized pattern search (GPS) Hooke-Jeeves MultiStart Algorithm was used. Indeed, it allows to look over some parameters and to avoid the result values to be obtained from a local minimum value. In addition, this algorithm is compatible with all building thermal insulation codes and gives reliable results. The cost function used is calculated by using a mathematical model tool based on the modified standard deviation:

$$s = \sqrt{\frac{1}{n}\sum_{i=1}^{n}(x_i - \bar{x})^2} \tag{5}$$

If several model outputs are used to optimize the model, the global cost function is the sum of the absolute values of costs functions for each output. The optimization sequence is stopped, if and only if, same minimum values of the cost functions were reached for 10 different optimization sequences. In our approach, several optimization sequences (1500 optimization sequences were reached approximately) and statistics studies were performed.

5.2. Results of the Optimization

Results from the optimization sequence are summarized in Table 4.

Table 4. Parameters before and after optimization sequences.

Parameters	Before Optimization	After Optimization	Unit
$h_{ci,floor}$	3.50	5.00	$W \cdot m^{-2} \cdot K^{-1}$
$h_{ci,w}$	1.00	1.00	$W \cdot m^{-2} \cdot K^{-1}$
$h_{ci,PCM}$	1.00	1.50	$W \cdot m^{-2} \cdot K^{-1}$
γ (when $T \neq T_m$)	0.04	0.01	-
$h_{ci,CI}$	3.50	1.75	$W \cdot m^{-2} \cdot K^{-1}$
$h_{ce,CI}$	25	5.7 V + 11.4 [15]	$W \cdot m^{-2} \cdot K^{-1}$
α_{CI}	0.85	0.76	-

Each value of an optimized parameter is found in agreement with the physical phenomena. Indeed, convective exchange coefficients are very small and correspond to the non-ventilated upper air layer as the building was kept closed during the experimental sequence. To apply the process for determining the γ parameter, this coefficient has been evaluated to 0.01 when $T \neq T_m$. The value of absorptivity coefficient (α_{CI}) is equal to the value that was determined by [30]. This value, lower than initially expected, can be justified by the fact that the cell was erected and covered a few years ago, time tending to decrease the absorptivity or dark color of the roof. Indeed, over time, the performances of buildings will not be the same than when the building was new [15].

5.3. Comparison between Optimized Thermal Model of Phase Change Material with Measurements

On the whole period, the dynamic behaviour of the components of the complex roof and the indoor air temperature are predicted well. Through the different curves (Figure 7), we can ensure that the given melting temperature for this application was correctly chosen because the temperatures on both sides of PCM allow the storage and the release of energy during phase change process.

The surface temperature of the suspended ceiling is predicted at ±1.3 °C (Figure 8). The used conditions show that PCM model is able to properly predict the behavior of the PCM panel. The comparison between numerical simulations and experimental data shows a good agreement. The indoor air temperature of LGI test cell is predicted with an accuracy of $\pm 0.5\,^\circ$C (Figure 9). The criteria for the validation of the thermal model are respected. According to Table 5, the mean errors between numerical solutions and experimental measurements do not exceed 5%.

Table 5. Standard deviations, maximum differences and errors after optimization sequences.

Localization	Standard Deviation σ ($^\circ$C)	Maximum Difference ($^\circ$C)	Maximum Error (%)	Mean Error (%)
Inside air temperature (zone 1)	0.2	− 0.5	− 6.2	2.1
Suspended ceiling inside surface temperature (plasterboard)	0.4	− 1.1	− 8.9	2.9
Interface PCM/plasterboard temperature	0.4	− 1.3	8.8	3.1
Suspending ceiling outside surface temperature (PCM)	0.5	− 1.7	− 9.8	2.3
Air-gap temperature (zone 2)	1.0	− 2.6	− 9.4	3.1
Corrugated iron temperature	1.3	3.5	9.4	3.2

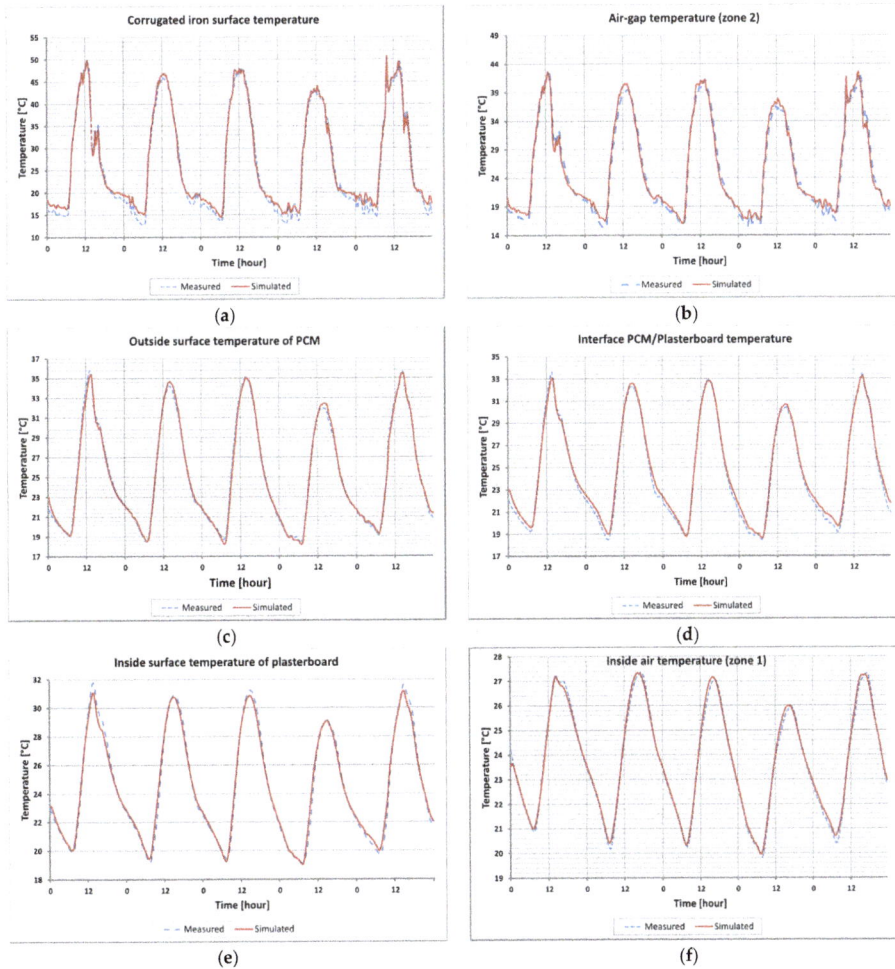

Figure 7. Comparison of model results with experimental measurements for all components of the complex roof (**a–e**) and for inside air temperature (**f**) (zone 1).

Difference between model and measurements

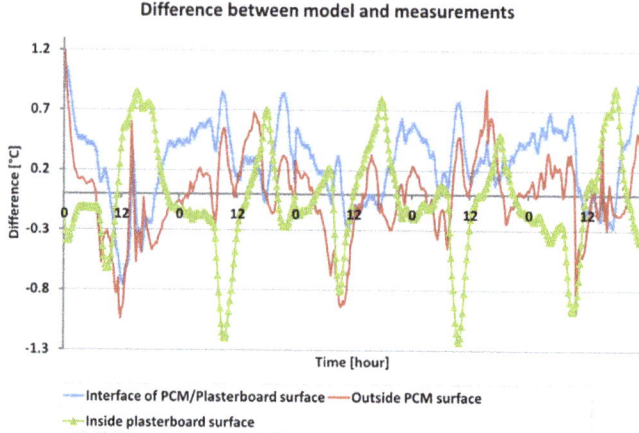

Figure 8. Errors between numerical solutions and experimental data for suspended ceiling temperatures.

Difference between model and measurements

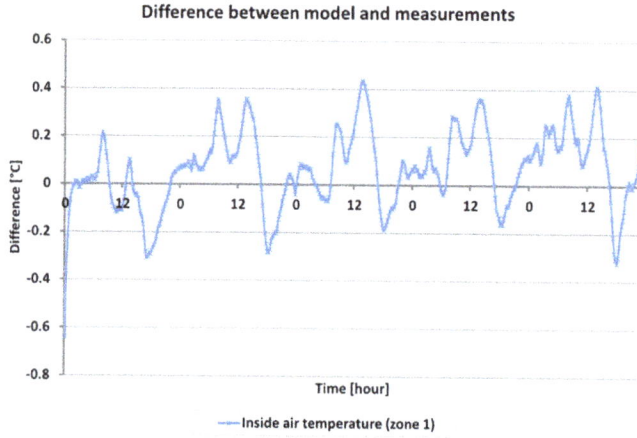

Figure 9. Errors between numerical solutions and experimental data for inside air temperature.

The most important errors result from the prediction of the air-gap temperature and the prediction of the metal sheet temperature in zone 2. For the air-gap, the model must be improved and models from the literature based on computational fluid dynamics (CFD) showed that the results are inconclusive due to the influence of boundary conditions. Thus, an empirical correlation from experiments will have to be determined. Nevertheless, the values of convective exchange coefficients ($h_{ce,CI}$ and $h_{ci,PCM}$) determined by the optimization sequences correspond with the empirical correlation developed by Alamdari and Hammond [31] (it has been verified later that the obtained results fit the correlation). The proposal of these authors is a combination of two correlations for taking into account both natural convection in laminar regime and turbulent regime:

$$h = \left[\left(1.51 \frac{|\Delta T|^{\frac{1}{4}}}{H} \right)^6 + \left(1.33 \, |\Delta T|^{\frac{1}{3}} \right)^6 \right]^{\frac{1}{6}} \tag{5}$$

where ΔT is the averaged difference temperature between the surface of the wall and the indoor air of the room. H corresponds to the height of the wall.

With the predicted air temperature in zone 2, the errors between the curves can also be explained by the prediction of the corrugated iron surface temperature. Indeed, the latter has a direct effect on the temperature of each component of the complex roof. To reduce these errors, the radiative model should be improved and the radiosity method should be used [15].

To conclude this step, the thermal model was fully-coupled with the ISOLAB code and the results of the PCM model are very encouraging. Indeed, for different type of walls with PCM or not, the model is able to predict temperatures in actual conditions. However, to definitely validate the PCM model implemented in the ISOLAB building simulation code, it is necessary to compare numerical solutions with another experimental data sequence as presented in the next part.

5.4. The Corroboration Step

The corroboration step consists in using another experimental period in order to verify if all parameters determined can be generalized and are not specific for a given period. Moreover, it also allows evaluating the efficiency of the proposed model. For this step of corroboration, the meteorological data used are from 2 to 5 October 2012. This period was characterized by [20]:

- an average maximal global radiation of 900 W$\cdot$m^{-2}
- an average wind speed of 3.50 m$\cdot$s^{-1}
- an average outdoor temperature of 26 °C
- an average rate humidity of 65%

The results from the corroboration step are summarized in Table 6. In this part, the validation criteria are also respected. For instance, comparison between numerical simulations and measurements for the predicted indoor air temperature is depicted in Figure 10. For more details, the interested reader may refer to [20].

Table 6. Standard deviations, maximum differences and errors for the corroboration period.

Localization	Standard Deviation σ (°C)	Maximum Difference (°C)	Maximum Error (%)	Mean Error (%)
Inside air temperature (zone 1)	0.2	0.6	5.6	2.1
Suspended ceiling inside surface temperature (plasterboard)	0.5	1.5	9.2	3.4
Interface PCM/plasterboard temperature	0.5	1.7	8.9	3.6
Suspending ceiling outside surface temperature (PCM)	0.5	1.7	9.9	2.8
Air-gap temperature (zone 2)	1.3	3.4	8.9	3.2
Corrugated iron temperature	1.6	4.7	9.8	3.3

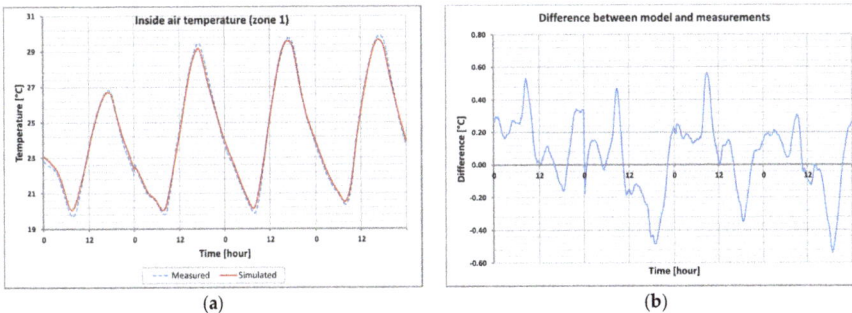

(a)   (b)

Figure 10. Comparison (a) and errors (b) between model results with experimental measurements for inside air temperature (zone 1) during the corroboration step.

Tables 5 and 6 show that maximum differences between predictions and measurements are a little more important than the results of the first experimental sequence. A possible explanation is a high setting of parameters for specific environmental conditions. Nevertheless, despite these differences the model can be considered as valid because the validation criteria have again been reached. Furthermore, the maximal standard deviation is approximately of 1.6 °C and all mean errors are below 5%.

6. Conclusions and Further Works

In this paper, a generic thermal model of PCM in buildings, validated with reliable experimental data, was presented. An actual building equipped with PCM in a complex roof was set up and each component's surfaces in contact with the indoor air temperature were measured. A detailed investigation was carried out to evaluate the efficiency of the thermal behavior of the model, with important steps included in experimental validation.

A mathematical model based on the apparent heat capacity has been presented. Moreover, an approach of nodal description of the complex wall and a finite difference method in one-dimensional were used. Thanks to parametric sensitivity analysis, the most influential factors on model outputs such as, all convective exchange coefficients, γ parameter and the absorptivity coefficient of corrugated iron, were determined by a generic optimization tool. Then, a new comparison between the optimized thermal model and experimental data was performed. With the corroboration step, the results showed that the validity criteria were respected. Finally, the validation process of model has been reached and the thermal model of PCM has been validated.

Despite the empirical validity of the PCM model, further work is necessary to improve the air-gap temperature prediction and the radiative model used in order to obtain better agreement between the numerical solutions and measurements. Moreover, a mathematical formulation of the γ parameter should be presented. Other experimental studies should be performed to confirm the use of these materials in tropical climates such as Reunion Island. Future works will also focus on the comfort study from the dedicated test cell.

Acknowledgments: The authors wish to thank Fonds Social Européen and La Région Réunion for their support and their fundings of the first author's thesis. In addition, this research received funding from the Ministère de l'Outre-Mer.

Author Contributions: Stéphane Guichard designed the experiments and developed PCM model for the experimental and numerical studies. Frédéric Miranville, Dimitri Bigot and Harry Boyer helped for the coupling model of PCM with ISOLAB code and performed the validation stages of this one. Bruno Malet-Damour and Teddy Libelle helped to monitor the data from the experimentation in order to ensure they were reliable for the empirical validation of the PCM model.

Conflicts of Interest: The authors declare no conflict of interest.

Nomenclature

Variables

c	Specific capacity ($J \cdot kg^{-1} \cdot K^{-1}$)
C	Apparent heat capacity ($J \cdot m^{-3} \cdot K^{-1}$)
f	Solid or liquid fraction
h	Convection heat transfer coefficient ($W \cdot m^{-2} \cdot K^{-1}$)
L	Latent heat ($J \cdot kg^{-1}$)
T	Temperature (°C)
t	Time (s)
x	Spatial variable (m)
U	Thermal conductance ($W \cdot m^{-2} \cdot K^{-1}$)
V	Wind speed ($m \cdot s^{-1}$)

Subscripts

ce	Exterior convection
ci	Interior convection
ext	Exterior
int	Interior
l	Liquid phase
m	Melting point
re	Exterior radiative
ri	Interior radiative
s	Solid phase

Greek Symbols

γ	Constant parameter
δT	Phase change interval ($^{\circ}$C)
ρ	Density (kg$\cdot$m^{-3})
λ	Thermal conductivity (W$\cdot$m$^{-1}\cdot$K^{-1})

Abbreviations

IEA BESTEST	International Energy Agency, Building Energy Simulation Test
ISOLAB	A building simulation software, which integrates the hygro-thermal and aeraulic phenomena
PCM	Phase change material

References

1. Pérez-Lombard, L.; Ortiz, J.; Pout, C. A review on buildings energy consumption information. *Energy Build.* **2008**, *40*, 394–398. [CrossRef]
2. Zhou, D.; Zhao, C.-Y.; Tian, Y. Review on thermal energy storage with phase change materials (PCMs) in building applications. *Appl. Energy* **2012**, *92*, 593–605. [CrossRef]
3. Zalba, B.; Marín, J.M.; Cabeza, L.F.; Mehling, H. Review on thermal energy storage with phase change: Materials, heat transfer analysis and applications. *Appl. Therm. Eng.* **2003**, *23*, 251–283. [CrossRef]
4. Farid, M.M.; Khudhair, A.M.; Razack, S.A.K.; Al-Hallaj, S. A review on phase change energy storage: Materials and applications. *Energy Convers. Manag.* **2004**, *45*, 1597–1615. [CrossRef]
5. Cabeza, L.F.; Castell, A.; Barreneche, C.; de Gracia, A.; Fernández, A.I. Materials used as PCM in thermal energy storage in buildings: A review. *Renew. Sustain. Energy Rev.* **2011**, *15*, 1675–1695. [CrossRef]
6. Seong, Y.-B.; Lim, J.-H. Energy Saving Potentials of Phase Change Materials Applied to Lightweight Building Envelopes. *Energies* **2013**, *6*, 5219–5230. [CrossRef]
7. Hasan, A.; McCormack, S.J.; Huang, M.J.; Norton, B. Energy and cost saving of a photovoltaic-phase change materials (PV-PCM) system through temperature regulation and performance enhancement of photovoltaics. *Energies* **2014**, *7*, 1318–1331. [CrossRef]
8. Medina, M.A.; King, J.B.; Zhang, M. On the heat transfer rate reduction of structural insulated panels (SIPs) outfitted with phase change materials (PCMs). *Energy* **2008**, *33*, 667–678. [CrossRef]
9. Sarı, A.; Karaipekli, A. Thermal conductivity and latent heat thermal energy storage characteristics of paraffin/expanded graphite composite as phase change material. *Appl. Therm. Eng.* **2007**, *27*, 1271–1277. [CrossRef]
10. Khudhair, A.M.; Farid, M.M. A review on energy conservation in building applications with thermal storage by latent heat using phase change materials. *Energy Convers. Manag.* **2004**, *45*, 263–275. [CrossRef]
11. Lo Brano, V.; Ciulla, G.; Piacentino, A.; Cardona, F. On the efficacy of PCM to shave peak temperature of crystalline photovoltaic panels: An FDM model and field validation. *Energies* **2013**, *6*, 6188–6210. [CrossRef]
12. Miranville, F.; Boyer, H.; Lauret, P.; Lucas, F. A combined approach for determining the thermal performance of radiant barriers under field conditions. *Solar Energy* **2008**, *82*, 399–410. [CrossRef]

13. Guichard, S.; Miranville, F.; Bigot, D.; Boyer, H. A thermal model for phase change materials in a building roof for a tropical and humid climate: Model description and elements of validation. *Energy Build.* **2014**, *70*, 71–80. [CrossRef]

14. Guichard, S.; Miranville, F.; Bigot, D.; Malet-Damour, B.; Boyer, H. Experimental investigation on a complex roof incorporating phase-change material. *Energy Build.* **2015**, *108*, 36–43. [CrossRef]

15. Miranville, F. *Contribution à l'étude des parois complexes en physique du bâtiment modélisation, expérimentation et validation expérimentale de complexes de toitures incluant des produits minces réfléchissants en climat tropical humide*; Université de la Réunion: Saint-Pierre, France, 2002. (In French)

16. Boyer, H.; Chabriat, J.P.; Grondin-Perez, B.; Tourrand, C.; Brau, J. Thermal building simulation and computer generation of nodal models. *Build. Environ.* **1996**, *31*, 207–214. [CrossRef]

17. Wetter, M. GenOpt-A generic optimization program. In Proceedings of the Seventh International IBPSA's Building Simulation Conference, Rio de Janeiro, Brazil, 13–15 August 2001.

18. Bigot, D.; Miranville, F.; Boyer, H.; Bojic, M.; Guichard, S.; Jean, A. Model optimization and validation with experimental data using the case study of a building equipped with photovoltaic panel on roof: Coupling of the building thermal simulation code ISOLAB with the generic optimization program GenOpt. *Energy Build.* **2013**, *58*, 333–347. [CrossRef]

19. *French Thermal Rules 2012*; Règles Th-U: Paris, France, 2012.

20. Guichard, S. *Contribution à l'étude des parois complexes intégrant des matériaux à changements de phase Modélisation, Expérimentation et évaluation de la performance thermique globale*; Université de La Réunion: Saint-Pierre, France, 2013. (In French)

21. Kuznik, F.; Virgone, J. Experimental investigation of wallboard containing phase change material: Data for validation of numerical modeling. *Energy Build.* **2009**, *41*, 561–570. [CrossRef]

22. David, D.; Kuznik, F.; Roux, J.-J. Numerical study of the influence of the convective heat transfer on the dynamical behaviour of a phase change material wall. *Appl. Therm. Eng.* **2011**, *31*, 3117–3124. [CrossRef]

23. Dutil, Y.; Rousse, D.R.; Salah, N.B.; Lassue, S.; Zalewski, L. A review on phase-change materials: Mathematical modeling and simulations. *Renew. Sustain. Energy Rev.* **2011**, *15*, 112–130. [CrossRef]

24. Guichard, S.; Miranville, F.; Boyer, H.; La Réunion, F. A Mathematical Model of Phase Change Materials (PCM) used in Buildings. In Proceedings of the Third IASTED African Conference, Gaborone, Botswana, 6–8 September 2010.

25. Dauvergne, J.L. Réduction et inversion de problèmes de diffusion thermique avec changement de phase. Ph.D. Thesis, University of Bordeaux, Bordeaux, France, 2008. (In French).

26. Miranville, F.; Lauret, P.; Medina, M.; Bigot, D. A simplified model for radiative transfer in building enclosures with low emissivity walls: Development and application to radiant barrier insulation. *J. Solar Energy Eng.* **2011**, *133*. [CrossRef]

27. Mara, T.A.; Boyer, H.; Garde, F. Parametric sensitivity analysis of a test cell thermal model using spectral analysis. *J. Solar Energy Eng.* **2002**, *124*, 237–242. [CrossRef]

28. Mara, T.A.; Garde, F.; Boyer, H.; Mamode, M. Empirical validation of the thermal model of a passive solar cell test. *Energy Build.* **2001**, *33*, 589–599. [CrossRef]

29. Lauret, A.J.P.; Mara, T.A.; Boyer, H.; Adelard, L.; Garde, F. A validation methodology aid for improving a thermal building model: Case of diffuse radiation accounting in a tropical climate. *Energy Build.* **2001**, *33*, 711–718. [CrossRef]

30. Bigot, D. *Contribution à l'étude du couplage énergétique enveloppe/système dans le cas de parois complexes photovoltaïques (pc-pv)*; Université de La Réunion: Le Tampon, France, 2011. (In French)

31. Alamdari, F.; Hammond, G.P. Improved data correlations for buoyancy-driven convection in rooms. *Build. Serv. Eng. Res. Technol.* **1983**, *4*, 106–112. [CrossRef]

3

Direct and Indirect Impacts of Vegetation on Building Comfort: A Comparative Study of Lawns, Green Walls and Green Roofs

Laurent Malys [1,2], Marjorie Musy [1,2,*,†] and Christian Inard [2,3,†]

Academic Editor: Nyuk Hien Wong

[1]   L'Université Nantes Angers Le Mans, ensa Nantes, UMR CNRS 1563, Centre de REcherche Nantais Architecture Urbanité, 6 quai F. Mitterrand, Nantes 44000, France; laurent.malys@crans.org
[2]   Institut de Recherche en Sciences et Techniques de la Ville, FR CNRS 2488, 1 rue de La Noé, Nantes 44000, France
[3]   Laboratoire des Sciences de l'Ingénieur pour l'Environnement, Université de La Rochelle, UMR CNRS 7356, Avenue M. Crépeau, 17042 La Rochelle Cedex 1, France; christian.inard@univ-lr.fr
*   Correspondence: marjorie.musy@crenau.archi.fr
†   These authors contributed equally to this work.

Abstract: Following development and validation of the SOLENE-microclimat tool, the underlying model was used to compare the impacts of various "greening strategies" on buildings' summer energy consumption and indoor comfort. This study distinguishes between direct and indirect impacts by successively implementing the test strategies on both the studied building and surrounding ones; it also considers insulated *vs.* non-insulated buildings. Findings indicate that green walls have a direct effect on indoor comfort throughout the entire building, whereas the effect of green roofs is apparently primarily confined to the upper floor. Moreover, the indirect effect of a green wall is greater, mainly due to the drop in infrared emissions resulting from a lower surface temperature. It has also been proven that the indirect effects of green walls and surrounding lawns can help reduce the loads acting on a non-insulated building.

Keywords: adaptive comfort; urban climate; lawn; building simulation; SOLENE-microclimat

1. Introduction

France's total energy consumption and the share generated by the country's residential and commercial sectors have stabilized since 2006 at values of 164 Mtoe and 68 Mtoe, respectively. These two sectors account for 43% of all energy consumption and 23% of CO_2 emissions. For the European building sector taken as a whole, the air conditioning of occupied spaces has been estimated at 57% of total energy demand and 33% of CO_2 generation. The building sector therefore is key to fulfilling the commitment made in 2003 by the French Government to the international community that calls for cutting by 75% the nation's greenhouse gas emissions by 2050. Technically speaking, such a reduction is much easier to achieve in new buildings than from retrofitting the large stock of older buildings, as part of a program that would need to be planned over a several-year period. Reducing summer heat stresses in an urban context may be partially and indirectly achieved by modifying the local climate through the introduction of adaptive techniques. The effectiveness of such solutions, however, requires extensive knowledge of the correlations existing between climate and the thermal behavior of buildings.

Techniques aimed at expanding the green spaces in cities have been approved by a large majority of interested parties because of their numerous benefits to the city ecosystem. Such techniques have been the focus of a wide array of studies over the past decade.

This is especially true for trees, whose effect has been broadly studied and shown to be the most efficient source for cooling cities and improving thermal comfort at the local scale [1–5].

Since roofs constitute a very high percentage of the exposed urbanized area, green roofs have received widespread attention for their direct impact on energy consumption, as well as for their impact caused by urban climate modification. Santamouris recently presented a state-of-the-art assessment of green roofs and proceeded to compare their effects with those of cool roofs [6].

It has been demonstrated that whenever vegetative, or green, roofs are installed in high- or even medium-rise buildings, their mitigation potential becomes almost negligible [7–9].

Nonetheless, the direct impacts of green roofs on building energy consumption have been extensively studied, and depending on climate conditions and building characteristics, results show a variable thermal efficiency derived from green roofs [10].

Walls also occupy a high fraction of the total urban surface, with total wall areas potentially greater than the space available for green roofs. Walls have been shown to reduce a building's cooling load and overall energy consumption, while improving indoor comfort [11–15], as the cooling potential is in fact influenced by the choice of plant species [16]. Temperatures on city streets are also cooler when green walls are installed on building facades [17,18]; their global impact on a city's climate and outdoor comfort however has not been widely investigated.

Due to their limited impact on climate, lawns have naturally received scant attention despite the fact that these surfaces, which are maintained at low temperatures compared to paved surfaces, do indeed exert a thermal impact on their environment [19].

Urban vegetation has direct impacts, namely those due to the building envelop modifications, such as decrease of solar heat gain by shading and a decrease of convective heat fluxes by lowering near-wall air velocity or increasing the envelop thermal resistance due to the substrate layer. Indirect impacts are due to the buildings' environment modifications as air and surface temperatures that influence long-wave (LW) radiation transfer, convective heat fluxes and ventilation and infiltration loads [20]. Due to the fact that their study requires using both a microclimate model and building energy simulation, they are more rarely studied: Djedjig [21] did this at the street scale using transient system simulation tool (TRNSYS) and a street canyon model and de Munck [9] at the city scale in town energy budget (TEB), where the building are averaged buildings and represented by a mono-zone mode.

As an extension of previous works that intended to developed new sub-models in SOLENE-microclimat and validate them [22–25], in this paper, through a case study conducted in Nantes (northwestern France), the direct and indirect effects will be compared from three greening configurations applicable to urban surfaces, namely: lawns, green roofs and green walls. The detailed simulations performed herein will enable analyzing how these configurations modify a city's thermal environment and then alter indoor building comfort.

2. Methodology

2.1. The SOLENE-Microclimat Model

The SOLENE-microclimat project was first developed for the purposes of an urban insolation assessment. Sub-modules have since been added, now making it possible to incorporate:

- LW radiation exchanges using the radiosity method;
- conduction heat transfer and thermal storage in walls and soil;
- airflow and convective exchanges through the coupling with a computational fluid dynamics (CFD) code.

The near-surface air velocity and temperature calculated by the CFD code allow evaluating the convective heat transfer of the surfaces used as the entries for the SOLENE software. On the other hand, the surface temperatures calculated with SOLENE software are used as entries for the conservation of energy equation of the CFD code, evapotranspiration from natural surfaces, like plants and water ponds or humidification systems and, lastly, the energy balance (*i.e.*, energy demand or indoor temperature) for a building within the simulated zone. Towards this end, a multi-zone building nodal network model is used. The temperature nodes correspond to each building floor, and the boundary conditions are the canopy air temperature and humidity calculated by the CFD code. In addition, the heat fluxes exchanged by radiation are computed with the thermo-radiative model [26]. It should be noted that the energy balance of each node includes ventilation and infiltration loads, both latent and sensible.

Only the SOLENE-microclimat thermo-radiative validation as a whole has been carried out, and its coupling with the Saturne CFD code is presented in [21]. Due to the difficulty in getting complete and accurate experimental datasets (airflows, canopy and surface temperatures, turbulent heat fluxes, *etc.*), it is very difficult to validate the SOLENE-microclimat model as a whole. Nevertheless, some validations have been carried out concerning sub-models of the software. For instance, walls, roofs and soil surface temperatures were compared successfully to measurements [22]. It is the same for green walls' behavior [23]. In the future, it is planned to increase these validation efforts. The SOLENE-microclimat thermo-radiative model and its validation have been presented in [22]. The coupling with the Saturne CFD code is also explained.

The modules added to study the direct and indirect impacts of adaptation and mitigation solutions are detailed in [23,25].

These latest developments introduced into SOLENE-microclimat models for lawns, green roofs and walls are now available for use. In all of these models, the vegetation layer acts on the following energy transfers:

- Evapotranspiration, *i.e.*, latent heat flux due to transpiration and evaporation of water contained in both the plants and substrate.
- Radiation: the foliage layer modifies both short-wave (SW) and LW radiation; moreover, it reflects, absorbs and transmits a portion of the SW radiation to the shaded surface area. In covering the surface, this layer exchanges LW radiation with the surrounding surfaces, as well as with the surface being covered.
- Convection: the foliage layer is rough and modifies the near surface airflow, thus implying a modification to the heat transfer.

The transfer by conduction is also changed whenever the green surface contains a substrate; this layer can then be included in the wall (or ground) model. The way in which these transfers have been taken into account in the SOLENE model is detailed in [23,25].

2.2. The Studied Urban District

Our study has been carried out in an actual urban district, called Pin Sec, in Nantes. Pin Sec underwent urbanization from 1954 through the end of the 1980s. The first buildings constructed here were called "large blocks" ("grands ensembles" in French), though those built in Nantes were not as high as in other French cities. The construction techniques employed in this district were typical of practices during France's post-war economic boom, *i.e.*, featuring non-insulated concrete.

This district was chosen because of the presence of comprehensive measurement instrumentation. Furthermore, it is representative of suburban residential districts that constitute land reserves due to their low density. As a countermeasure to urban sprawl, these districts were earmarked for increased density, as was the case for similar districts in the city of Nantes. The proposed Pin Sec densification scenario corresponds to an increase in density from 20% to 35%, with this latter figure having been attained in current projects. The block plans of this district, in both its present state and with a denser layout, are shown in Figure 1.

2.3. The Studied Building

The building under study is the so-called "Dunant block" (Figure 2); this choice was based on the fact that a meteorological station had been installed on the roof, though complementary measurements were conducted so as to verify that our model would correctly reproduce its thermal behavior [23]. Predominantly south facing, this building measures 99.2 m long by 9.6 m wide by 15 m high, corresponding to a total floor area of 4767 m², encompassing 120 housing units.

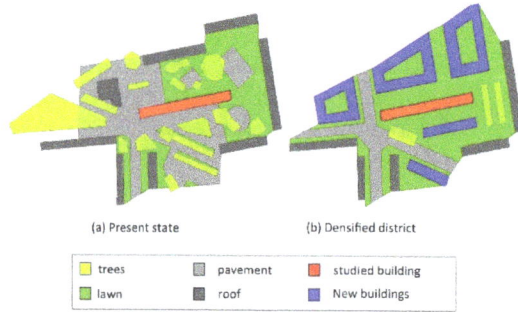

Figure 1. Block plans for the district, in both (a) its present state and (b) following densification.

Figure 2. The Dunant building: (a) southern facade; and (b) northern facade.

This building has never been insulated, although the original windows have been replaced by double-glazed panes, with the following characteristics: a glazing ratio of 17%, glass transmissivity of 0.7 and a low facade albedo (0.3). The walls and roof are made of a single layer of concrete (10 cm thick for walls and 20 cm for the roof).

For the simulation, the indoor boundary conditions applied are as follows:

- constant air change rate: $Q_r = 0.3$ vol· h^{-1};
- constant indoor heat gain: 30 kW.

Artificial soil surfaces are composed of a 10 cm-thick layer of asphalt and a 10 cm thick layer of concrete above a 2 m thick layer of soil. A description of the soil model is given in [27].

2.4. Greening Scenario

This study took place within the framework of the role of vegetation in sustainable urban development (VegDUD) Project [28], and the greening strategies studied were selected from among those identified in the project, *i.e.*,

- architectural greening: green roofs and walls;
- lawns;

- trees;
- sustainable urban drainage systems using vegetation;
- vegetation in deep soil *vs.* a soilless culture.

These strategies, or volume devices, were studied with respect to two families of parameters, namely:

- their management: extensive *vs.* intensive;
- their spatial distribution (city, district or building).

At the building or district scale, the use of SOLENE-microclimat allows handling the first three options. We have elected to focus on green roofs, green walls and lawns, because they all comprise surfaces and are thus more comparable, though admittedly, trees are volume devices, as well.

With the capability of explicitly representing urban geometry, it then becomes possible to explore the relative effect of various types of surfaces (roofs, soil, facades) based on their distribution across the urban landscape.

The presence of trees was included in the model of the district, and a description of the way to model them is given in [22]. Nevertheless, various tree layouts were not studied here. This will be done in the future. The green walls are living walls with an irrigated substrate. This model has been developed and validated by Malys *et al.* [25] from measurement data. For all configurations, we have assumed an absence of hydric stress; subsequently, results are to be analyzed in terms of water consumption.

The hypotheses adopted for vegetation modeling are as follows:

- foliar density: leaf area index (LAI) = 2;
- thickness: $L = 0.2$ m;
- extinction coefficient: $k_s = 0.8$;
- transmissivity: $\tau_f = 0.2$;
- an irrigation equal to potential evaporation (PTEc).

The goal here is to isolate the vegetation effect, neglecting both the insulation and inertia effects of the substrate. Only the solar mask and evapotranspiration of vegetation, which cannot be replaced by wall construction materials, have been studied. This set-up implies comparing a building whose living wall contains a substrate to one with walls featuring the same thermal conductivity, density and thermal capacity.

Similarly, albedo values have been set at 0.2 for facades and 0.3 for roofs. We have assumed that artificial pavement surfaces are covered with a paint whose reflectivity equals that of vegetation.

2.5. Adaptive Comfort

To compare the effects of these test strategies, we have used the indoor summer thermal comfort assessment (with no cooling system). An adaptive thermal comfort approach takes into account the ways people perceive their environment change. Their seasonal expectations of temperature and relative humidity, as well as on their capacity to control the spatial conditions are also considered [29,30].

A calculation method is presented in the European standard (NF-EN) 15251 Standard [31]. Four thermal comfort categories are defined on a predicted percentage dissatisfied-predicted mean vote (PPD-PMV) index that takes into account clothes, activity, mean radiant temperature, air temperature, velocity and humidity. In this standard [31], thermal comfort categories for naturally-ventilated spaces are differentiated by:

I   high expectations, used solely for spaces occupied by very sensitive and fragile individuals;
II   normal expectations, used for new buildings and renovated spaces;
III moderate expectations, used for existing buildings;
IV   expectation level for buildings outside the first 3 category levels. It is suggested that this category
     should be used for just a short period of the year.

With a relative humidity of 50% and a low air velocity, these categories are defined strictly from operating temperature ranges, as calculated by taking into account the variations in outdoor temperature during the previous days via the outdoor running mean temperature $\theta_{m,i}$. This parameter is formulated on the basis of its value on the previous day ($\theta_{m,i-1}$), the mean daily value of outdoor temperature on the previous day ($\theta_{e,i-1}$) and a coefficient α. Its recommended value equals 0.8:

$$\theta_{m,i} = (1 - \alpha)\,\theta_{e,i-1} + \alpha\theta_{m,i-1} \tag{1}$$

The limits of these recommended comfort categories for a residential building are expressed in the following and then presented in Figure 3 as a function of the outdoor running mean temperature:

$$\theta_{I,max} = 0.33\theta_m + 18.8 + 2 \tag{2}$$

$$\theta_{I,min} = 0.33\theta_m + 18.8 - 2 \tag{3}$$

$$\theta_{II,max} = 0.33\theta_m + 18.8 + 3 \tag{4}$$

$$\theta_{II,min} = 0.33\theta_m + 18.8 - 3 \tag{5}$$

$$\theta_{III,max} = 0.33\theta_m + 18.8 + 4 \tag{6}$$

$$\theta_{III,min} = 0.33\theta_m + 18.8 - 4 \tag{7}$$

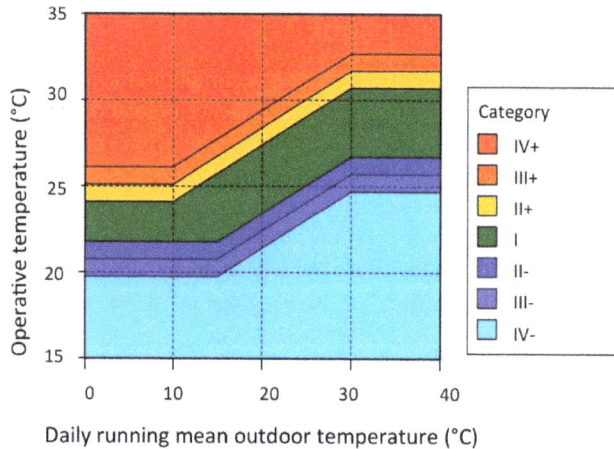

Figure 3. Thermal comfort categories for the operating temperature in naturally-ventilated buildings, according to the European standard (NF-EN) 15251 Standard [31].

3. Simulations

3.1. Weather Conditions

With respect to the weather conditions, we relied on *in situ* measurements for the year 2010 [32], consisting of:

- outdoor temperature (sensors/Vaisala HMP45C-L, Campbell Scientific, Logan, UT, USA);
- solar radiation (pyranometer Skye SKS 1110, Environmental Measurements Limited, North Shields, UK);
- infrared atmospheric flux (pyrgeometer Kipp&Zonen CGR3, Kipp & Zonen, Delft, The Netherlands);
- wind velocity and direction (wind monitor Young 05103, Young, Traverse City, MI, USA);

These data are acquired at a 3 s time step, but provided by the experimentalist after an average over a period of 15 min. Four our application, the average period is 1 h. They are then used as entries for all of the simulations. Wind velocity is used to calculate the airflow field (coupling with the Code-Saturne model) and the convective heat transfer coefficient. The LW energy balance within this urban landscape and exchanges with the sky are also calculated.

Our analysis focused on the hottest week recorded in 2010, *i.e.*, 25 June to 2 July. Meteorological data recorded on the building roof during this period are given in Figure 4. Air temperature ranged between 15 °C and 30 °C; and wind velocity, while highly variable, did not exceed 3 m/s. Just after the summer solstice, solar fluxes on a horizontal surface were near their maximum at nearly 1000 W·m^{-2} on the afternoon of 26 June. Cloud cover was prevalent on 27 and 28 June, as well as on 1 July, which also resulted in an increase of infrared thermal flux from the sky vault. With the exception of these periods, the infrared flux varied from 320 W·m^{-2} to 400 W·m^{-2}.

Even though it was not excessively hot, this week still gave rise to thermal stresses by virtue of following a rather cold week. Consequently, according to the principles of adaptive comfort, the temperatures used to ascertain indoor comfort will be low, *i.e.*, 6 °C less than the mean daytime temperature on 26 June.

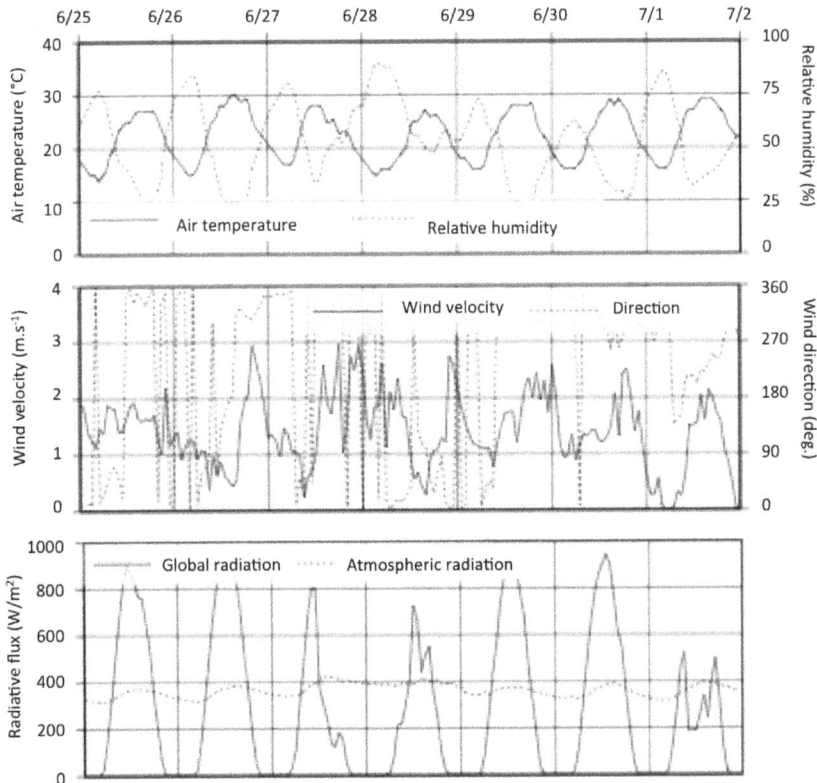

Figure 4. Weather conditions measured on the Dunant roof from 25 June to 2 July 2010.

3.2. The Targeted Greening Strategies

When choosing the applicable greening strategy, we considered two objectives:

- separating direct from indirect effects of the green envelopes;
- comparing the disposal rates.

This approach led us to distinguish five types of surfaces for potential greening, namely:

- facades of the studied building;
- roof on the studied building;
- facades of the other buildings;
- roofs on the other buildings;
- the soil.

Considering all combinations would yield 32 distinct case studies. The cases actually studied herein are presented in Table 1, *i.e.*,

- the reference case, without vegetation, or Case (0);
- Cases (1)–(3) are dedicated to assessing the direct effects (only the surfaces on the studied building are greened), with Case (1) being the facades, Case (2) being the roof and Case (3) being both facades and roof;
- Cases (4)–(7) consider just the indirect effects of greening, with Case (4) focusing on the facades, Case (5) focusing on the roofs and Case (6) focusing on the lawns; both the facades and soil are greened in Case (7);
- In Case (8), the entire envelop of the studied building is covered with vegetation, as are the other buildings;
- Cases (0+), (3+), (4+) and (8+) are similar to Cases (0), (3), (4) and (8), respectively, yet with an insulated building; for this insulated building, 10 cm of mineral wool have been added to the internal surface of three walls and the roof.

Table 1. The 13 selected simulation cases.

Simulation cases		0	0+	1	2	3	3+	4	4+	5	6	7	8	8+
Effect		Reference		Direct				Indirect					Direct + indirect	
Building greening	Facades	-	-	X	-	X	X	-	-	-	-	-	X	X
	Roof	-	-	-	X	X	X	-	-	-	-	-	X	X
Greening of the vicinity	Facades	-	-	-	-	-	-	X	X	-	-	X	X	X
	Roof	-	-	-	-	-	-	-	-	X	-	-	-	-
	Soil	-	-	-	-	-	-	-	-	-	X	X	-	-
+: with an insulated building		-	X	-	-	-	X	-	X	-	-	-	-	X

All types of surfaces, whose thermal characteristics remain variable, are shown in Figure 5, with their extreme cases being depicted in Figure 6. Table 1 presents the 13 cases simulated herein.

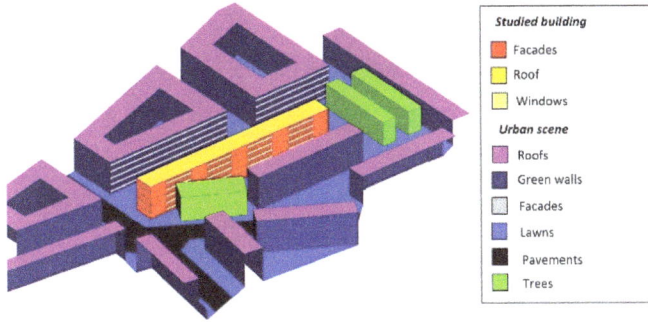

Figure 5. The various types of surfaces available for assignment.

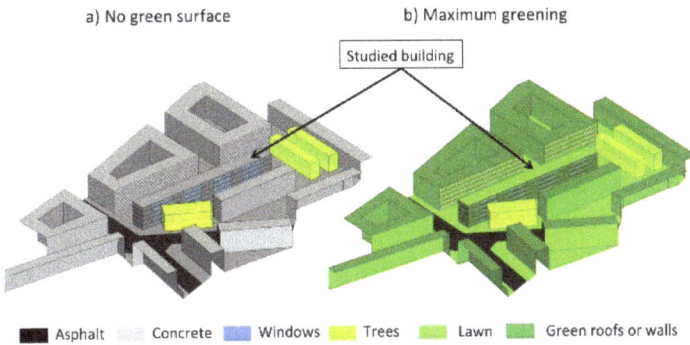

Figure 6. 3D models of the simulated configurations for the two extreme cases: (**a**) with no green surfaces; and (**b**) maximum greening.

4. Results

These results have been analyzed from four perspectives on compiling and comparing several simulation cases:

- Perspective 1, the effects of vegetation on the thermal insulation level: Cases (0), (3), (4) and (8) for the non-insulated building and Cases (0+), (3+), (4+) and (8+) for the insulated building;
- Perspective 2, comparison of the direct and indirect effects from green walls: Cases (0), (3), (4) and (8);
- Perspective 3, study of the direct effects of green envelopes: Cases (0)–(3);
- Perspective 4, study of the indirect effects of urban green surfaces: Cases (4)–(7).

These results simultaneously consider: indoor thermal conditions, modification of the urban microclimate and thermal exchanges between the building and its surroundings.

The large quantity of data generated has led to various analyses for each perspective. Comparisons will first be carried out for indoor conditions. In seeking to highlight the green wall effect, which until now has been studied the least, focus is placed at the mid-height of the building, *i.e.*, on the third floor.

Next, the energy exchanges between the building and its surroundings will be studied by analyzing heat fluxes at the outer surfaces.

Lastly, water requirements to irrigate the green envelopes for effect optimization will be provided.

4.1. Perspective 1: Impact of Green Surfaces on the Thermal Insulation Level

According to the results previously obtained [19,27], it may be assumed that a well-insulated building will be little affected by adding green envelopes. The results derived for both the insulated and non-insulated buildings are displayed in Figure 7.

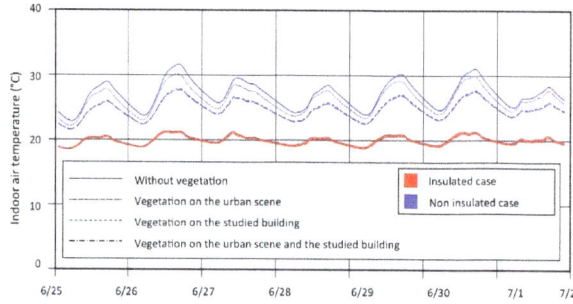

Figure 7. Indoor air temperatures (average of all five floors) for the insulated and non-insulated cases.

These findings simultaneously reveal the sensitivity of the building's thermal behavior to direct effects (*i.e.*, green strategies on the building), indirect effects (green strategies on surrounding surfaces) and a compounding of these two effects.

Let us note the major differences between cases relative to the insulated building and those relative to the non-insulated one. The amplitude and mean values of indoor temperature variations differ sharply. For the insulated cases, the amplitude of daily temperature variation is less than 3 °C, while this figure may exceed 6 °C for the non-insulated building without vegetation, where indoor temperatures can top 30 °C. In all insulated cases, indoor temperatures vary slightly in the vicinity of 20 °C.

To highlight the differences due to vegetation, it can be postulated that they become imperceptible when the building is insulated and remain capable of reaching 2 °C to 7 °C, depending on the greening configuration actually implemented. This finding confirms those from other studies in the literature [6,10,24] and leads us at this point to focus our study solely on the non-insulated building.

4.2. Perspective 2: Comparison of Direct and Indirect Effects of Green Facades

The focus here lies on a non-insulated building where facades have been greened either on the building itself, or on surrounding buildings, or on both the studied building and surrounding ones.

Figure 8 presents the variation in indoor temperatures on a floor at mid-height (*i.e.*, the third floor) of the building for four cases:

- the reference case, *i.e.*, without vegetation (0);
- green facades on just the studied building (3);
- green facades on the surrounding buildings, but not on the studied one (4);
- green facades on both the surrounding buildings and the studied one (8).

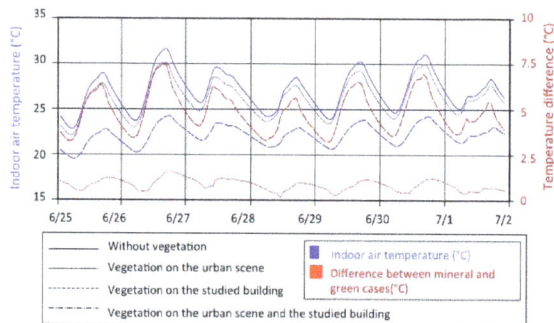

Figure 8. Indoor temperatures of the third floor (**in blue**) and the difference with the mineral case (**in red**). Nota bene (N.B.): the results of the cases "vegetation on the studied building" and "vegetation both in the urban environment and on the studied building" are superimposed.

4.2.1. Indoor Air Temperature

Implementing green facades on the surrounding buildings exerts a quite limited impact on the indoor temperature of the non-greened building. The greatest reduction obtained is 1.7 °C. The magnitude of the effect increases at the end of the afternoon, yet this effect does not fundamentally alter the thermal behavior of the building: the daily amplitude also lies between 5 °C and 7 °C.

The effect on indoor air temperature is correlated with modifications to the surrounding surface temperatures that influence LW radiation exchanges and convective exchanges with the air, thus modifying air temperature, as well.

Greened facades on the studied building lead to cooler indoor air, particularly during the daytime. The highest observed difference equaled 7.5 °C, and the lowest was 3 °C. Temperatures were capped at 25 °C, while in the reference case, the 30 °C threshold was often surpassed.

The amplitude in daily variations was also attenuated and did not exceed 4 °C. The fact that the impact was greater during the daytime suggests that the shading effect of vegetation is predominant.

The indirect effect of greening is practically zero when the studied building contains green facades. Implementing green facades on a building implies that it will be less sensitive to a modification of thermal boundary conditions, in particular LW radiation.

4.2.2. Indoor Thermal Comfort

Figure 9 shows the distributions of the occurrences among thermal comfort categories for the third floor. In the reference case, "hot" thermal comfort categories occur over 30% of the time; moreover, the IV+ category accounts for 5% of the occurrences. In contrast, this category tends to disappear under all green scenarios.

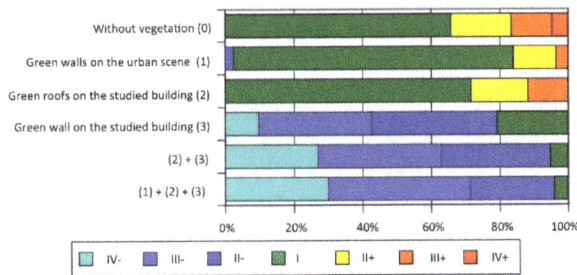

Figure 9. Distribution of the occurrences among thermal comfort categories on the third floor during the day.

The results when implementing a green roof or green facades on the studied building are also shown in Figure 9.

As regards the indirect effect, let us note that even though the difference in indoor temperature remained low, it was still sufficient to eliminate the occurrences of the IV+ comfort category. For the third floor, the indirect effect of green facades is greater than that of a green roof on the building, with this finding being somewhat counterintuitive.

The use of green facades on the studied building leads to significant changes in the comfort category distribution. Hot categories disappear and are offset by a sizable increase in cold categories. The appearance of these so-called "cold" categories might seem alarming, but it can still be verified that minimum temperatures lie around 20 °C at night (Figure 8) when the outdoor temperature is around 17 °C.

The indirect effect of implementing green facades on a studied building, in cases with green facades, as well, is perceptible, albeit quite small.

4.2.3. Water Balance

Green facades and green roofs represent passive solutions for cooling buildings. Both however require water not only to keep the plants alive, but also to obtain a satisfactory cooling effect [9]. When comparing the irrigation rate used in the samples by Malys *et al.* [25], which equaled $6 \text{ L} \cdot \text{d}^{-1} \cdot \text{m}^{-2}$, to what would be necessary to irrigate the entire Dunant building surface and to its number of inhabitants, a daily need of $53 \text{ L} \cdot \text{d}^{-1}$ per inhabitant is derived. This figure corresponds to over one-third of France's mean daily per capita water consumption. Fortunately, the simulated consumption levels are lower (Figure 10), due to the fact that conditions in the vicinity of the simulated green facades differ from those of the samples. More specifically, the samples were facing south, whereas according to the district simulations, they received varying levels of sun exposure. Similarly, the samples were recorded on a rooftop exposed to high wind speeds, while in our case, wind speeds were lower. The mean water consumption values obtained were in the range of $1 \text{ L} \cdot \text{d}^{-1}$.

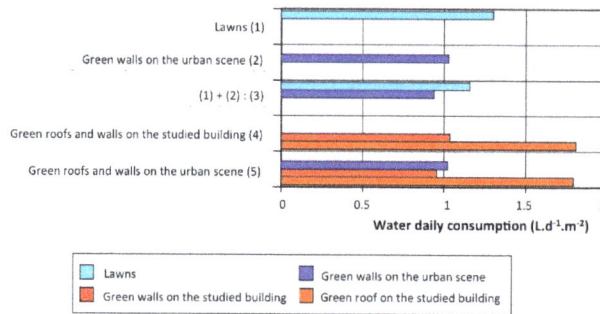

Figure 10. Mean daily water consumption of the various surfaces (in $\text{L} \cdot \text{d}^{-1} \cdot \text{m}^{-2}$).

The impact of solar heat fluxes on water consumption is depicted in Figure 10, with the surface requiring the largest quantity of water being the roofs. The second biggest consumer is the ground, whose insolation is attenuated by the buildings' shading effect. Facades need less water.

Let us also remark on the indirect effect of green surfaces on the quantity of evaporated water:

The addition of water consumption in Cases (1) and (2) (Figure 10) is 10% higher than in Case (3), which is accounted for in the implementation. The greater presence of green surfaces leads to a local increase in relative humidity and a decrease in air temperature, which in turn raises the evaporating power of air in the district. The decrease in LW radiation fluxes is another influential phenomenon.

These water needs have been compared to precipitation data. Table 2 presents the evaporated water quantities during a typical day, as correlated with the given ground coverage. As regards the facades, the reference area was set as the building roof area. The precipitation ratio is defined as the volume of evaporated water to that of the rainwater potentially stored on the reference. In Nantes, the mean June monthly precipitation equals approximate 46 mm, or 1.5 mm a day.

Table 2. Quantities of water evaporated by the various surfaces relative to both their ground coverage and precipitation.

Configuration	Total Quantity (L)	Footprint (m^2)	Total Height of Evaporated Water (mm·m^{-2})	Evaporation Ratio (%)
Green walls on the studied building	2594	954	2.72	177
Green roof on the studied building	1723	954	1.80	118
Green walls and roof on the studied building	4317	954	4.53	195
Green walls in the urban environment	18,527	8012	2.31	151
Lawns	15,568	13,444	1.16	75
Total	38,412	23,350	1.65	107

The water needs of the green roof and facades of the studied building correspond to nearly twice the mean daily rainwater falling on the roof in June. The facades require two-thirds of this water amount.

As for the lawns, the precipitation ratio exceeds one.

This approach is too simplistic to adequately address the hydrological stresses of the surfaces and to mitigate their evaporation rates, which would require a dynamic water balance and include an irrigation system in the vertical wall facade.

This crude analysis does however make it possible to highlight that evaporative cooling has a water consumption effect and that implementing green strategies must be considered in paying special attention to water availability, as this will also become a challenge relative to climate change.

This approach has also demonstrated that the opportunity to implement green facades must take into account the wall orientation, as it influences water needs.

4.3. Perspective 3: Direct Effect of Green Envelopes

4.3.1. Variation of Temperature Inside the Building

Our thermal model has yielded the temporal variations of both internal surface temperatures and indoor air temperature. The mean radiant temperature (MRT) can be calculated on each floor, along with the various heat fluxes between indoors and outdoors or between two contiguous floors.

Temporal variations in the internal surface temperature of windows and walls, in MRT and in indoor air temperature, are all given in Figure 11 for three simulation cases, namely:

-   building without any vegetation;
-   building with a green roof and green facades;
-   an insulated building without vegetation.

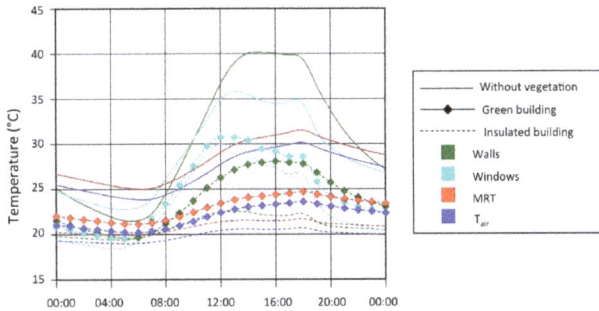

Figure 11. Internal surface temperature of windows and walls, mean radiant temperature (MRT) and indoor air temperature (T_{air}) of the third floor on 26 June 2010.

It can be observed that the internal surface temperature is high in the absence of any greening or insulation, allowing it to reach 40 °C, thus providing a sense of the extent of heat transmission by conduction. This finding can be explained by the low albedo values (0.25) and thermal resistance of the walls. The maximum internal surface temperature of the windows is lower because less solar flux is being absorbed by these surfaces. The variation in amplitude of MRT and T_{air} is smaller, and the difference between these two variables remains constant at around 1 °C. The green wall effect is especially noticeable on the internal surface temperature of walls, which can be as much as 12 °C lower than in the case without a green wall. The impact of insulation however is greater, with the decrease capable of reaching 25 °C. With insulation, the amplitude of variation is reduced to 3 °C.

4.3.2. Thermal Fluxes at the External Building Surface

Figure 12 indicates the temporal variation of heat fluxes at the external surface of the south-facing wall, on the third floor. The heat fluxes are positive when entering the building. The solar flux, while not shown here, is much greater than the other fluxes and reaches a maximum of 400 W·m^{-2} for the facade directly facing the Sun's rays. This flux is reduced by a factor of 0.2 when vegetation is added to the surface, thus declining to 80 W·m^{-2}. This decrease is taken into account when plotting the balance curves that represent the thermal flux either transmitted to the internal surface or stored in the wall:

$$Q_{Balance} = Q_{SW} + Q_{conv} + Q_{LW} + Q_E \qquad (8)$$

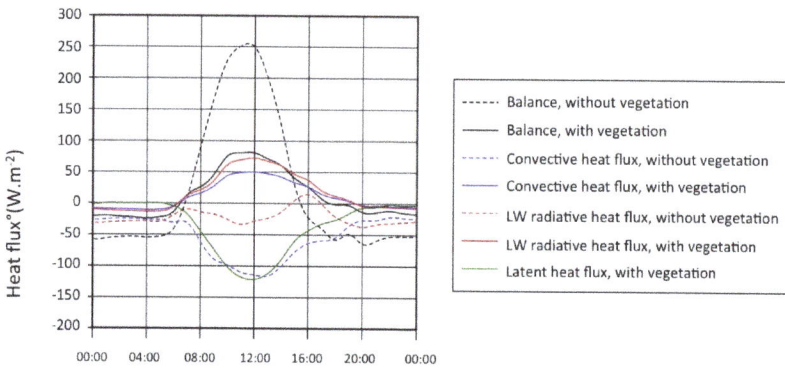

Figure 12. Heat fluxes on external surfaces of the south facing wall, third floor, 26 June 2010.

The difficulty of comparing these heat fluxes stems from the difference in calculation protocol depending on whether a green facade is being implemented or not. In the absence of a green wall, the convective heat flux is calculated directly between the outdoor ambient air and the external wall surface, while the LW radiative flux is between the surfaces and the sky. When a facade is present, the convective flux is calculated between the wall surface and the air confined in the foliage layer, with the LW radiative flux between the wall and the foliage cover. The latent heat flux must also be considered, along with the SW transmission coefficient of the foliage cover.

Let us note that the thermal behavior of the external wall surface differs. Without vegetation, the convective and LW radiative fluxes are the main contributors to its cooling. The LW radiative flux depends on the insolation of the facing surfaces, which explains its irregular variation.

With vegetation, the confined air layer and foliage cover exhibit a higher temperature than that of the external wall surface during the day; hence, both convective and LW radiative exchanges are positive.

The variation in these fluxes seems to be mainly due to incoming SW radiation. The nocturnal cooling is then reduced. Figure 12 clearly highlights the importance of the effect due to solar mask and evaporation at the substrate surface.

Lastly, the maximum flux transmitted to the indoor air or stored in the wall is reduced by two-thirds when a green wall has been implemented.

Figure 13 displays the mean daily heat fluxes calculated for two facades, one facing south, the other north, at the external surface of the building wall, both with and without a green wall. For the sake of comparison, the fluxes calculated for the non-insulated and mineral case are also given.

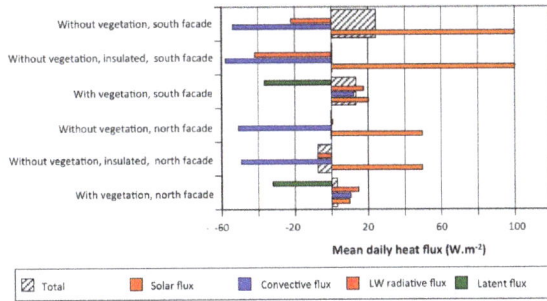

Figure 13. Mean daily heat fluxes on the external walls of the third floor, on 26 June 2010.

Averaged over the day, the thermal fluxes transmitted through the wall and stored in it are reduced by 50% for the south-facing facade due to greening. For the north-facing facade, the effect is reversed: without vegetation, the thermal balance is negative, *i.e.*, this facade contributes to cooling the building, but with vegetation, it is slightly positive. This finding is consistent with previously-observed trends: the large decrease in incoming solar flux and evaporation flux limits the contribution of the surface to the heating of indoor air.

In the case of an insulated building, the high thermal resistance of the walls leads to an increase in the external surface temperature of the walls and subsequently increases the convective and LW heat fluxes, which contribute more to microclimate heating and exert a negative impact on the surrounding buildings.

It is difficult to establish this kind of reasoning with regard to green walls, since the fluxes at the external surface are not directly exchanged with the outdoor air, but instead remain within the confined air layer. These effects are indirect and will be studied in Perspective 4 below.

4.4. Perspective 4: Indirect Effects of Vegetation

The visibility of green roofs from the urban space is very limited. One can therefore presume that their effect on the thermal comfort of residents in the street is also limited, as is their impact on the thermal behavior of surrounding buildings. It has been shown however that green roofs can have a major impact on the thermal behavior of the buildings where they have been installed, as well as on the urban microclimate. In considering the high level of roof insolation during the summer, the direct impact of green roofs is an obvious one. For this same reason, plus the fact that roofs occupy a large proportion of a dense urban surface, it can be anticipated that the effect of these surfaces, if greened, on the urban climate would also be strong. However, at the smaller scale of the urban block or street, on which green facades and ground vegetation is considered to be more efficient, further examination is still required. Studying the indirect effect of a green surface should help to answer the following questions:

- To what extent do green roofs modify air temperature in the district and affect the thermal behavior of surrounding buildings?
- What are the impacts of lawns and green walls on the thermal behavior of surrounding buildings?
- Are the direct and indirect effects of these techniques comparable? In particular, what is the predominant effect: convective or radiative?

To better answer these questions, the following configurations have been compared:

- the reference case, without vegetation (0);
- a green roof on the studied building (2);
- lawns on the uncovered ground (6);
- lawns and green facades on the studied building and surrounding buildings, (7) + (3).

Simulation results have been analyzed in considering both the impact on indoor thermal comfort and the modification of radiative and convective heat fluxes.

4.4.1. Indoor Thermal Comfort

The temporal variations in indoor temperatures on the third floor for the various configurations are given in Figure 14. Three distinct effects can be distinguished:

-   The effect of green roofs is practically negligible;
-   The effect of lawns and green walls leads to a decrease in indoor temperature, ranging from 1 °C to 2 °C. This difference slowly varied during the week, with some differences appearing between the configurations after 28 June.

The temperature decrease is maximized with both green walls and lawns, reaching 3 °C on the afternoon of 26 June.

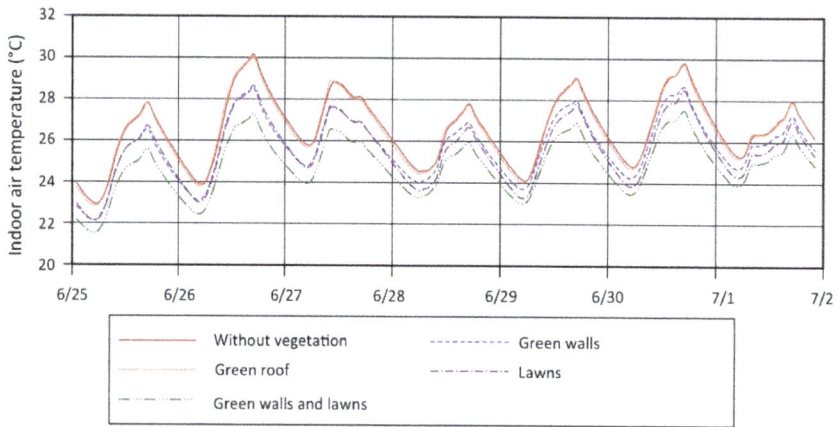

Figure 14. Indoor temperatures on the third floor for various greening configurations.

The three previous impact levels are once again observed, as applied to the distribution of thermal comfort categories (Figure 15). The effect of green walls is still imperceptible. Lawns and green walls together lead to a 50% decrease in the number of hours of discomfort and serve to eliminate the occurrence of Category +IV. A possible cold discomfort appears, which corresponds to night periods.

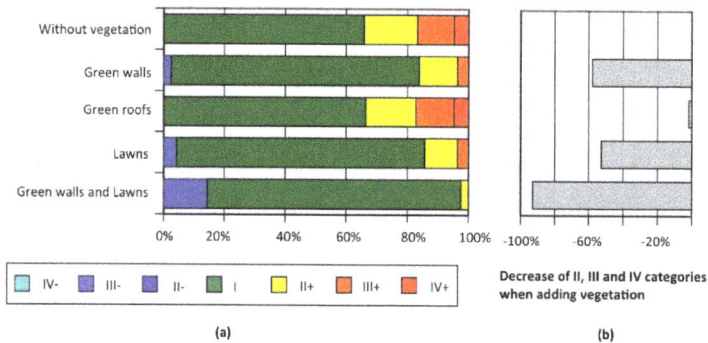

Figure 15. (a) Distribution of occurrences among thermal comfort categories for the third floor; and (b) relative decrease in the occurrences of hot categories compared to the reference case.

Covering all surfaces visible from the studied building with vegetation makes it possible to nearly eliminate all periods of discomfort.

4.4.2. Convective and Radiative Heat Fluxes

When analyzing the thermal balance at the external wall surface, it becomes clear how the various fluxes act upon both the thermal behavior of the studied building and the urban microclimate. The temporal variations of mean flux values (as calculated over the entire external surface of the third floor) on 26 July are shown in Figure 16 for the studied cases, using the same conventions as before.

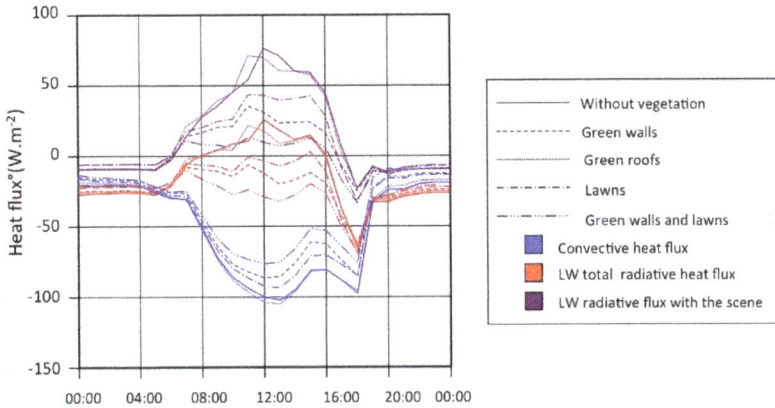

Figure 16. Mean values of convective and radiative heat fluxes over the entire third floor external surface. LW, long wave.

The net radiative LW flux exchange with the other urban surfaces is also provided so as to better highlight the contribution of these specific surfaces. The differences between total LW radiative flux and this latter value yield the exchanges with the sky.

Regardless of the configuration, a change in the fluxes can be noticed every day around 5:00 pm: the LW radiative flux becomes negative, as does the absolute value of the convective heat flux (Figure 16). This outcome is certainly due to the fact that at this moment in time, sun rays are reaching the north facade.

The convective heat flux is the main negative flux (cooling the facades) not only during the day, but at night, as well. LW radiation with the other surfaces makes a slight contribution to cooling the building, yet the effect on indoor comfort is negative overall due to strong contributions during the day. This radiation is partially offset by the LW exchanges with the sky, though the net LW balance remains positive over the course of the day (Figure 17).

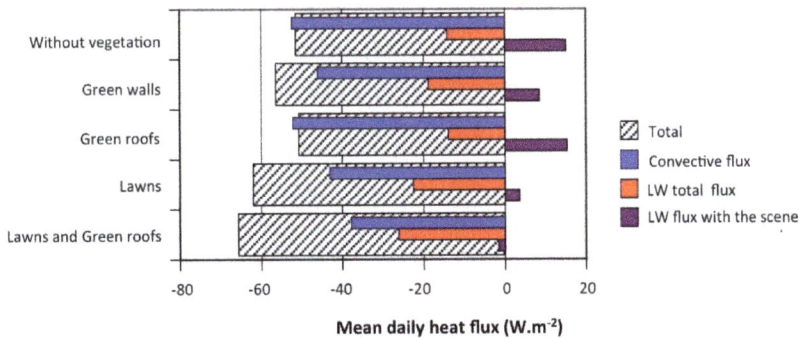

Figure 17. Convective and long-wave (LW) radiative heat fluxes at the external third floor surface, cumulated throughout the day.

Except in the case of green roofs, the effect of green surfaces on heat fluxes is significant. The positive effect of LW radiation is reinforced, ranging from 17 W·m^{-2} to 25 W·m^{-2}. With green facades or lawns, the net LW fluxes are almost always negative (Figures 16 and 17). When implementing green surfaces on all buildings and on the ground, the net LW flux with the environment is negative. The impact of ground vegetation is most significant with respect to the LW radiative flux, as the result of a higher view factor (compared to facade intervisibility) and of the fact that the ground is more exposed to sun rays.

Implementing vegetation on the surfaces leads to an overall decrease in convective heat fluxes, due at first to the decrease in surface temperature and, then, to the temperature differential with the air.

5. Discussion and Conclusions

The various configurations studied herein have demonstrated that introducing vegetation on buildings and on the ground has a positive summer effect on comfort in a non-insulated building, when considering the temperate climate of the city of Nantes. We decomposed the effect of the multiple heat fluxes in order to show how vegetation acts. It has been revealed that the direct effect resulting from the shading of leaves is predominant. The surface temperature of a green surface, due to its evaporation capacity, is much less than that of a mineral surface. This effect has contributed to improving the building's thermal conditions in the summer and reducing LW radiative fluxes between surfaces in a district, which also exerts a significant impact on a building's energy balance.

The results obtained for a non-insulated building were not applicable to an insulated one. They can however be of great interest when considering the proportion of buildings that are impossible to refurbish for structural or other reasons. It has been shown that for such buildings, acting on the surroundings can lead to an improvement in the level of summer comfort.

Green roofs appear to be effective only for the upper floors of the building where they have been installed. Green facades have both direct and indirect impacts, while lawns only provide an indirect effect, which deserves to be considered in the case of a non-greened and non-insulated building.

Even though such was not the stated purpose of this paper, it can still be deduced from our calculations that green surfaces, due to their lower temperature, will also have an improvement effect on outdoor thermal comfort, as a result of their ability to decrease the LW radiative flux.

Many complementary studies could be carried out, for various insulation and glazing levels, as performed by [24], and by taking into account in greater detail the kind of plants (e.g., climbing plants, which prove to be the least expensive green facades). This additional research should be conducted for different climate conditions, as well.

Acknowledgments: This research work was carried out within the scope of the VegDUD Project, funded by the French Research Agency (ANR) under Contract No. ANR-09-VILL-0007. The authors are grateful to the Pays de la Loire Regional Council and ADEME (French Environment and Energy Management Agency) for their financial support of this study, as well as to the Laboratory Energy, Environment & architecture (LEEA) at Haute Ecole du Paysage, d'Ingénierie et d'Architecture de Genève (HEPIA) facility for providing us with experimental data.

Author Contributions: This paper is part of a Ph.D work that has been carried out by Laurent Malys and co-supervised by Marjorie Musy and Christian Inard.

Conflicts of Interest: The authors declare no conflict of interest.

References

1.   Nasir, R.A.; Ahmad, S.S.; Zain-Ahmed, A.; Ibrahim, N. Adapting Human Comfort in an Urban Area: The Role of Tree Shades Towards Urban Regeneration. *Proced. Soc. Behav. Sci.* **2015**, *170*, 369–380. [CrossRef]

2.   De Abreu-Harbich, L.V.; Labaki, L.C.; Matzarakis, A. Effect of tree planting design and tree species on human thermal comfort in the tropics. *Landsc. Urban Plan.* **2015**, *138*, 99–109. [CrossRef]

3.   Hong, B.; Lin, B. Numerical studies of the outdoor wind environment and thermal comfort at pedestrian level in housing blocks with different building layout patterns and trees arrangement. *Renew. Energy* **2015**, *73*, 18–27. [CrossRef]

4. Vailshery, L.S.; Jaganmohan, M.; Nagendra, H. Effect of street trees on microclimate and air pollution in a tropical city. *Urban For. Urban Green.* **2013**, *12*, 408–415. [CrossRef]

5. Shashua-Bar, L.; Tsiros, I.X.; Hoffman, M.E. A modeling study for evaluating passive cooling scenarios in urban streets with trees. Case study: Athens, Greece. *Build. Environ.* **2010**, *45*, 2798–2807. [CrossRef]

6. Santamouris, M. Cooling the cities—A review of reflective and green roof mitigation technologies to fight heat island and improve comfort in urban environments. *Sol. Energy* **2014**, *103*, 682–703. [CrossRef]

7. Chen, H.; Ookab, R.; Huang, H.; Tsuchiyab, T. Study on mitigation measures for outdoor thermal environment on present urban blocks in Tokyo using coupled simulation. *Build. Environ.* **2009**, *44*, 2290–2299. [CrossRef]

8. Smith, K.; Roeber, P. Green roof mitigation potential for a proxy future climate scenario in Chicago, Illinois. *J. Appl. Meteorol. Climatol.* **2011**, *50*, 507–522. [CrossRef]

9. De Munck, C. Modélisation de la végétation urbaine et stratégies d'adaptation pour l'amélioration du confort climatique et de la demande énergétique en ville. Ph. D. Thesis, University de Toulouse, Toulouse, France, 2013. (In French).

10. Ascione, F.; Bianco, N.; de' Rossi, F.; Turni, G.; Vanoli, G.P. Green roofs in European climates. Are effective solutions for the energy savings in air-conditioning? *Appl. Energy* **2013**, *104*, 845–859. [CrossRef]

11. Susorova, I.; Angulo, M.; Bahrami, P.; Stephens, B. A model of vegetated exterior facades for evaluation of wall thermal performance. *Build. Environ.* **2013**, *67*, 1–13. [CrossRef]

12. Gros, A. Modélisation de la demande énergétique des bâtiments à l'échelle d'un quartier. Ph.D. Thesis, University de La Rochelle, La Rochelle, France, 2013. (In French).

13. Chen, Q.; Li, B.; Liu, X. An experimental evaluation of the living wall system in hot and humid climate. *Energy Build.* **2013**, *61*, 298–307. [CrossRef]

14. Wong, N.H.; Tan, A.Y.K.; Tan, P.Y.; Wong, N.C. Energy simulation of vertical greenery systems. *Energy Build.* **2009**, *41*, 1401–1408. [CrossRef]

15. Hoyano, A. Climatological uses of plants for solar control and the effects on the thermal environment of a building. *Energy Build.* **1988**, *11*, 181–199. [CrossRef]

16. Cameron, R.W.F.; Taylor, J.E.; Emmett, M.R. What's "cool" in the world of green facades? How plant choice influences the cooling properties of green walls. *Build. Environ.* **2014**, *73*, 198–207. [CrossRef]

17. Alexandri, E.; Jones, P. Temperature decreases in an urban canyon due to green walls and green roofs in diverse climates. *Build. Environ.* **2008**, *43*, 480–493. [CrossRef]

18. Djedjig, R.; Bozonnet, E.; Belarbi, R. Analysis of thermal effects of vegetated envelopes: Integration of a validated model in a building energy simulation program. *Energy Build.* **2015**, *86*, 93–103. [CrossRef]

19. Armson, D. The Effect of Trees and Grass on the Thermal and Hydrological Performance of an Urban Area. Ph.D. Thesis, University of Manchester, Manchester, UK, 2012.

20. Akbari, H. Shade trees reduce building energy use and CO2 emissions from power plants. *Environ. Pollut.* **2002**, *116*, S119–S126. [CrossRef]

21. Djedjig, R. Impacts des enveloppes végétales à l'interface bâtiment microclimat urbain. Ph.D. Thesis, University de La Rochelle, La Rochelle, France, 2013. (In French).

22. Malys, L.; Musy, M.; Inard, C. Microclimate and building energy consumption: Study of different coupling methods. *Adv. Build. Energy Res.* **2015**, *9*, 151–174. [CrossRef]

23. Musy, M.; Malys, L.; Morille, B.; Inard, C. The use of *SOLENE-microclimat* model to assess adaptation strategies at the district scale. *Urban Clim.* **2015**, *14*, 213–223. [CrossRef]

24. Morille, B.; Musy, M.; Malys, L. Preliminary study of the impact of urban greenery types on energy consumption of building at a district scale: Academic study on a canyon street in Nantes (France) weather conditions. *Energy Build.* **2015**. [CrossRef]

25. Malys, L.; Musy, M.; Inard, C. A hydrothermal model to assess the impact of green walls on urban microclimate and building energy consumption. *Build. Environ.* **2014**, *73*, 187–197. [CrossRef]

26. Bouyer, J.; Inard, C.; Musy, M. Microclimatic coupling as a solution to improve building energy simulation in an urban context. *Energy Build.* **2011**, *43*, 1549–1559. [CrossRef]

27. Bouyer, J. Modélisation et simulation des microclimats urbains—Étude de l'impact de l'aménagement urbain sur les consommations énergétiques des bâtiments. Ph.D. Thesis, Ecole Polytechnique de l'Université de Nantes, Nantes, France, 2009.

28. Musy, M. VegDUD project: Role of vegetation in sustainable urban development. In Proceedings of the 8th International Conference on Urban Climate and 10th Symposium on the Urban Environment, Dublin, Ireland, 6–10 August 2012.

29. Nicol, F.; Humphreys, M.A.; Roaf, S. *Adaptive Thermal Comfort: Principles and Practice*; Routledge: London, UK; New York, NY, USA, 2012.

30. Nicol, J.F.; Humphreys, M.A. Adaptive thermal comfort and sustainable thermal standards for buildings. *Energy Build.* **2002**, *34*, 563–572. [CrossRef]

31. *NF-EN 15251: Critères d'ambiance intérieure pour la conception et l'évaluation de la performance énergétique des bâtiments couvrant la qualité de l'air intérieure, la thermique, l'éclairage et l'acoustique*; AFNOR: La Plaine Saint-Denis, France, 2007. (In French)

32. Mestayer, P.; Rosant, J.-M.; Rodriguez, F.; Rouaud, J.-M. The experimental campaign FluxSAP 2010: Climatological measurements over a heterogeneous urban area. *Int. Assoc. Urban Clim.* **2011**, *40*, 22–30.

4

Phase Change Materials-Assisted Heat Flux Reduction

Hussein J. Akeiber [1,*], Seyed Ehsan Hosseini [1,*], Mazlan A. Wahid [1], Hasanen M. Hussen [2]
and Abdulrahman Th. Mohammad [3]

Academic Editor: Nyuk Hien Wong

[1]   High-Speed Reacting Flow Laboratory, Faculty of Mechanical Engineering, Universiti Teknologi Malaysia,
      81310 UTM Skudai, Johor 81310, Malaysia; mazlan@fkm.utm.my
[2]   Machine and Mechanical Department, University of Technology, Baghdad 35023, Iraq;
      drhasanen_hvac@yahoo.com
[3]   Baqubah Technical Institute, Middle Technical University, Baghdad 06800, Iraq; abd20091976@gmail.com
*   Correspondence: husseinutm@yahoo.com.my (H.J.A.); seyed.ehsan.hosseini@gmail.com (S.E.H.)

Abstract: Phase change materials (PCM) in the construction industry became attractive because of several interesting attributes, such as thermo-physical parameters, open air atmospheric condition usage, cost and the duty structure requirement. Thermal performance optimization of PCMs in terms of proficient storage of a large amount of heat or cold in a finite volume remains a challenging task. Implementation of PCMs in buildings to achieve thermal comfort for a specific climatic condition in Iraq is our main focus. From this standpoint, the present paper reports the experimental and numerical results on the lowering of heat flux inside a residential building using PCM, which is composed of oil (40%) and wax (60%). This PCM (paraffin), being plentiful and cost-effective, is extracted locally from waste petroleum products in Iraq. Experiments are performed with two rooms of identical internal dimensions in the presence and absence of PCM. A two-dimensional numerical transient heat transfer model is developed and solved using the finite difference method. A relatively simple geometry is chosen to initially verify the numerical solution procedure by incorporating in the computer program two-dimensional elliptic flows. It is demonstrated that the heat flux inside the room containing PCM is remarkably lower than the one devoid of PCM.

Keywords: phase change material (PCM); wax; melting temperature; heat flux

1. Introduction

Over the years, it has been established that thermal energy storage systems (TESSs) with phase change materials (PCMs) can efficiently reduce the excessive usage of fossil fuels and subsequent global warming [1,2]. Thermal energy storage (TES) is found to play a vital role in a broad range of industrial and residential applications. The use of TESSs with PCMs in buildings enhances the human comfort by reducing the internal air temperature fluctuation. Consequently, the indoor air temperature remains near the required temperature over an extended time period [3]. Sensible heat TES, which stores heat in fluid or solid form, latent heat TES, which uses latent heat during the phase change process, and thermoelectric devices are the common examples of different types of TES technologies. TESSs can rapidly discharge or store huge amounts of heat with solar energy and heat as alternating sources [4]. Recent research reveals the advancements of extensive building architecture and management based on TES with PCMs. TES is attractive due to its absolute suitability to reduce the gap between energy demand and supply [5].

Several commercial PCMs have been developed with varying melting temperatures. Considering the cost of various PCMs, Rezaei et al. [6] examined their influence on energy and exergy efficiencies. Based on the lumped-parameter method, Li et al. [7] established an analytical temperature model. They calculated the enthalpy difference of inorganic salts derived from two types of composite materials. It is shown that the melting point and the enthalpy difference of the binary eutectics ($LiNO_3$-$NaNO_3$, LiCL-NaCL and Li_2CO_3-Na_2CO_3) are consistent with the ones obtained from standard methods. Meanwhile, many efforts are made to produce high performance PCM integration in building walls. Romero-Sanchez et al. [8] evaluated the thermal performance of PCMs by incorporating in natural stone. Experiments and numerical simulation are carried out to improve the thermal properties of natural stone, where concrete pilot houses are constructed. These pilot houses are covered with trans-ventilated facade designs via Spanish Bateigazul natural stone. An improvement in human comfort with the reduction in energy consumption is evidenced upon implementing PCMs.

Izquierdo-Barrientos et al. [9] inspected the effects of PCMs on external building walls with various configurations by altering the PCMs' layer position, ambient conditions, wall orientation and phase transition temperature. A 1D transient heat transfer numerical model is developed and solved using a finite difference method. Results revealed no significant reduction in the total heat lost during winter irrespective of the variation of the PCM wall orientation or the transition temperature. Moreover, during the summer time, a significant difference in the heat gain is evidenced, which is attributed to the elevated solar radiation fluxes.

The thermal performances of a PCM based co-polymer composite wallboard were experimentally assessed by Kuznik and Virgone [10], where two identical enclosures, called Test Cells 1 and 2, were constructed. The volume of each test cell is (3.10 m × 3.10 m × 2.50 m) and is bounded on five sides by a fixed temperature regulated air mass. The sixth side is a glazed face that separates the test cell from a climate compartment. The air temperature in the room containing PCM is found to decrease up to 4.2 °C without any thermal stratification as compared to the room without PCM composite.

Employing PCM in a representative Mediterranean building, De Gracia et al. [11] evaluated its environmental impact in terms of warming. Three scenarios, such as different temperature control systems, different PCM types or different weather conditions, are emphasized based on the life cycle assessment (LCA) process. It is shown that the presence of PCM in the building envelope decreased the energy consumption without considerable reduction of the global impact. The LCA for the real rooms exhibited an impact reduction of 37% upon incorporating polyurethane (PU) into the reference room (REF). Kuznik et al. [12] optimized the PCM wallboard thickness in lightweight buildings with reduced air temperature fluctuations inside the room, where the in-house numerical code CODYMUR is used to calculate the optimal thickness.

In another experiment, Navarro et al. [13] evaluated the PCM performance in terms of internal thermal gains. Three different rooms with the same internal dimensions (2.4 m × 2.4 m × 2.4 m) are considered. These rooms are labelled as (1) the REF (built using traditional two-layered brick with an air gap and without insulation), (2) the PU (constructed by a traditional brick with of spray foam thickness of 5 cm (walls) and 3 cm (roof)) and (3) the PCM (made with a PCM layer in the southern and western walls and on the roof). It is found that during the summer season the PCM room stored the heat produced by the internal loads, thus limiting the heat dissipation to the outer environment. The REF is found to possess higher temperature fluctuations in its envelope (27.5–24 °C) than other rooms with insulation (28–26 °C).

Pasupathy and Velraj [14] analyzed (theoretically and experimentally) the thermal performance of an inorganic eutectic PCM-based thermal storage system (TSS) for energy conservation in buildings. In the design, one room contained PCM on the roof, and the other room is devoid of the PCM panel. The inner walls, except the ceiling of the rooms, are insulated by a 6 mm-thick plywood on all sides to determine the sole influence of the PCM panel on the roof. The PCM panel is made of (2 m × 2 m) stainless steel having a thickness of 2.54 cm. The stainless steel accommodated an inorganic salt hydrate (48% $CaCl_2$ + 4.3% NaCl + 0.4% KCl + 47.3% H_2O) as the PCM. The measured room temperatures are observed to vary ~27 ± 3 °C during the experiment. Despite many dedicated efforts, a comprehensive understanding of the PCM-mediated reduction in heat flux inside the building is far from being achieved.

In this paper, experimental and numerical investigations were performed using PCMs in building architecture to determine their impact on lowering the heat flux inside the building. Two identical rooms, one without and the other with PCM, were considered for experiments. Numerical simulation was done based on the transient heat transfer model. The heat flux inside the room is determined to assess the thermal performance of such PCMs. Results are discussed, analyzed, compared and validated.

2. Numerical Scheme

The boundary condition on the inner surface of the aluminum frame follows natural convection. Most of the previous researchers considered the bottom wall as insulated, because the temperature difference between the room and the wall was very small. The heat transfer coefficient (h) inside the room is calculated (FORTRAN programming). Figures 1 and 2 display the schematics for the numerical model formulation, which assumes the following:

(i)   One dimensional heat conduction in the composite wall is considered, and the end impacts are not taken into account.

(ii)   The thermal conductivity of the aluminum frame and the roof top slab are constant irrespective of temperature variation.

(iii)   The PCM is uniform and isotropic.

(iv)   The convection impact in the molten PCM is not considered.

(v)   The interfacial resistances are negligible.

(vi)   The value of C_p for the PCM panel is considered as follows:

$$T < T_m - \Delta T, \ C_p = C_{ps} \tag{1}$$

$$T > T_m + \Delta T, \ C_p = C_{pl} \tag{2}$$

$$T_m - \Delta T < T < T_m + \Delta T, \ C_p = h_{sl}/2\Delta T \tag{3}$$

where C_p is the specific heat capacity, h_{sl} is the enthalpy change of solid-liquid, ΔT is half of the temperature range over which the phase change occurs and T_m is the phase transition temperature.

(vii)   The latent heat being highly sensitive to the phase transition process of the PCM is modeled over a range of temperatures, where C_p is considered to be uniform during the phase conversion. Although, in reality, C_p varies with temperature.

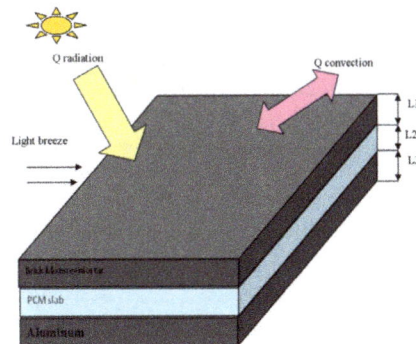

Figure 1. Schematic diagram of phase change material (PCM) incorporated ceiling.

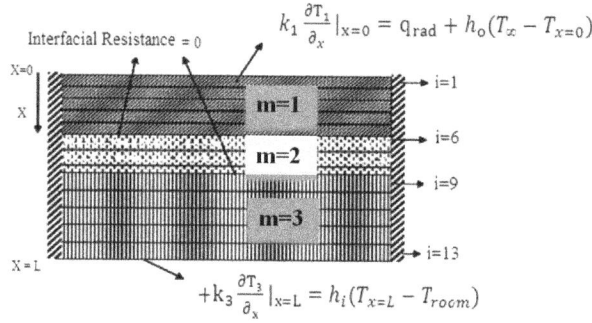

Figure 2. Finite volume grid for the analysis.

Following these assumptions, the governing equation and the boundary conditions are written as:

$$k_m \frac{\partial^2 T_m}{\partial x^2} = \frac{\rho_m C_{pm} \partial T_m}{\partial t} \quad \{0 < x < L\}; m = 1, 2, 3 \tag{4}$$

where $m = 1$ for the roof top slab, $m = 2$ for the PCM panel and $m = 3$ for the bottom aluminum frame. The same equation holds for all three material regions and takes different values of k, ρ and C_p.

In the outer walls ($x = 0$), where the floor is exposed to solar radiation, the boundary condition is expressed as:

$$k_1 \frac{\partial T_1}{\partial x}\Big|_{x=0} = q_{rad} + h_o \left(T_\infty - T_{x=0}\right) \tag{5}$$

The radiation effect is considered only during the sunshine hours. The boundary condition ($x = L$) in the bottom layer of the aluminum frame is:

$$k_3 \frac{\partial T_3}{\partial x}\Big|_{x=L} = h_i \left(T_{x=L} - T_{room}\right) \tag{6}$$

The instantaneous continuity of heat flux and temperature at the interfaces $x = L_1$ and L_2 is preserved.

The equation for the top volume cell is written as:

$$\left(\frac{\rho_1 c_1 \Delta x_1}{\Delta t} + \frac{f k_1}{\delta x_1} + h_o f\right) T_1 - \frac{f k_1}{\delta x_1} T_2$$
$$= h_o f T_\infty + (1-f)\left[\frac{k_1 (T_2 - T_1)}{\delta x_1} - h_o (T_1 - T_\infty)\right] + \frac{\rho_1 c_1 T_1^0}{\Delta t} \Delta x_1 + \alpha q_s \tag{7}$$
$$+ \sigma \left[\alpha T_{sky}^4 - \epsilon T_s^4\right]$$

The equation for the volume cells located in between the top and bottom volume cells yields:

$$-\frac{f k_m}{\Delta x_m} T_{i+1} + \left[\frac{\rho_m c_m \Delta x_m}{\Delta t} + \frac{f k_m}{\Delta x_m} + \frac{f k_m}{\Delta x_m}\right] T_i - \frac{f k_m}{\Delta x_m} T_i$$
$$= (1-f)\left[\frac{k_m (T_{i+1} - T_{-i-1})}{\delta x_m} - \frac{k_m (T_i - T_{i-1})}{\delta x_m}\right] + \frac{\rho_m c_m T_i^0 \Delta x_m}{\Delta t} \tag{8}$$

The above-mentioned discretized equations are applicable to volume cells for 2–4, 7 and for 10–12 of the roof top slab, PCM panel and aluminum frame, respectively, with: $m = 1$, $i = 2, 3, 4$; $m = 2$, $i = 7$; and $m = 3$, $i = 10, 11, 12$.

The equation for the interface volume cell 5 is written as:

$$-\frac{fk_1}{\delta x_1}T_4 + \left[\frac{\rho_1 c_1 \Delta x_1}{\Delta t} + \frac{f}{\delta x_1/2k_1 + \delta x_2/2k_2} + \frac{fk_1}{\delta x_1}\right]T_5 - \left[\frac{f}{\frac{\Delta x_1}{2k_i} + \frac{\Delta x_2}{2k_2}}\right]T_4$$
$$= (1-f)\left[\frac{k_1(T_6 - T_5)}{\delta x_2} - \frac{k_1(T_5 - T_4)}{\delta x_1}\right] + \frac{\rho_1 c_1 T_5^0 \Delta x_1}{\Delta t} \tag{9}$$

where Δx_1 and Δx_2 are the cell thickness of the roof top slab and PCM panel, respectively. Similarly, the equation can be written for volume cell 6. A similar process is extended to control Volumes 8 and 9, which involve cell thicknesses Δx_2 and Δx_3 corresponding to the PCM panel and the bottom aluminum frame, respectively.

The equation for the bottom volume cell 13 is given by:

$$\frac{-fK_3}{\delta x_3}T_{12} + \left[\frac{\rho_3 c_3 \Delta x_3}{\Delta t} + \frac{fK_3}{\delta x_3}\right]T_{13}$$
$$= f[h_i(-2)] + (1-f)\left[2h_i - k\frac{T_{13} - T_{12}}{\delta x_3}\right] + \frac{\rho_3 c_3 T_{13}^0 \Delta x_3}{\Delta t} \tag{10}$$

3. Development of the Numerical Model

3.1. One-Phase Solution

For either completely solid or liquid PCM layers, the temperature distribution on the n-th layer of the wall follows the one-dimensional diffusion equation given by:

$$K_n \frac{\partial^2 T}{\partial x^2} = \rho_n c_n \frac{\partial T}{\partial t} \tag{11}$$

with the boundary condition:

$$T(0,t) = T_o(t) \tag{12}$$

$$T(L,t) = T_l(t) \tag{13}$$

The numerical solution for a single material node yields:

$$T_i^{j+1} = T_i^j + \frac{Kn\Delta t}{\rho_n c_n (\Delta X)^2}\left(T_{i+1}^j - 2T_i^j + T_{i-1}^j\right) \tag{14}$$

The finite difference equation for the inner interface is written as:

$$T_i^{j+1} = T_i^j + \frac{2\Delta t\left[K_{ins}\left(T_{i-1}^j - T_i^j\right) + K_{pcm}\left(T_{i+1}^j - T_i^j\right)\right]}{\left(\rho_{pcm}C_{pcm} + \rho_{ins}C_{ins}\right)(\Delta X)^2} \tag{15}$$

The finite difference equation for the outer surface is written as:

$$T_i^{j+1} = T_i^j + \frac{2\Delta t\left[K_{pcm}\left(T_{i-1}^j - T_i^j\right) + K_{ins}\left(T_{i+1}^j - T_i^j\right)\right]}{\left(\rho_{pcm}C_{pcm} + \rho_{ins}C_{ins}\right)(\Delta X)^2} \tag{16}$$

Depending on whether the PCM is completely solid or completely liquid upon entering the one-phase subroutine, this solution is run until the maximum nodal temperature within the PCM layer increases above the melting point or until the minimum temperature within the material drops below the freezing point. The temperature distribution at this particular time instant becomes the initial condition for the two-phase subroutine.

3.2. Two-Phase Solution

The boundary points between the solid and liquid are held constant at the melting temperature of the material when the PCM layer consists of more than one phase and is expressed as:

$$T = T_\mathrm{m} \tag{17}$$

The movement of the solid-liquid boundary is given by:

$$K_{\mathrm{pcm}}[\lim_{x \to xsl+} \left(\frac{\partial T}{\partial x}\right) - \lim_{x \to xsl-} \left(\frac{\partial T}{\partial x}\right)] = \frac{\rho_{\mathrm{pcm}} \alpha d_{\mathrm{xsl}}}{dt} \tag{18}$$

where $xsl+$ and $xsl-$ denote the limits approaching the solid-liquid interface from the right and left, respectively.

An explicit numerical solution for a single material two-phase node is given by:

$$T_i^{j+1} = T_i^{j} = T_m \tag{19}$$

$$\lambda_i^{j+1} = \lambda_i^{j} + \frac{K_{\mathrm{pcm}} \Delta t}{\rho_{\mathrm{pcm}} \alpha (\Delta x)^2} (T_{i+1}^{j} - 2T_m + T_{i-1}^{j}) \tag{20}$$

where λ is the volume fraction of the i-th node, which is melted at the j-th time step. For the two PCM-insulation interface points, the finite difference equation is modified to consider that the node is half insulation and half PCM. For these two nodes, $\lambda = 1/2 = \lambda_{\max}$ indicates the fully-melted state.

For the inner interface node, the two-phase finite difference equation takes the form:

$$\lambda_i^{j+1} = \lambda_i^{j} + \frac{\Delta t \left[K_{\mathrm{ins}}(T_{i+1}^{j} - T_\mathrm{m}) - K_{\mathrm{pcm}}(T_{i+1}^{j} - T_\mathrm{m}) \right]}{\rho_{\mathrm{pcm}} \alpha (\Delta x)^2} \tag{21}$$

For the outer interface node, the two-phase finite difference equation yields:

$$\lambda_i^{j+1} = \lambda_i^{j} + \frac{\Delta t \left[K_{\mathrm{pcm}}(T_{i+1}^{j} - T_\mathrm{m}) - K_{\mathrm{ins}}(T_{i+1}^{j} - T_\mathrm{m}) \right]}{\rho_{\mathrm{pcm}} \alpha (\Delta x)^2} \tag{22}$$

Two-phase subroutine prepares a kind of switching between Equations (13)–(15) and Equations (15)–(18), based on the nodal state. Figure 3 shows a typical temperature for exiting the one-phase solid solution. In this case, the material is completely solid ($l = 0$) for all nodes with the outermost PCM node at some temperature slightly above T_m.

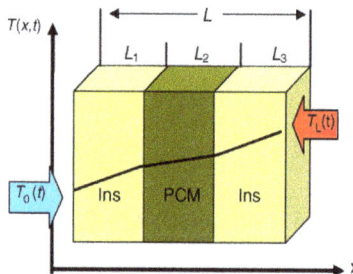

Figure 3. Typical temperature distribution for exiting the one-phase solid sub-routine.

The temperature of this node is shifted back to the PCM melting temperature and kept constant. Equation (18) is used to determine the fraction of the melted node. This is continued until l returns to zero, meaning the complete re-freezing of node or $\lambda = \lambda_{max}$, implying the complete melting of nodes. In such an instance, the temperature is again allowed to alter according to Equation (12). For the nodes that are below T_m, the temperatures are governed by the one-phase equations until they exceed T_m. At this time, the node is in two phases, and the method of calculation is toggled to Equations (15)–(18).

4. Experimental Scheme

Experiments are set up using two full-scale rooms. The test room with PCM is constructed to determine its effect on the roof and wall for thermal management of a residential building. The thermal performance of this organic eutectic PCM having a melting temperature within 40–44 °C is computed. The following criteria are satisfied for precise measurements:

(1) The PCM heat transfer is one-dimensional.
(2) The heat flow from uncontrolled outside influences is negligible compared to the applied heat.
(3) The rate of the temperature rise and fall of the PCM is comparable to reality.

The general properties of the PCM are listed in Table 1.

Table 1. PCM properties.

Properties	Value	Unit
T_m	40–44	°C
C_{ps}	2.21	kJ/kg·°C
C_{pL}	2.3	kJ/kg·°C
K_s	0.51	W/m·K
K_L	0.22	W/m·K
ρ_S	830	kg/m^3
ρ_L	878	kg/m^3
α_L	9.59×10^{-8}	m^2/s
α_S	7.92×10^{-8}	m^2/s
H	146	kJ/kg

4.1. Full-Scale Test Rooms

Two identical full-scale outdoor test rooms with dimension (3 m × 2.5 m × 2 m) as shown in Figure 4 are built at the University Technology site of Iraq. The thermal performance (internal thermal gains) of these two test rooms (PCM incorporated wall and roof) is analyzed to determine their impact under the Iraqi climate. The wall and roof of one room is covered with PCM and the other without any PCM. The basic structures (windows and supports) of both rooms are identical and made of a metallic insulated door positioned on the north wall. The thermal behavior of these rooms is compared to obtain the performance and effect of PCM. The detailed structure of these rooms is as follows:

(1) The REF is a test room without PCM. The walls (without insulation) are built using a commune brick system, cemented plaster and gypsum board. Conversely, the roof contains two layers, where the bottom slab (thickness 12 cm) is made from concrete and the top slab (thickness 10 cm) is made using a brick mixture plus mortar (Figure 4).
(2) Room 1: The structure is the same as the REF together with the incorporation of a PCM layer in the southern, eastern and western walls, as well as in the roof. The PCM (paraffin) layer is 2.5 cm thick and contained in aluminum panels (Figure 4).

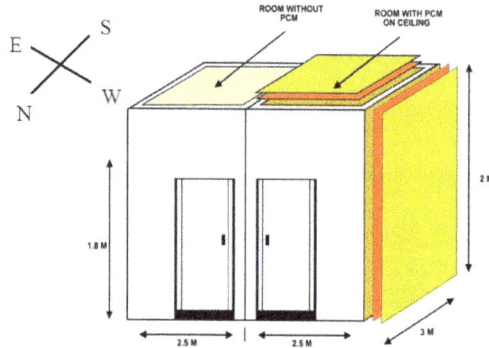

Figure 4. Geometry of the constructed test rooms, reference room (REF) (**left**, without PCM) and PCM2 (**right**).

5. Results and Discussion

Figure 5 illustrates the average variation of ambient temperature and the temperature inside the test rooms with and without PCM at a 1.5 m height in the month of August. As shown in the figure, the temperature of the room without PCM is found to increase at $\tau = 12$ h, reaches the maximum (35 °C) between 16 h and 17 h and then decreases. This is due to the fact that initially (up to 12 h), the heat is absorbed by the room walls and roof exposed to sun and then slowly released, which caused a greenhouse-like effect of an inner increase of the temperature, as expected. Needless to say, the sun sign (solar irradiance) reaches the maximum at 12 h and remains there for a few hours before gradually dropping in later in the afternoon (beyond 17 h). On the other hand, the implementation of the PCM in the building structures reduced the room temperature peak load by 5 °C for the same time period (between 16 h and 17 h). This lowering in the peak temperature inside the room and the creation of thermal comfort is attributed to the effect of the heat storage capacity of the PCM on the temperature variation. Actually, during the first few hours of the day (up to 12 h), paraffin absorbed the latent heat from the room environment and then underwent phase transformation, thereby reducing the peak temperature. However, with the decrease of the sun sign (beyond 17 h), the latent heat is slowly released into the atmosphere. The temperature fluctuation in this case is much weaker than the one without PCM incorporated in the building. It is evident that the greenhouse effect was much lowered in the presence of PCM. This verifies the environmental friendliness of PCM when used in building management and architecture.

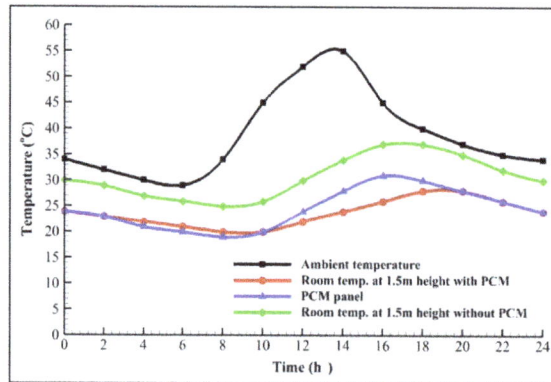

Figure 5. Temperature variation of the test rooms with and without PCM (REF) at a 1.5 m height in the month of August.

Figure 6 depicts the average variation of ambient temperature and the temperature inside the test rooms with and without PCM at a 1.5 m height in the month of January. The temperature of the room without PCM started to increase at 10 h and reached its peak load at 20 h. The temperature of the roof top reaches the maximum between 12 h and 16 h before dropping. The higher value of the roof top temperature compared to the ambient and inside ones clearly indicates the role of both radiation and convection throughout the day, as long as the sun signs; while the temperature inside the room with PCM in the winter month of January is stable all day under ambient temperature. This stability is due to the effect of using the PCM, which acts as a heat storage capacitor and diminishes the rapid temperature fluctuation via phase transformation. During phase conversion in paraffin, a huge amount of latent heat exchange occurs at a constant temperature. This elevated absorption of heat by the PCM materials is indeed responsible for the reduction of the heat flux inside the room and the maintenance of thermal comfort. Specifically, the use of PCM in the building walls and roof increased their thermal resistance and thereby reduced the overall heat transfer through the walls with much lowered heat load compared to the building without PCM.

The temperature variation for the PCM and non-PCM walls is theoretically examined. The east, south and west walls are used to calculate the heat flux through the external and internal walls' surface and the amount of heat storage in the walls. In Figures 7–9 the external heat flux, heat storage and internal heat flux on the east wall of both types of rooms have been compared. Figure 9 (heat flux through the external wall) clearly reveals that the heat flux for the PCM incorporated room is slightly higher than the one without PCM on the walls. This observation is attributed to the low thermal conductivity of the liquid PCM, which reduced the heat transmission to the room and thereby increased the wall top surface temperature. The maximum value of heat flux through the PCM including and excluding the external wall at 9 h is found to be 312 W/m^2 and 162 W/m^2, respectively. Conversely, the heat flux (at 21.5 h) through the internal east wall of the room with and without PCM is discerned to be 26 W/m^2 and 44 W/m^2, respectively. Furthermore, heat storage is observed to be maximum at 12 h with values of 240 W/m^2 and 135 W/m^2 for the PCM integrated and non-integrated east walls, respectively.

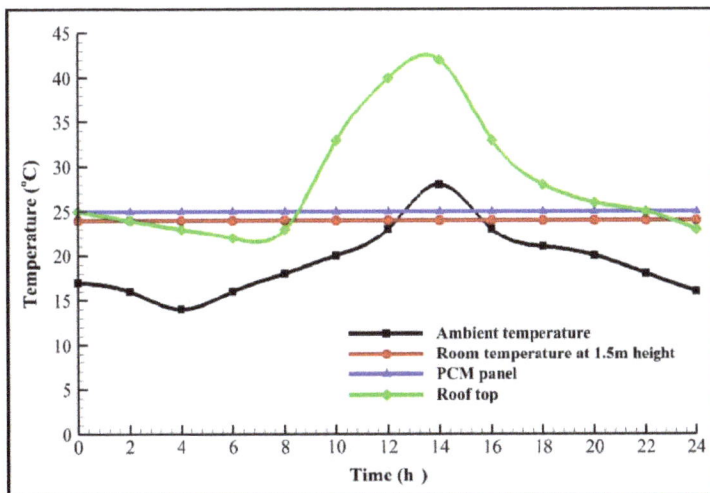

Figure 6. Temperature variation of the test room with and without PCM (REF) at a height of 1.5 m in the month of January.

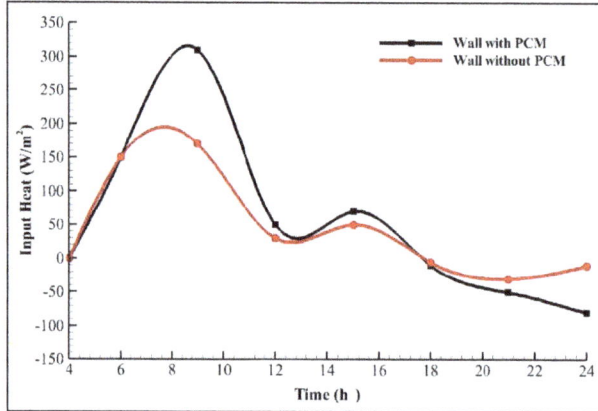

Figure 7. Heat flux through the external surface of the non-PCM- and PCM-treated east walls.

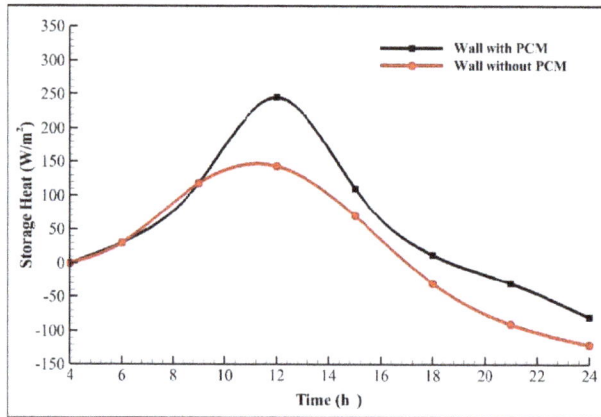

Figure 8. Amount of heat storage in the non-PCM- and PCM-treated east walls.

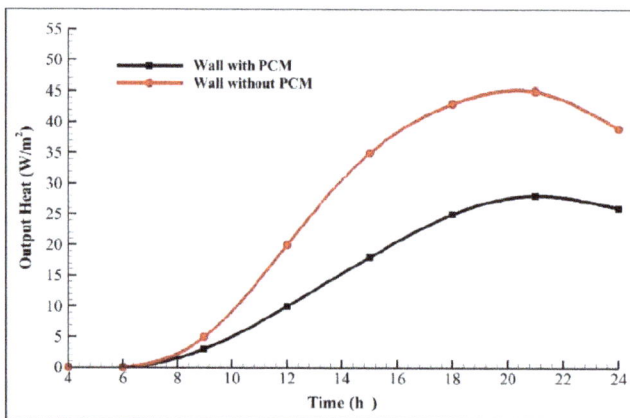

Figure 9. Heat flux through the internal surface of the non-PCM- and PCM-treated east walls.

The variation of heat flux and storage for the south and west walls for the rooms with and without the incorporation of PCM is demonstrated in Figures 10–15. These variations in heat and storage profiles are similar to those of the east wall.

The value of thermal flow and heat storage through a selected wall is computed using finite volume method (FVM), where the thermal flow (through the wall for both the internal and external surface) is calculated. A fixed value of the internal and external heat transfer coefficient is used, which is 7.73 W/m^2· °C and 23.3 W/m^2· °C for the internal and external wall surface, respectively. The amount of stored heat in the walls containing PCMs is found to be larger than the traditional (REF) walls. This elevated storage in PCM walls is ascribed to the high heat capacity and heat-retaining susceptibility without connecting any storage space to the air conditioner. The observed higher thermal storage for the western walls is related to the longest exposure to the solar radiation.

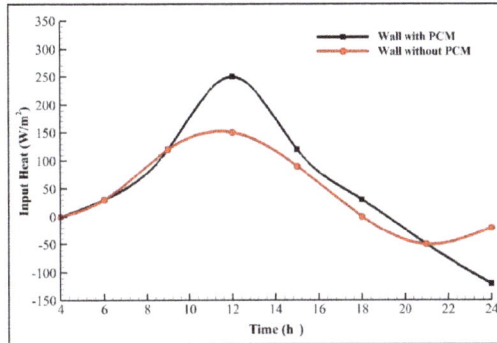

Figure 10. Heat flux through the external surface of the non-PCM- and PCM-treated south walls.

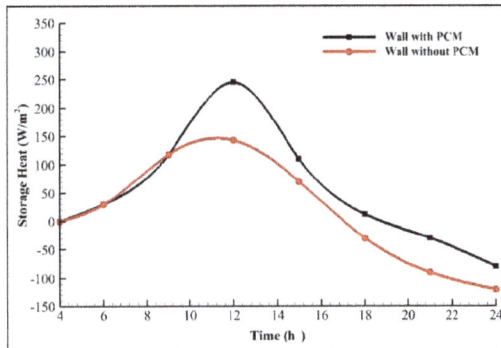

Figure 11. Amount of heat storage of the non-PCM- and PCM-treated south walls.

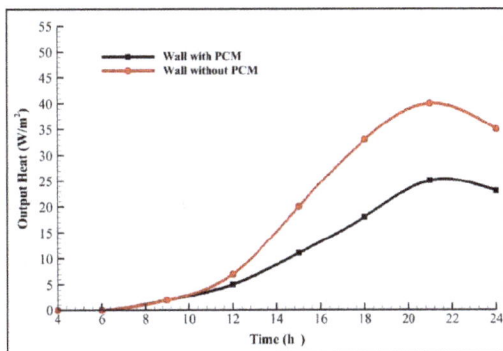

Figure 12. Heat flux through the internal surface of the non-PCM- and PCM-treated south walls.

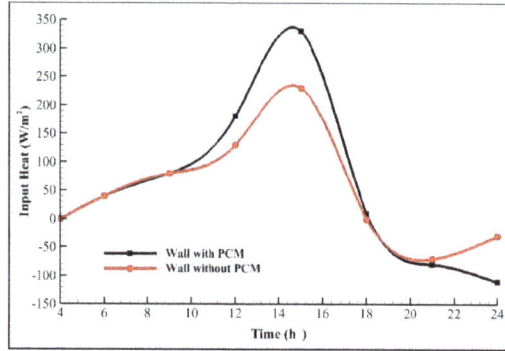

Figure 13. Heat flux through the external surface of the non-PCM- and PCM-treated west walls.

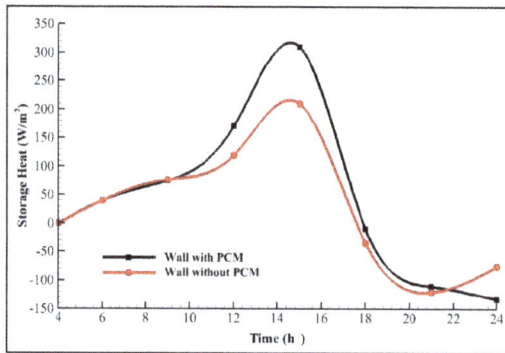

Figure 14. Amount of heat storage in the non-PCM- and PCM-treated west walls.

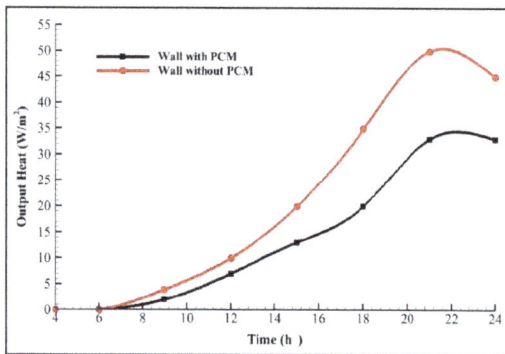

Figure 15. Heat flux through the internal surface of the non-PCM- and PCM-treated west walls.

The simulated temperature contours for REF and PCM2 at different times over the day are illustrated in Figures 16–19. It is evident that the room with PCM integration in the roof achieves a superior temperature distribution throughout the day than the one without PCM. However, with the increase of sun sign, the temperature fluctuation became more prominent (at 12 h) for the REF room, as shown in Figure 20a. The temperature fluctuation kept on increasing at 16 h and 18 h, as depicted in Figures 21a and 22a.

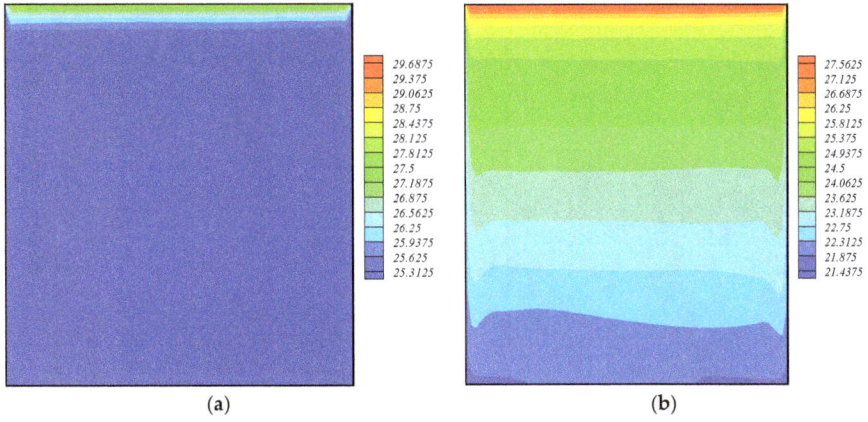

Figure 16. Temperature counters of the roof at 8 h for the room: (**a**) without and (**b**) with PCM.

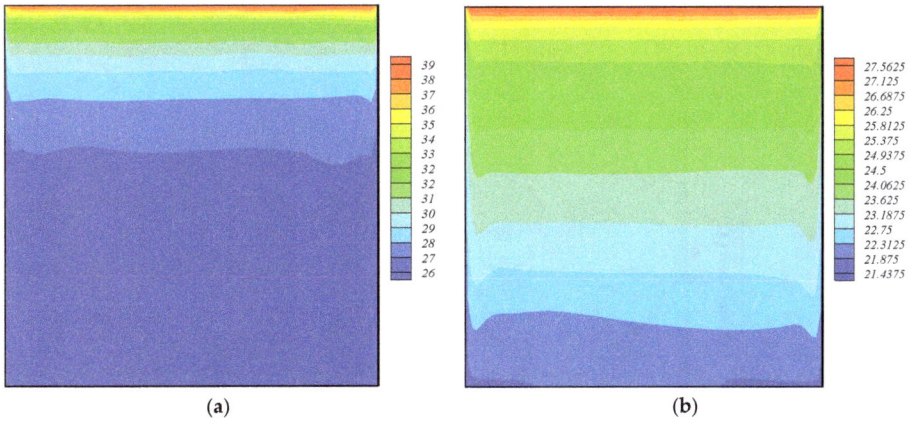

Figure 17. Temperature counters of the roof at 12 h for the room: (**a**) without and (**b**) with PCM.

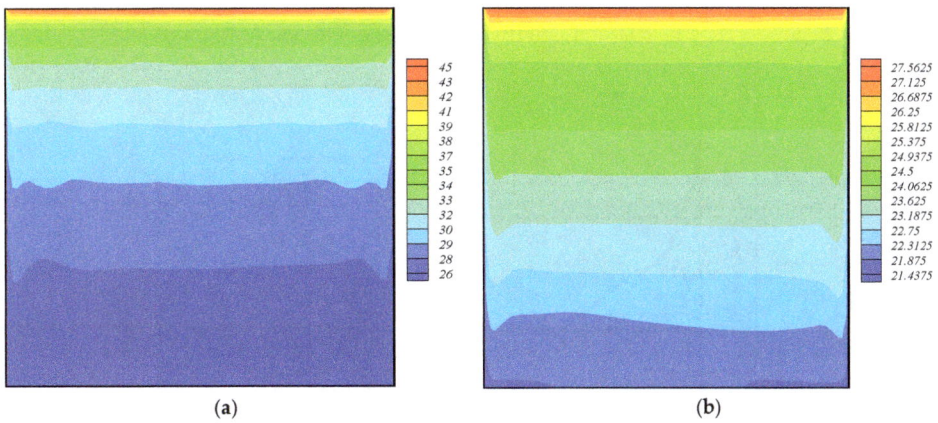

Figure 18. Temperature counters of the roof at 16 h for the room: (**a**) without and (**b**) with PCM.

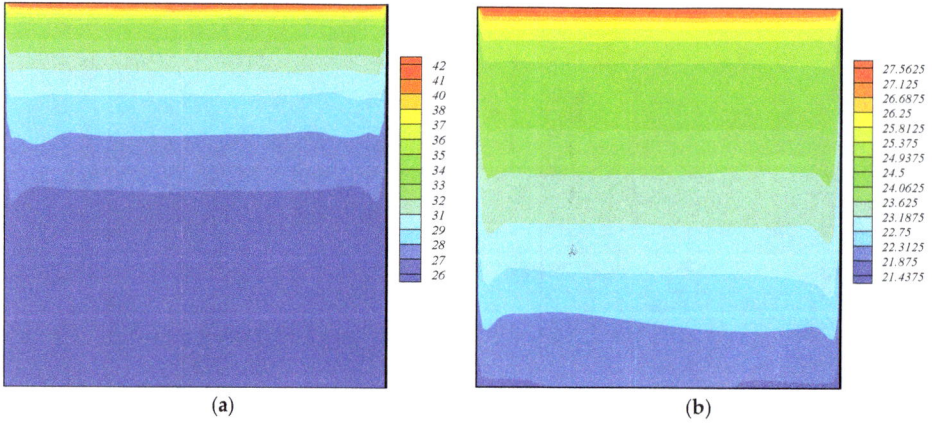

Figure 19. Temperature counters of the roof at 18 h for the room (**a**) without and (**b**) with PCM.

Figures 20 and 21 compare the simulation and experimental results of the roof temperature for the rooms with and without PCM, respectively.

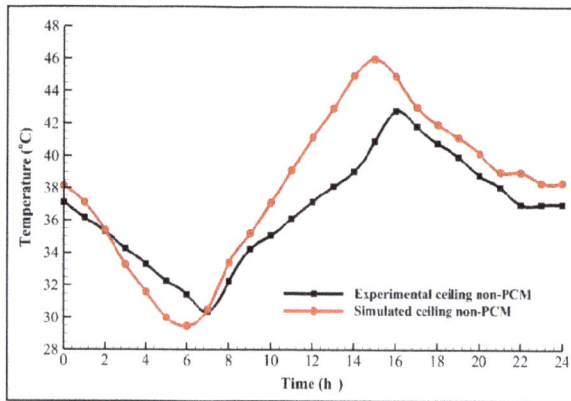

Figure 20. Experimental and simulated temperature of the PCM integrated roof.

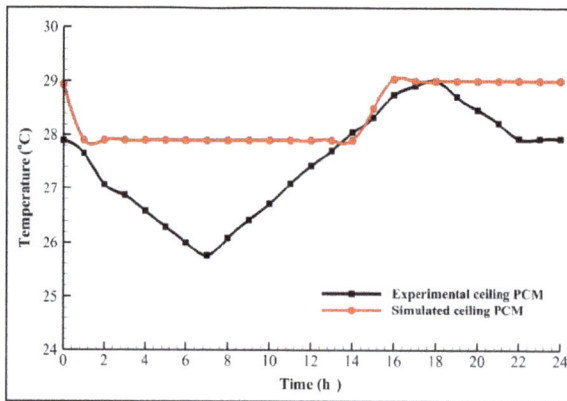

Figure 21. Experimental and simulated temperature of the REF roof.

It is evident that the ceiling (concrete) temperature of the room containing PCM maintained the temperature constant (27 °C) throughout the day as compared to that of the conventional room. This demonstrates that the environment insignificantly affects the inner surface of the ceiling, because all of the heat energy is absorbed by the PCM installed in the roof. Conversely, a considerable temperature fluctuation is observed in the ceiling of the REF (without PCM), because the outside environment immediately influenced its ceiling. Furthermore, the experimental results for PCM2 revealed a small reduction in ceiling temperature during the day time and slight augmentation during the night time. This diminished temperature fluctuation of PCM2 arose from the large heat storage capacity of the PCM. The occurrence of the observed temperature differences between the simulation and experimental results is ascribed to the following reasons:

(1) The room ceiling is influenced by interior condition, where an actual temperature variation occurred.
(2) The effective thermal conductivity of the PCM in the experiment is higher due to the presence of uniformly-distributed high conductivity heat exchanger material in the PCM panel.
(3) The actual phase change may not occur during the phase change temperature as prescribed in the theory.

Figure 22 shows the simulated heat flux entering the room. It is clear that the PCM incorporated roof is better than the one without PCM. The implementation of PCM in the building structure remarkably reduced (more than two-thirds) the heat entry compared to the room without PCM integration. Moreover, the presence of PCM reduced the heat transfer by 46.71% which is directly proportional to reduction in the electricity consumption to maintain the room at 25 °C. Hence, the incorporation of PCM in the building architecture of Iraq is recommended because of thermal comfort, cost-effectiveness and environmental friendliness.

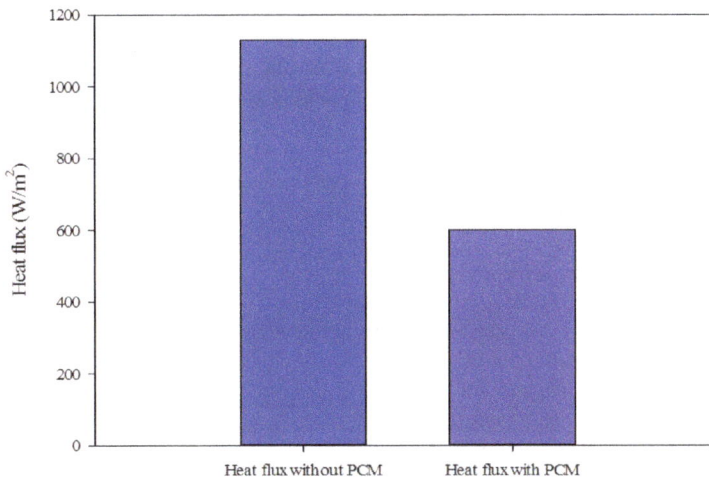

Figure 22. The simulated heat flux entering the room.

6. Conclusions

Thermal management of locally-extracted PCM (paraffin with 40% oil and 60% wax) incorporated building structures in the specific climatic condition of Iraq is reported. Experiments and numerical investigations are made to examine the heat flux reduction inside a residential building using such PCMs. These inexpensive PCMs are obtained from waste petroleum products in Iraq. Two rooms of identical internal dimensions are built, one with PCM and the other without PCM in the roof and walls. The thermal performances of these full-scale test rooms are evaluated. The experimental results are complemented using two-dimensional numerical transient heat transfer (laminar and turbulent flows) model simulation, where the finite difference method is used to solve the discretized equations. The computations are conducted to obtain the solution of heat transfer in a square cavity with differentially-heated side walls. Simulation and experimental results revealed a good agreement. The heat flux inside the PCM integrated room is demonstrated to be considerably lower than the one without PCM. The admirable features of the results suggest that our systematic theoretical and simulation studies may contribute towards the development of thermal management of PCM integrated buildings in Iraq.

Acknowledgments: The authors would like to thank the Ministry of Science, Technology and Innovation (MOSTI) and the Universiti Teknologi Malaysia for supporting this research activity under a science grant under Grant Research No. R.J130000.7924.4S080. The authors are grateful to the staff of the Department of Mechanical Engineering, University of Technology-Iraq, for their help. Hussein is grateful to the government of Iraq for providing the financial support and study leave to complete the doctoral degree.

Author Contributions: The contributions of each author are as follows: Hussein J. Akeiber and Seyed Ehsan Hosseini provided the impetus for this work and analyzed the experimental and numerical results and drafted the manuscript. Mazlan A. Wahid, Hasanen M. Hussen and Abdulrahman Th. Mohammad provided insights that led to highlighting some of the distinctions between equations and worked on rewrites and clarifications. All authors have read and approved the final manuscript.

Conflicts of Interest: The authors declare no conflict of interest.

Nomenclatures

C_1, C_3	Specific heat of roof top slab and concrete slab (kJ/kg·K)
C_{pl}	Specific heat of liquid PCM (kJ/kg·K)
C_{ps}	Specific heat of solid PCM (kJ/kg·K)
f	Implicit factor
Gr_L	Grashof number
h_i	Inside heat transfer coefficient (W/m²·K)
h_o	Outside heat transfer coefficient (W/m²·K)
k_1, k_2, k_3	Thermal conductivity of roof top slab, PCM panel and bottom concrete slab (W/m·K)
L_1, L_2, L_3	Thickness of roof top slab, PCM panel and bottom concrete slab (m)
Nu_L	Nusselt number
P_r	Prandtl number
q_{rad}	Radiation flux (W/m²)
Re	Reynolds number
T	Temperature
T_∞	Ambient temperature
T_i^0	Previous time step temperature at i-th volume cell
T_i	Current time step temperature at i-th volume cell
T_{in}	Initial temperature
T_{room}	Room temperature
T_s	Surface temperature
T_{sky}	Sky temperature
α	Absorptivity
ε	Emissivity
h_{sl}	Solid-liquid enthalpy change (kJ/kg)
σ	Stefan-Boltzmann constant
ρ_1, ρ_2, ρ_3	Density of roof top slab, PCM panel and bottom concrete slab (kg/m³)
Δt	Time step (s)
$\delta x_1, \delta x_2, \delta x_3$	Nodal distances (m)
$\Delta x_1, \Delta x_2, \Delta x_3$	Control volume length of roof top slab, PCM panel and bottom concrete slab (m)

References

1. Tenorio, J.A.; Sánchez-Ramos, J.; Ruiz-Pardo, Á.; Álvarez, S.; Cabeza, L.F. Energy Efficiency Indicators for Assessing Construction Systems Storing Renewable Energy: Application to Phase Change Material-Bearing Façades. *Energies* **2015**, *8*, 8630–8649. [CrossRef]

2. Lo Brano, V.; Ciulla, G.; Piacentino, A.; Cardona, F. On the efficacy of PCM to shave peak temperature of crystalline photovoltaic panels: An FDM model and field validation. *Energies* **2013**, *6*, 6188–6210. [CrossRef]

3. Seong, Y.B.; Lim, J.H. Energy Saving Potentials of Phase Change Materials Applied to Lightweight Building Envelopes. *Energies* **2013**, *6*, 5219–5230. [CrossRef]

4. Normura, T.; Tsubota, M.; Oya, T.; Okinka, N.; Akiyama, T. Heat storage in direct-contact heat exchanger with phase change material. *Appl. Therm. Eng.* **2013**, *50*, 26–34. [CrossRef]

5. Oro, E.; De Gracia, A.; Castell, A.; Farid, M.M.; Cabeza, L.F. Review on phase change materials (PCMs) for cold thermal energy storage applications. *Appl. Energy* **2012**, *99*, 513–533. [CrossRef]

6. Rezaei, M.; Anisur, M.R.; Mahfuz, M.H.; Kibria, M.A.; Saidur, R.; Metselaar, I.H.S.C. Performance and cost analysis of phase change materials with different melting temperatures in heating system. *Energy* **2013**, *5*, 173–178. [CrossRef]

7. Li, Y.; Zhang, Y.; Li, M.; Zhang, M. Testing method of phase change temperature and heat of inorganic high temperature phase change materials. *Exp. Therm. Fluid Sci.* **2013**, *44*, 697–707. [CrossRef]

8. Romero-Sanchez, M.D.; Guillem-Lopez, C.; Lopez-Buendia, A.M.; Stamatiadou, M.; Mandilaras, I.; Katsourinis, D.; Founti, M. Treatment of natural stones with phase change materials: Experiments and computational approaches. *Appl. Therm. Eng.* **2012**, *48*, 136–143. [CrossRef]

9. Izquierdo-Barrientos, M.A.; Belmonte, J.F.; Rodríguez-Sánchez, D.; Molina, A.E.; lmendros-Ibáñez, J.A. A numerical study of external building walls containing phase change materials (PCM). *Appl. Therm. Eng.* **2012**, *47*, 73–85. [CrossRef]

10. Kuznik, F.; Virgone, J. Experimental assessment of a phase change material for wall building use. *Appl. Energy* **2009**, *86*, 2038–2046. [CrossRef]

11. De Gracia, A.; Rincón, L.; Castell, A.; Jiménez, M.; Boerb, D.; Medrano, M.; Cabez, L.F. Life Cycle Assessment of the inclusion of phase change materials (PCM) in experimental buildings. *Energy Build.* **2010**, *42*, 1517–1523. [CrossRef]

12. Kuznik, F.; Virgone, J.; Noel, J. Optimization of a phase change material wallboard for building use. *Appl. Therm. Eng.* **2008**, *28*, 1291–1298. [CrossRef]

13. Navarro, L.; de Garcia, A.; Solé, C.; Castell, A.; Cabeza, L.F. Thermal loads inside buildings with phase change materials: Experimental results. *Energy Proced.* **2012**, *30*, 342–349. [CrossRef]

14. Pasupathy, A.; Velraj, R. Effect of double layer phase change material in building roof for year round thermal management. *Energy Build.* **2008**, *40*, 193–203. [CrossRef]

5

A Numerical Study on System Performance of Groundwater Heat Pumps

Jinsang Kim [1] and Yujin Nam [2,*]

Academic Editors: Vincent Lemort and Samuel Gendebien

[1] Blue Economy Strategy Institute, Suite 911, 55 Digital-Ro 34-Gil, Seoul 08378, Korea; kim6755@nate.com
[2] Department of Architectural Engineering, Pusan National University, Jangjun-2 Dong, Busan 609-735, Korea
* Correspondence: namyujin@pusan.ac.kr

Abstract: Groundwater heat pumps have energy saving potential where the groundwater resources are sufficient. System Coefficients of Performance (COPs) are measurements of performance of groundwater heat pump systems. In this study, the head and power of submersible pumps, heat pump units, piping, and heat exchangers are expressed as polynomial equations, and these equations are solved numerically to determine the system performance. Regression analysis is used to find the coefficients of the polynomial equations from a catalog of performance data. The cooling and heating capacities of water-to-water heat pumps are determined using Energy Plus. Results show that system performance drops as the water level drops, and the lowest flow rates generally achieve the highest system performance. The system COPs are used to compare the system performance of various system configurations. The groundwater pumping level and temperature provide the greatest effects on the system performance of groundwater heat pumps along with the submersible pumps and heat exchangers. The effects of groundwater pumping levels, groundwater temperatures, and the heat transfer coefficient in heat exchanger on the system performance are given and compared. This analysis needs to be included in the design process of groundwater heat pump systems, possibly with analysis tools that include a wide range of performance data.

Keywords: groundwater heat pump; coefficient of performance; regression analysis; UA value; heat exchanger

1. Introduction

A large portion of energy use in buildings goes towards heating and cooling. Operating buildings contributes more to energy use and climate change than either transportation or industry, according to an estimate of global energy consumption [1]. On the other hand, buildings offer the largest global potential to reduce greenhouse gas consumption [2]. Geothermal heat pumps are the most environmentally friendly and efficient method for heating and cooling buildings, and thus could have the largest potential to mitigate greenhouse gas emissions.

Generally, ground heat source systems utilize the annually stable underground temperature. In particular, Groundwater Heat Pumps (GWHPs) directly use groundwater that has a huge amount of potential as a heat source. The Equitable Building—located in Portland (OR, USA)—initiated the use of groundwater as a heat source for heating and cooling in 1948; in recognition, the American Society of Mechanical Engineers recorded it as a National Mechanical Engineering Landmark. It is possibly the oldest geothermal heat pump and the oldest commercial heat pump.

In Korea, the use of renewable energy systems, including ground source heat pump systems, has been spreading widely in the building sector since it became mandatory in public buildings in

2004. Figure 1 shows applications of ground source heat pump systems in Korea [3,4]. According to a market analysis [5], the market for ground source heat pump systems in Korea may grow to 500 billion dollars. Although the market has been growing dramatically, most systems were closed loop system using a U tube pipe as a ground heat exchanger. GWHP systems are used in large-scale buildings, but it is difficult to say that the systems have spread widely.

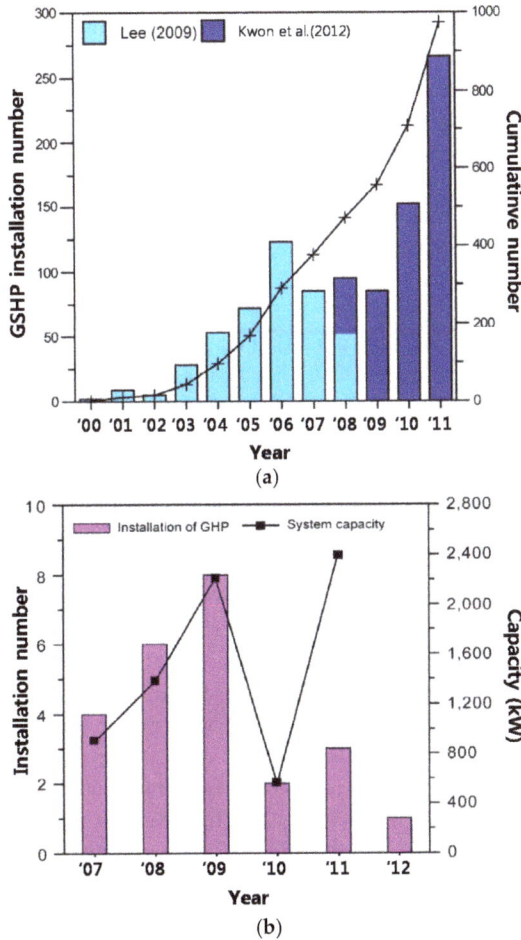

Figure 1. Current status of ground source heat pump system use in Korea. (a) Application of ground source heat pump systems; (b) Application of groundwater heat pump systems [3,4].

There are several barriers to the installation of a GWHP system, such as the limitations of groundwater condition, pumping power, and local regulations. In particular, most building designers did not recognize the system as a heating and cooling system, and did not understand the proper design methods. Although the Korean government has suggested design and installation guidelines for ground source heat pump systems [6], the explanations for GWHP systems are not sufficient.

Proper design of GWHP systems can realize both low installation and operation costs [7]. While ground-source heat pumps can be installed almost anywhere, GWHPs can be successfully constructed and operated only where an aquifer can sustainably provide groundwater to the heat pump units. Adequate site characterization and system design are essential for stable operation and efficiency. Several studies have been conducted on the design factors of GWHP systems. Nam and Ooka conducted a numerical simulation of ground heat and water transfer for GWHP systems compared with real-scale experiments [8]. The results confirmed that the groundwater flow and the well position are important for the optimum design of GWHP systems, and the electricity consumption of a pump must be reduced significantly to achieve a higher system coefficient of performance (COP). Zhou *et al.* also conducted numerical and experimental investigations of the thermal behavior in the well of a GWHP [9]. They compared the changes in the pumping temperature in different aquifer advection configurations and found that horizontal downstream cross-flow can improve the system efficiency in heating and cooling. ASHRAE (American Society of Heating, Refrigerating and Air-Conditioning Engineers) Research Report 1119 focused on the standing column well (SCW) design, and a design method for SCW systems was developed [10]. In that research, a simplified simulation tool for SCW systems was developed, and a parametric study for the design guidelines was conducted. Through the simulation tool, the groundwater temperature can be calculated for various conditions of the specific system and operation. Although the ground temperature or circulation temperature is important for calculating the performance of a heat pump, from the perspective of total system performance, the design and selection of the pumping system is also important.

Pumping systems and their operating conditions greatly influence GWHP system performance. Groundwater usually provides favorable operating conditions for heat pump units as compared to ground-source heat pumps, with higher COPs for heat pump units in GWHP systems. However, open-loop groundwater pumping power consumption increases rapidly as the groundwater level drops.

This theory is very simple and well known, but, in real system applications, it is difficult to quantitatively consider the relationship between the system performance and the groundwater conditions. Most system designers did not consider the COP of the total system, but only the lift of the pump for groundwater level and the friction factor. Of course, numerical simulations with ground heat and groundwater transfer models can support optimum designs, but they also have the limitation of practical calculation time. In order to develop a simple and practical design guide, it is necessary to utilize available performance data and individual component models.

In this study, the efficiencies of GWHP systems are analyzed based on the system heating and cooling COPs. The system COPs are calculated by employing polynomial models for heat pump units and pumps. The coefficients of the polynomial models are found by applying regression analysis to the manufacturers' catalog data. The effects of groundwater level and temperature, as well as heat exchanger UA (overall heat transfer coefficient) value (*i.e.*, the rate of heat transfer per temperature difference), on the system performance are analyzed. The analysis will be helpful in determining system performance during the design process of GWHP systems, possibly with analysis tools that include a wide range of heat pump units and pump performance data.

2. Groundwater Heat Pump (GWHP) System

GWHP systems may consume a smaller amount of energy in operating heat pump units than ground source heat pump (GSHP) systems because their heat pump units usually operate under more favorable operating conditions, *i.e.*, groundwater temperatures that remain stable throughout the year. However, in some instances, GWHP systems could consume more energy for pumping groundwater. The pumping energy consumption increases greatly as the groundwater level drops or the pumping flow rate increases.

The water table is maintained in wells and unconfined aquifers when no water is pumped from them. The static water level is the distance from the ground surface to the water level in the well. When water is pumped from a well, the water level in the well drops from the static water level. The water level in the well while pumping is called the pumping water level. The distance between the static water level and the pumping water level is called the drawdown [11,12]. The pumping water level is used in calculating the GWHP system performance.

GWHP systems are composed of one or more production water wells, heat pump units, submersible and circulation pumps, and heat exchangers. Their layouts can vary significantly with design intents and the number of production wells, heat exchangers, and other components. Some GWHP systems do not include heat exchangers between the heat pump units and the submersible pumps, allowing the groundwater to pass through the heat pump unit. This configuration is called a direct GWHP system, and is sometimes used in smaller capacity applications. Indirect GWHP systems, which include a heat exchanger between the groundwater submersible pump and the heat pump units, are often used in large commercial systems.

In this study, the indirect GWHP system is composed of one groundwater production well, one submersible pump, one water-to-water heat pump unit, one circulation pump, and one heat exchanger, as shown in Figure 2. Air handling units (AHUs) or fan coil units (FCUs) that are connected to the water-to-water heat pump unit are not included in calculating the system performance of the current GWHP system. The flow rates and inlet temperature to the AHUs are predefined and assumed to be unchanged. The system configuration is the same as that from the authors' preceding publication [13].

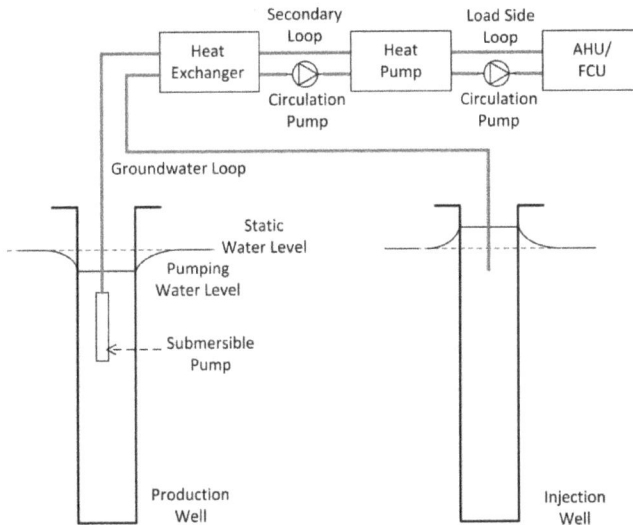

Figure 2. Groundwater heat pump system layout.

The submersible pump installed in the production well pumps the water, transfers it through the heat exchanger, and generally sends it to the injection well (unless there is another application). When the heat pump unit is in heating mode, the heat of the groundwater is transferred to the circulation water in the heat exchanger. The circulation fluid from the heat exchanger is sent to the heat pump unit by a circulation pump in the secondary loop. The heat of the circulation water is transferred to the refrigerant loop inside of the heat pump unit. The circulation water from the heat pump then returns to the heat exchanger in the secondary loop. The heat pump unit produces hot water with the capacity including the absorbed heat from the circulation water in the secondary loop and the electricity consumed by the compressor. The hot water produced by the heat pump units is sent to the AHUs or FCUs to heat the space, or for another process. In heating mode, the heat from the groundwater moves to the space or the process through the heat exchanger, heat pump unit, and the AHUs.

In cooling mode, the heat of the space or the process is transferred to the refrigerant loop inside of the heat pump unit via the AHUs and the circulation water in the load side loop. The circulation water in the secondary loop is chilled by the heat pump unit. Inside the heat exchanger, the heat travels from the circulation water to the groundwater, and the circulation water thus becomes cooler and the groundwater becomes hotter. The groundwater from the injection well is heated in the heat exchanger, and returns to the injection well (unless it is sent to another application).

3. Component Models

The GWHP system of this study is composed of one submersible pump installed in the groundwater production well, one water to water heat pump unit, one circulation pump, and one heat exchanger. In order to calculate the power consumption and the heating and cooling capacities of a GWHP system, the power consumption, head, and heating and cooling capacities of the components need to be modeled as functions of flow rate and temperature. When the flow rate varies in a GWHP system, the temperature distributions in the heat exchanger and heat pump unit vary. Accordingly, the heating and cooling capacities of the GWHP have different values.

3.1. Submersible Pump and Circulation Pump Model

Submersible pumps generally consist of multiple stages of identical impellers and housings installed in series with electric motors installed at the ends. The head and power consumption of submersible pumps are calculated by multiplying the number of stages n of the pump and the associated values for a single stage:

$$H_p = n \cdot h_p \tag{1}$$

$$W_p = n \cdot w_p \tag{2}$$

The head and power consumption for a single stage of a submersible pump are modeled in cubic polynomial equations in terms of flow rate V (in LPM (liters per minute)):

$$w_p = A_0 + A_1 V + A_2 V^2 + A_3 V^3 \tag{3}$$

$$W_p = B_0 + B_1 V + B_2 V^2 + B_3 V^3 \tag{4}$$

The eight coefficients in Equations (3) and (4) can be obtained through regression analysis from performance data typically provided in the manufacturer's catalog. For the single stage, h_p and w_p are the head (in m) and power consumption (in kW), respectively. The groundwater production well is assumed to be capable of providing the necessary flow rate to the heat exchanger or the heat pump unit, and it maintains the pumping water level throughout the operation. The circulation pumps in the secondary loops have similar characteristics as the submersible pumps. The same equation forms for the head and power consumption are used.

3.2. Heat Pump Model

When the circulation fluid of the secondary loop passes through the heat exchanger installed inside the heat pump unit to exchange heat with the refrigerant, head loss and temperature change occurs in the circulation fluid. The head loss in the secondary circulation loop of the heat pump unit H_{hp} is expressed in the following equation. The same equation is used for the head loss that occurs in the load side loop of the heat pump unit:

$$H_{hp} = C_{hp} \cdot V^2 \tag{5}$$

Temperature variations inside the heat exchanger and flow rate variations in the secondary loop cause the heat pump unit to produce different cooling and heating capacities—Q_c and Q_f (in kW), respectively—and consume different amounts of electricity—W_c and W_f (in kW), respectively:

$$\frac{Q_c}{Q_{c,ref}} = D_0 + D_1 \left(\frac{T_{S,in}}{T_{ref}} \right) + D_2 \left(\frac{T_{L,in}}{T_{ref}} \right) + D_3 \left(\frac{V_S}{V_{ref}} \right) + D_4 \left(\frac{V_L}{V_{ref}} \right) \tag{6}$$

$$\frac{Q_h}{Q_{c,ref}} = E_0 + E_1\left(\frac{T_{S,in}}{T_{ref}}\right) + E_2\left(\frac{T_{L,in}}{T_{ref}}\right) + E_3\left(\frac{V_S}{V_{ref}}\right) + E_4\left(\frac{V_L}{V_{ref}}\right) \tag{7}$$

$$\frac{W_c}{W_{c,ref}} = F_0 + F_1\left(\frac{T_{S,in}}{T_{ref}}\right) + F_2\left(\frac{T_{L,in}}{T_{ref}}\right) + F_3\left(\frac{V_S}{V_{ref}}\right) + F_4\left(\frac{V_L}{V_{ref}}\right) \tag{8}$$

$$\frac{W_h}{W_{h,ref}} = G_0 + G_1\left(\frac{T_{S,in}}{T_{ref}}\right) + G_2\left(\frac{T_{L,in}}{T_{ref}}\right) + G_3\left(\frac{V_S}{V_{ref}}\right) + G_4\left(\frac{V_L}{V_{ref}}\right) \tag{9}$$

Subscripts S and L indicate the secondary loop and the load side loop, and subscript ref is the reference value for the corresponding variable. The performance data of the water-to-water heat pump unit can be used to find the coefficients through regression analysis. The performance table includes the cooling and heating capacities and power consumption for source and load side water temperatures and flow rates [14].

3.3. Piping Model

Head loss occurs when the fluid passes inside the pipes and piping accessories. Pipe diameters, pipe lengths, and fluid properties influence the heat loss that occurs in the secondary and groundwater loops. The equation of the head losses expressed in terms of flow rate is as follows:

$$H_{piping} = C_{piping} \cdot V^2 \tag{10}$$

3.4. Heat Exchanger Model

Plate frame plate heat exchangers are normally used as heat exchangers installed between heat pump units and groundwater submersible pumps in indirect GWHP systems. The heat transfer occurs between the groundwater and the circulation fluid of a secondary loop inside the heat exchanger. When the flow rate of the groundwater or circulation fluid changes, the output temperatures vary in the following way:

$$H_{hx} = C_{hx} \cdot V^2 \tag{11}$$

$$Q_{hx} = \rho c_p \cdot \Delta T \cdot V \tag{12}$$

$$Q_{hx} = UA \cdot \Delta T_{LM} \tag{13}$$

$$\Delta T_{LM} = \frac{\Delta T_1 - \Delta T_2}{\ln\left(\dfrac{\Delta T_1}{\Delta T_2}\right)} \tag{14}$$

Subscript hx indicates the heat exchanger, UA is a heat exchanger characteristic, and ΔT_{LM} is the logarithmic temperature difference of the heat exchanger as defined in Equations (13) and (14). U value is overall heat transfer coefficient, which considers the heat transfer coefficients of circulation fluid and the resistance of the pipe material.

4. GWHP System Configuration

The indirect GWHP system layout in this study is composed of one submersible pump installed inside the groundwater production well, one water to water heat pump unit, one circulation pump, and one heat exchanger.

A 4 inch diameter submersible pump SP 75S made by Grundfos (Bjerringbro, Denmark) was considered here [15]. All pump models from single stage through sixteen-stage are assumed to be available in this series of products. The eight coefficients of Equations (3) and (4) for the single stage are obtained from regression analysis of the performance data as shown in Table 1.

The McQuay GRW360 model (McQuary, Minneapolis, MN, USA) which uses R407a refrigerant and has a nominal capacity of 30 refrigeration tons is used for the water-to-water heat pump [16]. The twenty coefficients from Equations (6) to (9) are calculated from regression analysis of the manufacturer's data as shown in Table 2. The reference values used in Equations (6) to (9) and the head loss coefficients of Equation (5) are also given in Table 2. Figure 3 shows performance curves of the heat pump according to heat source temperature, which is based on the catalog data.

Table 1. Pump model coefficients for the single stage.

A_0	A_1	A_2	A_3
1.022×10	-1.351×10^{-2}	2.5313×10^{-5}	-1.1589×10^{-7}
B_0	B_1	B_2	B_3
2.07403×10^{-1}	8.9077×10^{-4}	3.467×10^{-6}	-9.95602×10^{-9}

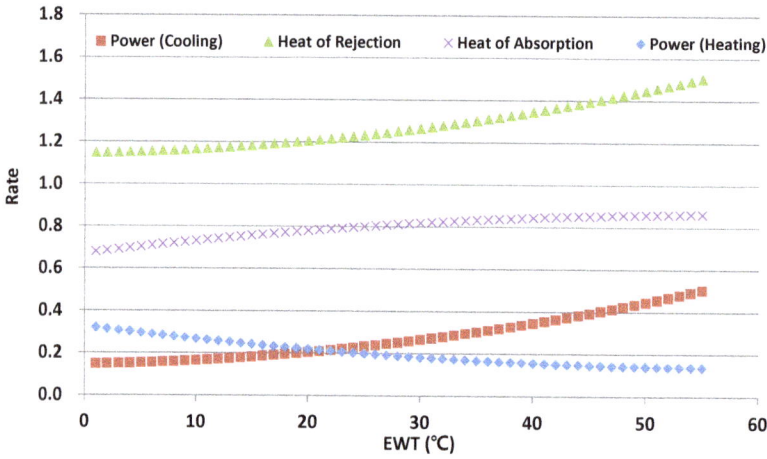

Figure 3. Performance curve of heat pump.

Table 2. Heat pump model coefficients.

D_0	D_1	D_2	D_3	D_4
-3.70415	-2.26533	6.75804	3.322×10^{-2}	7.447×10^{-2}
E_0	E_1	E_2	E_3	E_4
-6.28243	7.83147	-6.663×10^{-1}	9.162×10^{-2}	3.253×10^{-2}
F_0	F_1	F_2	F_3	F_4
-6.42	5.66874	1.32662	-1.106×10^{-1}	1.483×10^{-2}
G_0	G_1	G_2	G_3	G_4
-8.35467	9.0202×10^{-1}	7.81833	2.878×10^{-2}	6.669×10^{-2}
$Q_{c,ref}$	$Q_{h,ref}$	$W_{c,ref}$	$W_{h,ref}$	C_{hp}
70 kW	70 kW	15 kW	15 kW	5.08245×10^{-5}
T_{ref}	$T_{S,in}$	$T_{L,in}$	V_{ref}	V_S
273 K	288 K	313 K	284 LPM	250 LPM

The head loss that occurs in the piping of the groundwater loop is approximated by Equation (10), and the head loss coefficient in the piping H_{piping} is determined to be 3.25×10^{-5}. The submersible pumps installed in the wells supply the groundwater with the head to cover the system head of the groundwater loop, which includes the pumping water level and the head losses in the heat exchanger and piping. The system head equation is shown in Equation (15):

$$H_{SYS} = H_{GW} + H_{hp} + H_{hx} + H_{piping} \tag{15}$$

The head of the submersible pump and the system head vary as the groundwater flow rate varies. The residual head of the groundwater loop is defined as the difference between the groundwater loop system head and the submersible pump head. At the operating flow rate, the submersible pump head H_p matches the system head of the groundwater loop, meaning that the residual head becomes zero:

$$H_{res} = H_{SYS} - H_p \tag{16}$$

The Newton-Raphson method is used to find the operating flow rate. Iterations on the groundwater loop system head in terms of the groundwater flow rate are performed until the residual head vanishes. In the secondary loop, the chosen circulation pump is a Grundfos CR 15 single stage model (Manufacturer, Bjerringbro, Denmark). The operating circulation water flow rate of the circulation pump is 300 LPM with a head of 17.4 m. The power consumption of the pump is 1.17 kW, based on the performance data from the manufacturer's catalog.

5. Results

The performance of the GWHP system can be expressed in terms of system COPs, which are the ratios of the heating or cooling capacities to the corresponding electricity consumptions of the GWHP system. Meanwhile, the COPs of the heat pump units are the ratios of the heating and cooling capacities to the corresponding electricity consumption for only the heat pump unit, excluding the pumps. System COPs are always lower than heat pump unit COPs, since the denominators of the ratios always become larger with the additional pump power consumptions. In calculating the GWHP system performance, the power consumption of the AHUs or FCUs and the pump in the load side loop are not included. The performance and characteristics of the components in the load side loop varies greatly for various types of building uses and designs. To evaluate the system performance of the GWHP without the effects caused by load side designs, the load side loop is assumed unchanged in the analysis.

5.1. Groundwater Level

The pumping groundwater levels are assumed to be obtained from a series of pumping tests. Pumping tests are experimental tests for finding the aquifer transmissivity or the soil permeability. In the test, a well is pumped at a controlled rate, and water-level response (drawdown) is measured in one or more surrounding observation wells. The response data from pumping tests is used to estimate the hydraulic properties of aquifers and evaluate well performance. The drawdown is a function of various parameters including pumping flow rate, aquifer transmissivity, time since the start of pumping, and distance from the well [17]. However, the water levels in the well during pumping tests include the drawdown effects, which are thus not being treated separately in this study.

System COPs of GWHP systems are presented for three different constant pumping groundwater levels: 15, 25 and 35 m. At each groundwater level, groundwater flow rates for the selected submersible pumps are found, and performance results of the GWHP system are calculated for the flow rates. The submersible pumps of eleven different stages are considered at each groundwater level. Performance results include heat pump COP and GWHP system COP.

System heating and cooling COPs are shown in Figures 4 and 5 respectively. In Figure 4, as the groundwater level drops, the optimum flow rate for the GWHP heating system COP becomes more recognizable. In Figure 5, as the groundwater level drops, the slope of the curve in the low flow rate region becomes smaller in magnitude. However, in the high flow rate region, the system heating and cooling COPs drop rapidly as the flow rate increases.

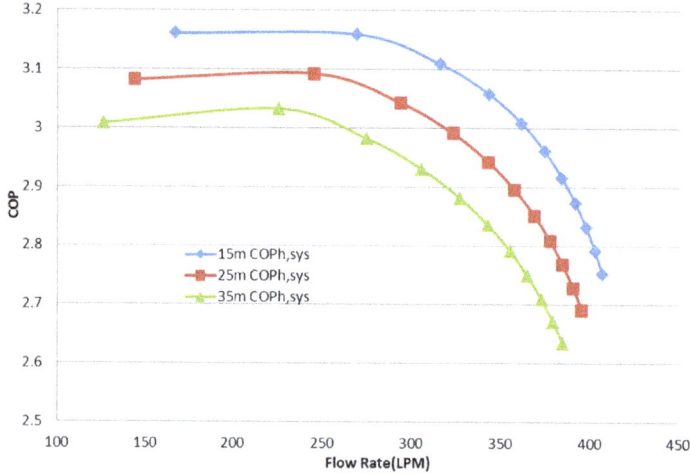

Figure 4. Heating system COP variation with the flow rate.

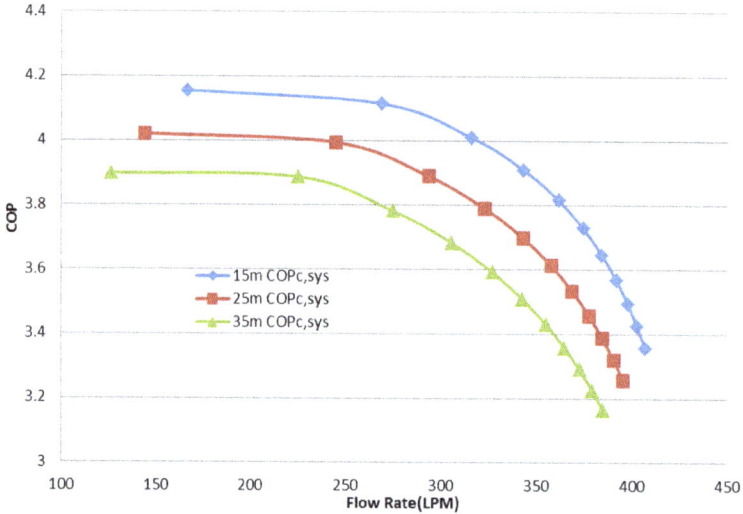

Figure 5. Cooling system COP variation with the flow rate.

The percentage of reduction of the COP of the GWHP system compared with that of the heat pump unit increases as the flow rate increases or the groundwater level drops. Figure 6 shows the system heating COP value variation with flow rates and groundwater levels. A 10%–15% reduction is shown for the 250 LPM groundwater flow rate. When the groundwater level is 35 m below the ground, the system heating COP shows the largest reduction (15%) as compared to the heat pump COP. At the higher flow rate of around 350 LPM, the system heating COP decreases by 15%–22% from the heat pump unit COP.

Figure 7 shows that the reduction percentage of system cooling COP as compared to the heat pump unit COP is larger than the reduction percentage of the system heating COP. A similar trend is found with the variation of flow rate and groundwater level. A 15%–22% reduction is observed for the 250 LPM groundwater flow rate. When the groundwater level is 15 m below the ground, the system heating COP shows the smallest reduction (15%) as compared to the heat pump COP. At the higher flow rate of around 350 LPM, the system heating COP shows a 22%–29% reduction from the heat pump unit COP.

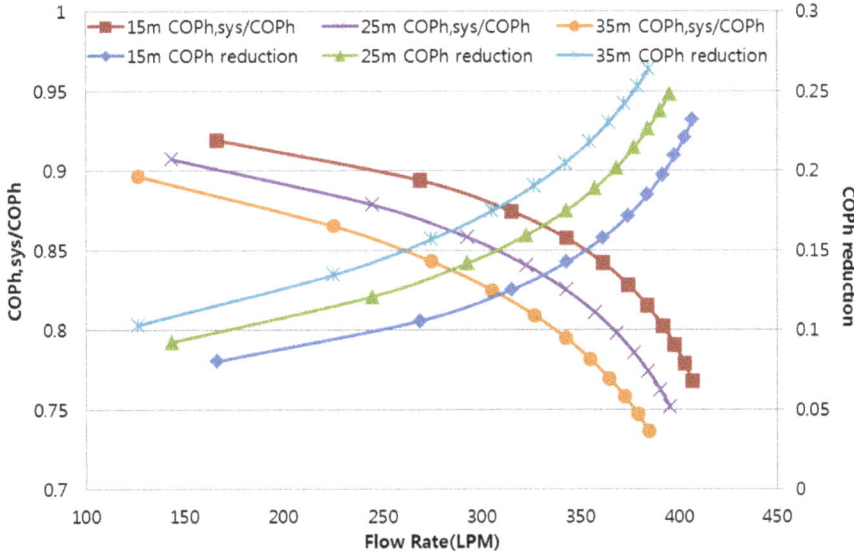

Figure 6. System heating COP reduction compared to heat pump unit COP.

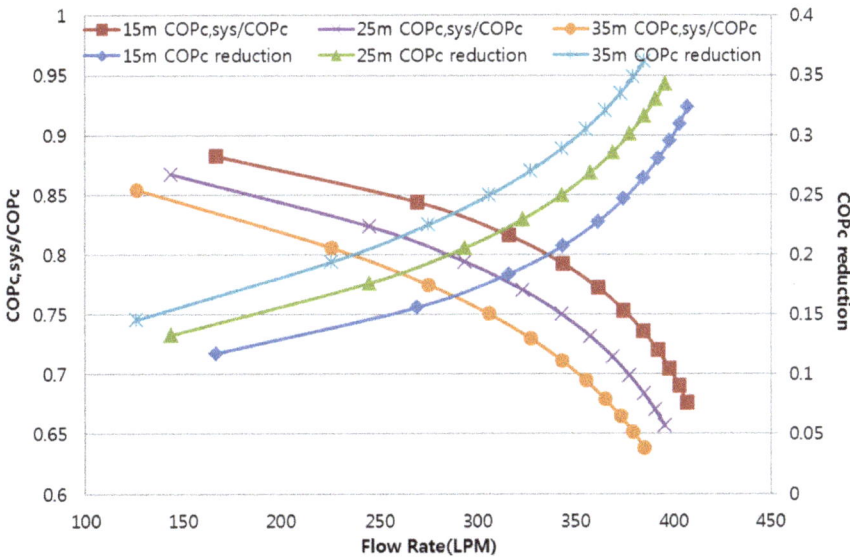

Figure 7. System cooling COP reduction compared to heat pump unit COP.

5.2. Groundwater Temperature

Groundwater temperature influences the system COPs of the GWHP system, as the heat pump capacities and COPs are dependent on the groundwater temperature. As the groundwater temperature increases, the heating COPs also increase, but the cooling COPs decrease. Conversely, as the groundwater temperature decreases, the heating COPs decrease, but the cooling COPs increase. While studying the effect of groundwater temperature on the system COPs of the GWHP, the groundwater level is set at a constant 25 m. The system COPs are calculated and shown in Figure 8 for groundwater temperatures of 12, 15 and 18 °C.

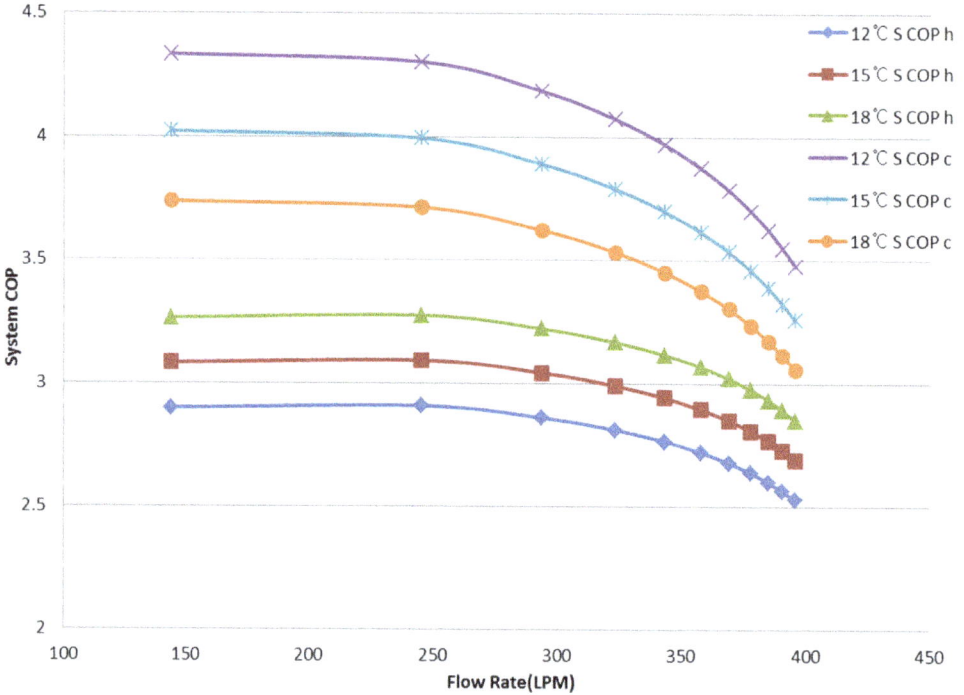

Figure 8. System Cooling and Heating COP variation to Groundwater Temperature.

The plot of the system cooling COPs for 12 °C groundwater shows the largest values; however, the system heating COPs show the smallest values. The system COPs for 18 °C groundwater show the opposite trend as compared to the case for 12 °C. The COPs for 15 °C groundwater lie between those for 12 and 18 °C. As the flow rate varies from 144.0 to 395.3 LPM, the system cooling COPs drop by approximately 19.0% (19.8% for 12 °C, 19.0% for 15 °C, and 18.3% for 18 °C), and the system heating COPs drop by approximately 12.7% (12.7% for 12 °C, 12.7% for 15 °C , and 12.6% for 18 °C).

At the flow rate of 245 LPM, the system heating COPs drop by 11.2% when compared with the COPs between 12 and 18 °C groundwater temperature, whereas the system cooling COPs drop by 13.7%. At a higher flow rate of 343 LPM, the system heating COPs drop by 11.2%, and the system cooling COPs drop by 13.1%

5.3. Heat Exchanger UA Value

Figure 9 shows the calculation results of system COP according to the change of heat exchanger UA values. The heat exchanger installed between the groundwater pump and the heat pump unit influences the system COP of the GWHP system. The temperature of the water circulated between the heat exchanger and the heat pump unit influence the capacity and power consumption of the heat pump unit.

The UA value of the heat exchanger determines the temperature differences between the hot and cold streams. In this study, a UA value of 12 W/K has been used for the previous results. A positive and negative variation by 25% from 12 W/K are considered, and three UA values of 9, 12 and 15 W/K are used to study its effect on the system COPs.

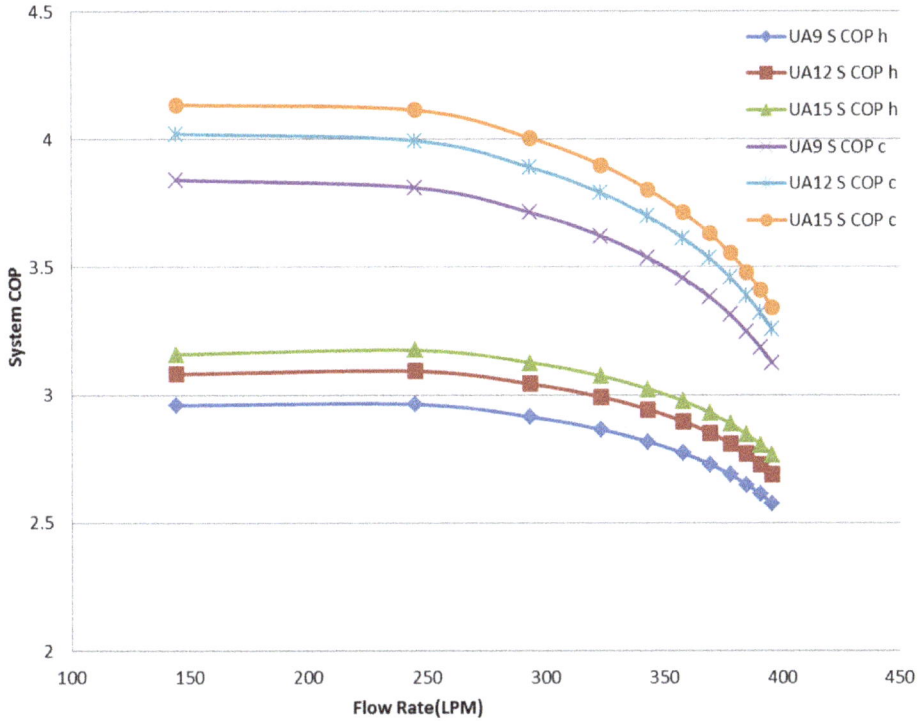

Figure 9. System cooling and heating COP variation to heat exchanger UA values.

When the UA value decreases from 12 to 9 W/K, the system heating COP drops by 4.2% and the system cooling COP drops by 4.6%. When the UA values increases from 12 to 15 W/K, the system heating COP rises by 2.7% and the system cooling COP rises by 3.0%. It is expected that the low UA values of the small heat exchanger may deteriorate the system COP values by a significant amount.

6. Conclusions

In this study, in order to develop an installation and operation guideline for a GWHP system, a numerical study was conducted using individual performance models and available system catalog data. The system COPs, as measures of performance for GWHP systems, were obtained by combining the component mathematical models. The system heating and cooling COPs were calculated for different groundwater levels, flow rates, groundwater temperatures, and UA values. The system COPs were almost unchanged or gradually changed in the low flow rate region as the flow rate increases. However, in the higher flow rate region, the system heating and cooling COPs decreased rapidly as the flow rate increased. The system heating COPs increased with warmer groundwater temperatures; similarly, the system cooling COPs increased with colder groundwater temperatures. Increasing the UA value of the heat exchanger universally improved the system heating and cooling COPs.

The system heating COPs were reduced 10%–15% at around 250 LPM and 15%–22% at around 350 LPM when compared to heat pump unit COPs. The system cooling COPs were reduced by a larger rate, 15%–22% at around 250 LPM, and 22%–29% at around 350 LPM. This result indicates that, if the flow rate is above 350 LPM at 30 refrigerant ton of heat pump, GWHP systems would be not proper due to low system COP.

At the flow rate of 245 LPM, the system heating COP dropped by 11.2% when 12 °C groundwater temperatures are compared to 18 °C, and the system cooling COP rose by 13.7%. At a higher flow rate of 343 LPM, the system heating COP dropped by 11.2%, and the system cooling COP rose by 13.1%.

When the groundwater temperature rose, the system heating COP of GWHP systems increased, and the system cooling COP decreased. These results show that the effect of groundwater temperature on system performance is as significant as the effect of groundwater level or flow rate. Generally, the groundwater temperature at shallow depths is determined by the annual regional air temperature, but it could be different according to the well depth. Therefore, well designers should decide the well depth considering the groundwater temperature and building load, such as deeper wells for heating dominant buildings.

The system heating COP dropped by 4.2% and the system cooling COP dropped by 4.6% when comparing UA value of 9 to 12 W/K. When the UA value of 15 W/K is compared to a 12 W/K one, the system heating COP rose by 2.7% and the system cooling COP rose by 3.0%. Selection of proper UA values of the heat exchanger may ensure proper system COP values.

GWHP system design engineers need to find the groundwater flow rates at which GWHP systems maintain high system COPs with a proper selection of heat exchanger. The GWHP systems need to be modeled with detailed information on components, such as submersible pumps, heat pump units, and heat exchangers. In order to accurately predict system performance, proper GWHP system models need to be used in conjunction with the building heating and cooling load variations.

Acknowledgments: This work was supported by the New and Renewable Energy (No. 20123040110010) of the Korea Institute of Energy Technology Evaluation and Planning (KETEP) grant funded by the Korea Government Ministry of Knowledge Economy.

Author Contributions: All authors contributed equally to this work. All authors designed the simulations, discussed the results and implications and commented on the manuscript at all stages. Jinsang Kim performed the numerical simulations and the result analysis. Yujin Nam led the development of the paper and performed discussion.

Conflicts of Interest: The authors declare no conflict of interest.

References

1. Mendler, S. *Buildings: An Essential Sector in Climate Change Mitigation*; Civil Society Institute: Newton, MA, USA, 2012.
2. Bernstein, L.; Bosch, P.; Canziani, O.; Chen, Z.; Christ, R.; Davidson, O.; Hare, W.; Huq, S.; Karoly, D.; Kattsov, V.; *et al.* Intergovernmental Panel on Climate Change. In *Climate Change 2007: Synthesis Report. Contribution of Working Groups I, II and III to the Fourth Assessment Report of the Intergovernmental Panel on Climate Change*; Core Writing Team, Pachauri, R.K., Reisinger, A., Eds.; IPCC: Geneva, Switzerland, 2007.
3. Kwon, K.-S.; Lee, J.-Y.; Mok, J.-K. Update of current status on ground source heat pumps in Korea (2008–2011). *J. Geol. Soc. Korea* **2012**, *48*, 193–199.
4. Park, Y.; Kwon, K.-S.; Kim, N.; Lee, J.-Y.; Yoon, J.-G. Change of geochemical properties of groundwater by use of open loop geothermal cooling and heating system. *J. Geol. Soc. Korea* **2013**, *49*, 289–296.
5. Korea Energy Agency. New and Renewable Energy Center of Korea, Report of Geothermal Energy. Available online: https://www.renewableenergy.or.kr/ (accessed on 18 December 2015).
6. Korea Energy Agency. New and Renewable Energy Center of Korea, Legislation 2015-153. Available online: http://www.knrec.or.kr/knrec/14/KNREC140110.asp?idx=505&page=1&num=68&Search2=&Search=&SearchString= (accessed on 18 December 2015).
7. Kavanaugh, S.P.; Rafferty, K. *Ground Source Heat Pumps, Design of Geothermal Systems for Commercial and Institutional Buildings*; American Society of Heating Refrigeration and Air-Conditioning Engineers: Atlanta, GA, USA, 1997.
8. Nam, Y.; Ooka, R. Numerical simulation of ground heat and water transfer for groundwater heat pump system based on real-scale experiment. *Energy Build.* **2010**, *42*, 69–75. [CrossRef]
9. Zhou, X.; Gao, Q.; Chen, X.; Yu, M.; Zhao, X. Numerically simulating the thermal behaviors in groundwater wells of groundwater heat pump. *Energy* **2013**, *61*, 240–247. [CrossRef]
10. Spitler, J.D.; Rees, S.J.; Deng, Z. *R&D Studies Applied to Standing Column Well Design*; ASHRAE Report 1119-RP; American Society of Heating Refrigeration and Air-Conditioning Engineers: Atlanta, GA, USA, 2002.

11. Banks, D. *An Introduction to Thermogeology: Ground Source Heating and Cooling*, 2nd ed.; John Wiley & Sons: Oxford, UK, 2012.

12. Sachs, H.M. *Geology and Drilling Methods for Ground-Source Heat Pump Installations: An Introduction for Engineers*; American Society of Heating, Refrigeration and Air-Conditioning Engineers: Atlanta, GA, USA, 2002.

13. Kim, J.; Nam, Y. Study on groundwater heat pump system performance with various groundwater level. In Proceedings of the 11th IEA Heat Pump Conference, Montreal, QC, Canada, 12–16 May 2014.

14. *EnergyPlus Engineering Reference: The Reference to EnergyPlus Calculations*; U.S. Department of Energy: Washington, DC, USA, 2012.

15. *Grundfos Product Guide: SP Submersible Pumps, Motors, and Accessories*; L-SP-PG-001; Grundfos: Bjerringbro, Denmark, 2012.

16. *Water to Water Source Heat Pumps 3 to 35 Tons R-22 and R407C Refrigerant/60 Hz*; Catalog1111-1; McQuary: Minneapolis, MN, USA, 2007.

17. Park, B.-H.; Bae, G.-O.; Lee, K.-K. Importance of thermal dispersivity in designing groundwater heat pump (GWHP) system: Field and numerical study. *Renew. Energy* **2015**, *83*, 270–279. [CrossRef]

6

Intermittent Very High Frequency Plasma Deposition on Microcrystalline Silicon Solar Cells Enabling High Conversion Efficiency

Mitsuoki Hishida [1,*]**, Takeyuki Sekimoto** [2]**, Mitsuhiro Matsumoto** [3] **and Akira Terakawa** [3]

Academic Editor: Alessio Bosio

[1] Automotive & Industrial Systems Company, Panasonic Corporation, Kadoma, Osaka 571-8506, Japan
[2] Advanced Research Division, Panasonic Corporation, Seika, Kyoto 619-0237, Japan;
    sekimoto.takeyuki@jp.panasonic.com
[3] Eco Solution Company, Panasonic Corporation, Kaizuka, Osaka 597-0094, Japan;
    matsumoto.mi@jp.panasonic.com (M.M.); terakawa.akira@jp.panasonic.com (A.T.)
* Correspondence: hishida.mitsuoki@jp.panasonic.com

Abstract: Stopping the plasma-enhanced chemical vapor deposition (PECVD) once and maintaining the film in a vacuum for 30 s were performed. This was done several times during the formation of a film of i-layer microcrystalline silicon (μc-Si:H) used in thin-film silicon tandem solar cells. This process aimed to reduce defect regions which occur due to collision with neighboring grains as the film becomes thicker. As a result, high crystallinity (X_c) of μc-Si:H was obtained. Eventually, a solar cell using this process improved the conversion efficiency by 1.3% (0.14 points), compared with a normal-condition cell. In this paper, we propose an easy method to improve the conversion efficiency with PECVD.

Keywords: thin-film silicon tandem solar cell; microcrystalline silicon; amorphous silicon; crystallinity; plasma-enhanced chemical vapor deposition; conversion efficiency

1. Introduction

Stable high quality hydrogenated microcrystalline silicon (μc-Si:H) films which are used in thin-film silicon tandem solar cells have been difficult to make since the quality of μc-Si:H varies depending on its underlying shape or film thickness [1]. This is because a defective component in the μc-Si:H film is affected by transparent conductive oxide (TCO) texture, as the film thickness of the microcrystalline silicon increases [2]. To avoid a defective region being formed and achieve high conversion efficiency (*Eff*), the surface morphology of TCO was modified to have a blunt angle or rounded shape [3]. Recently, École Polytechnique Fédérale de Lausanne has developed thin-film silicon solar cells with 12.63% efficiency [4]. Also, as the current maximum value of *Eff*, the National Institute of Advanced Industrial Science and Technology (AIST) reported the *Eff* of 12.69% with a triode plasma-enhanced chemical vapor deposition (PECVD) reactor [5].

As a model of a defective region generated, when μc-Si:H is grown from the slope of the texture shape of TCO, a defective region ingenerates at the collided point of each μc-Si:H [6,7]. A defective region is affected by the underlying shape or the underlying film quality. It has been reported that the defective region consists of vacancies and a low-density amorphous region [2]. Previously, we proposed another factor for the formation of a defective region due to the shadowing effect on incident radicals [2], which should reduce the X_c [8].

Beyond that, a defective region also occurs with the continuous deposition film after becoming thicker. This defect region was caused by colliding with the neighboring grains after the film become thicker [9]. This is because μc-Si:H grows perpendicular to the textured substrate. The reduction in X_c was revealed by micro-crystallinity depth profiling measurements [8,10]. An X_c profiling technique was reported by Choong *et al.* who measured the X_c profiling of the early-stage (<800 nm) of μc-Si:H. Therefore, in this report, we aim to reduce the collisions with the neighboring grains by stopping the PECVD once to make the epitaxial μc-Si:H film thinner (>1000 nm). Finally, a high conversion efficiency solar cell can be expected. Since the beginning period of discharge is an unstable process, the value of X_c can be expected to decrease if epitaxial growth of μc-Si:H is halted by temporarily stopping the PECVD. Impurities which come from the deposition chamber also influence the quality of μc-Si:H.

As a comparative experiment to reduce the defective region correlating to the film thickness gain, an experiment was carried out. The experimental method is very simple; the plasma-discharge of very high frequency (VHF) PECVD should be stopped once and then the chamber with a vacuum is held for a few seconds (although this process stopped the epitaxial growth of μc-Si:H, this experimental method is referred to as "*SEG*"). In the actual experiment, *SEG*, which was done twice or three times, produces an equal thickness during the i-layer deposition. For ease of evaluating *SEG*, all samples were deposited with pulsed-PECVD [11]. This is because the pulsed-PECVD can be expected to reduce the ion-bombardment and short lifetime species [12,13]. Using a different *SEG* method, characteristic improvement was reported by Urbain *et al.* [14]. Even though they used a different way of profiling the controlled SiH$_4$, the aim of suppressing the reduction of X_c was the same as that of the *SEG* experiments. In only the experiment where *SEG* was done twice using the thin-film silicon tandem solar cells, this X_c of test cells became higher than that of normal-condition cells, and it did not decrease during the process of forming a thick film as well as the case using X_c-adjustment layer [8]. Eventually, *Eff* improved by 1.3% (0.14 points), V_{oc} by 0.011 V, and *FF* by 0.002 compared with those for the normal-condition cell.

2. Experimental Methods

In this study, actual thin-film silicon tandem solar cells were used for the evaluation of *SEG*. The structure of these cells was glass/tin oxide (SnO$_2$)/a-Si:H (p-i-n)/μc-Si:H (p-i-n)/back electrode [15]. For TCO, a commercial SnO$_2$-based TCO substrate (Asahi VU, Asahi Glass, Tokyo, Japan) was adopted. μc-Si:H and a-Si:H were deposited by plasma-enhanced chemical vapor deposition (PECVD).

Typically, when the μc-Si:H is deposited with PECVD, the plasma-discharge power is not stopped before reaching the target thickness. However, in this study, when the thickness reached one-third of t (t/3) and two-thirds of t (2t/3), *SEG* was done twice, if the total thickness of i-μc-Si:H is t (=2300 nm). The evacuation times of *SEG* were carried out in two modes: 30 s and 600 s, and these were labeled cells A and B, respectively. This degree of vacuum was under 5×10^{-3} (pa). The vacuum time of 30 s was determined by the time required for the deposition gases (SiH$_4$/H$_2$) to me almost completely eliminated from the chamber. It is due to a return to the default deposition position. The impact of long time exposure of the cell to the vacuum chamber was compared with that for the cell with the 600 s exhaust time (B cells). Only the difference between cell A and cell B influences the impurities while discharge is halted. The structures of a normal-condition cell and a cell made using the *SEG* cell (cells A and B) are shown in Figure 1.

Additionally, different numbers of *SEG* experiments were carried out simultaneously. Since X_c is correlated with the film thickness, the value of X_c changes according to the number of *SEG*. When the thickness reached one-fourth of t (t/4), two-fourths of t (2t/4), and three-fourths of t (3t/4), *SEG* was done three times, if the total thickness of i-μc-Si:H is t (=2300 nm). One evacuation time of *SEG* was 30 s, which is the same as for cell A. This was labeled cell C. The degree of vacuum was under 5×10^{-3} (pa), which is the same as for cells A and B. Table 1 summarizes these conditions.

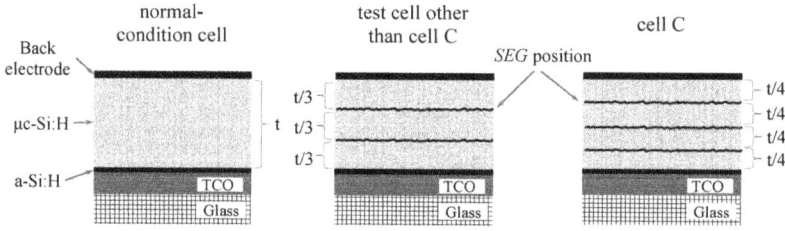

Figure 1. Cell structure comparison between normal-condition, *SEG* (other than cell C) and a *SEG* cell (cell C). The i-μc-Si:H thicknesses were t. *SEG* was done when the thicknesses reached t/3 and 2t/3, t/4, 2t/4, and 3t/4.

Table 1. Comparison of i-μc-Si:H thickness constitution with normal-condition cells A, B and C; as well as normal-condition film and films A. One time of evacuations times of *SEG*, and total evacuations times of that with films A and cells A, B and C.

Description	Normal Condition Cell/Film	Cell A/Film A	Cell B	Cell C
i-layer constitution	t	t/3 + t/3 + t/3		t/4 + t/4 + t/4 + t/4
total i-later thickness		t (=2300 nm)		
exhaust time between i-layer (s)	0	30	600	30
total exhaust time (s)	0	60	1200	90

The cell position and temperature of the heater were maintained during *SEG*. After the *SEG*, the same gases as the deposition conditions were supplied for 30 s, and then plasma-dischargewas restarted again. The total discharge time of μc-Si:H was the same for the whole duration of these experiments, although the discharge was stopped during the deposition. All the μc-Si:H treatment times with PECVD became longer by only around 5 min, except for cell B.

Furthermore, to verify the reproducibility of the *SEG* process, two other types of *SEG* tests were carried out. These test conditions were different from the conditions applied to cells A to C, such as the top layer and i-μc-Si:H. However, the *SEG* processes were carried out twice, the same as for cell A. The total thicknesses of i-μc-Si:H were 1500 and 1750 nm: these were labeled cells D and E, respectively. For cell D, the deposition conditions (thickness, gases, power and pressure) of i-μc-Si:H and the top layer condition were varied compared to other cells. Cell E was subjected to conditions different from those applied to cells A to D. The no-*SEG* cells were fabricated simultaneously, and were labeled as having been fabricated under the same normal conditions as those for cell D and cell E. Table 2 summarizes these conditions.

Table 2. Comparison of different i-μc-Si:H thickness (1500/1750 nm) *SEG* processes with cell D and normal-condition cell D, and cell E and normal-condition cell E. Single evacuation times of *SEG* and total exhaust time of cell.

Description	Normal Condition (D)	Cell D	Normal Condition (E)	Cell E
i-layer constitution	t	t/3 + t/3 + t/3	t	t/3 + t/3 + t/3
total i-later thickness	t (=1500 nm)		t (=1750 nm)	
exhaust time between i-layer (s)	0	30	0	30
tolal exhaust time (s)	0	60	0	60

The thickness of the i-layer (a-Si:H) in the top layer was sufficient at 240 nm. This was to ensure that the top layer current would have no effect on the bottom layer current, since the bottom layer current has a rate-controlling effect on I_{sc}.

For each condition test, five to eight cells were fabricated. The active area of each cell was 1 cm^2. The I–V characteristics of open circuit voltage (V_{oc}), short circuit current (I_{sc}), fill factor (FF), and conversion efficiency (Eff) were measured using a solar simulator under an air mass of 1.5 (AM 1.5) at 25 °C.

Pulsed PECVD was used at all times during the i-µc-Si:H process. As the pulsed PECVD conditions, a PECVD frequency of 40 MHz, a pulse modulation frequency of 30 kHz, and a duty ratio of 50% were adopted for i-layer deposition.

For the comparison of X_c, actual normal-condition and cell A cells were used in the analysis with a polishing method. This method was reported in [8]. The actual cell A was analyzed by secondary ion mass spectrometry (SIMS) [16]. Nitrogen and oxygen were analyzed through to the glass substrate from the back electrode side of the cell. SIMS measurementing conditions were as follows: Equipment: CAMECA ims5f (CAMECA, Gennevilliers Cedex, France) Ion conditions: Cs+, 14.5 keV, approximately 30 nA; Measuring area: approximately 30 µm diameter; Depth: calibration by Si sputtering.

In addition to the evaluation of cells, µc-Si:H films have been analyzed simultaneously. In the film evaluations, Eagle XG (Corning, NY, USA) was used as the substrate. A 20 nm p-layer was first deposited on a bare substrate under the same conditions as the cell. Then, two types of process (normal-condition and SEG) were carried out at the same thicknesses as for the cell, and these were labeled as normal-condition film and film A. The total µc-Si:H thickness of the normal-condition film and film A were "t", in the same manner as for the cell. For film A, when the thickness reached one-third of t (t/3) and two-thirds of t (2t/3), SEG processes were done in the same manner as for cell A. The evacuation time of the SEG of film A was 30 s. The i-µc-Si:H films of the normal-condition film and the film A were compared. Table 1 summarizes the conditions.

These two types of film (normal-condition film and film A) were analyzed by Raman spectrometry and Fourier transform infrared spectroscopy (FTIR). The quality of the µc-Si:H film was different from that of a cell because it is affected by the underlying shape. This evaluation was a relative comparison among the films. To distinguish the crystallinity of the cell expressed by "X_c" the crystallinity of the film is referred to as "X_c'". The crystallinity of µc-Si:H was evaluated using a 514 nm wavelength and a Via Reflex Raman microscope (Renishaw, Gloucestershire, UK). The crystallinity evaluated by Raman spectroscopy was defined as $X_c = I_{520}/I_{480}$, where the peak heights of I_{480} and I_{520} are the wavenumbers of 480 and approximately 520 cm^{-1}, respectively [17,18]. Fourier transform infrared spectroscopy (FTIR) measurements using the attenuated total reflection (ATR) method were carried out using a Spectrum 100 instrument (PerkinElmer, Waltham, MA, USA) [19]. The following stretching modes (SM) of Si–H bonds were used to fit the IR spectra: oxide silicone (O_ySiH_x) at about 2250 cm^{-1}, high SM (HSM) at 2100 cm^{-1}, middle SM (MSM) at 2040 cm^{-1}, low SM (LSM) at 2000 cm^{-1}, and extremely low SM (ELSM) at 1925 and 1895 cm^{-1} [20–22]. Each SM was analyzed using the Gaussian mode [23], and each SM fraction was defined as the ratio of the total area of all SMs divided by that SM's area. The LSM was used in the discussion section of this study. The two ELSM values were also simply expressed as ELSM.

3. Results

3.1. Comparison of Conversion Efficiency with the SEG Process Cell

Test results for the cells A–C are shown in Figure 2 and Table 3. Each of the values of Eff, V_{oc}, I_{sc}, and FF in Table 3 was the average for each condition. The error bars are also the standard deviation (δ) calculated for every condition. All results for the cells (five cells) are plotted in Figure 2. For cell A, Eff improved by 1.3% (0.14 points), V_{oc} by 0.011 V, I_{sc} by 0.0021, and FF by 0.002 compared with those for the normal-condition cell. The improved I_{sc} and FF were within a standard deviation range. Simply performing SEG thus improved the Eff, V_{oc}, I_{sc}, and FF.

For cell B, which has the extended vacuum time, Eff improved by 1.1% (0.12 points), V_{oc} by 0.008 V, I_{sc} by 0.08 mA compared with those for the normal-condition cell. This result shows that SEG has similar effect to that of cell A. However, no added benefit of Eff is obtained by extending the vacuum time. However, I_{sc} was higher than that for cells prepared under other conditions. The longer the vacuum time is, the higher the I_{sc} available is, but the value of the changed I_{sc} resembled the standard deviation range. The result of FF was within the margin of error.

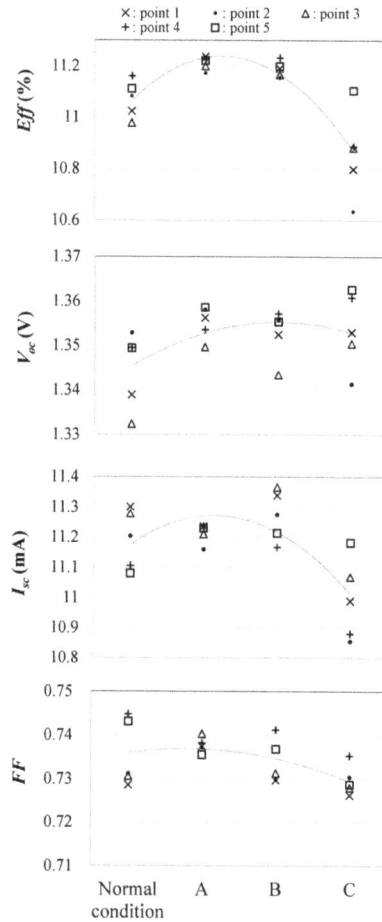

Figure 2. Results of each test cell: normal-condition cell, cell A, cell B, and cell C. For each cell type, five cells were fabricated. Points 1-5 were cells location of substrate.

Table 3. Results for each test cell. Each of the values of Eff, V_{oc}, I_{sc}, and FF was averaged for each condition. Error bars are standard deviation (δ) calculated for each condition.

	Normal-Condition Cell	Cell A	Cell B	Cell C
V_{oc} (V)	1.345 ± 0.008	1.355 ± 0.003	1.353 ± 0.005	1.354 ± 0.008
I_{sc} (mA)	11.194 ± 0.089	11.214 ± 0.029	11.272 ± 0.074	10.996 ± 0.121
FF	0.736 ± 0.007	0.738 ± 0.002	0.734 ± 0.005	0.730 ± 0.003
Eff (%)	11.07 ± 0.06	11.21 ± 0.02	11.19 ± 0.03	10.86 ± 0.15

For cell C, which increased the number of SEG treatments to three, Eff decreased by 1.8% (0.21 points), I_{sc} by 0.20 mA compared with those for the normal-condition cell. Although V_{oc} improved by 0.008 V, it was not better than for cell A. At this time of the experiment, two SEG trials are better than three trials.

Test results for cells D and E are shown in Table 4 and Figure 3. For cell D, Eff increased by 13.1% (1.14 points), I_{sc} by 1.33 mA, FF by 0.005 compared with the normal-condition cell D. However,

V_{oc} decreased by 0.027 V. For cell E, *Eff* increased by 2.1% (0.21 points), V_{oc} by 0.006, I_{sc} by 0.31 mA compared with the normal-condition cell E. But increases of V_{oc} and I_{sc} were within a standard deviation range. However, *FF* decreased by 0.009.

Table 4. Results for each test of cell D and cell E. Each of the values of *Eff*, V_{oc}, I_{sc}, and *FF* was averaged for each condition. Normal condition of D is without *SEG*. Normal condition of E is without *SEG*. Error bars are standard deviation (δ) calculated for each condition.

	Normal Condition (E)	Cell D	Normal Condition (D)	Cell E
V_{oc} (V)	1.337 ± 0.015	1.309 ± 0.021	1.325 ± 0.009	1.331 ± 0.005
I_{sc} (mA)	9.099 ± 0.660	10.424 ± 0.380	10.289 ± 0.320	10.597 ± 0.306
FF	0.718 ± 0.01	0.724 ± 0.00	0.7335 ± 0.00	0.724 ± 0.007
Eff (%)	8.72 ± 0.49	9.87 ± 0.167	10.00 ± 0.213	10.21 ± 0.278

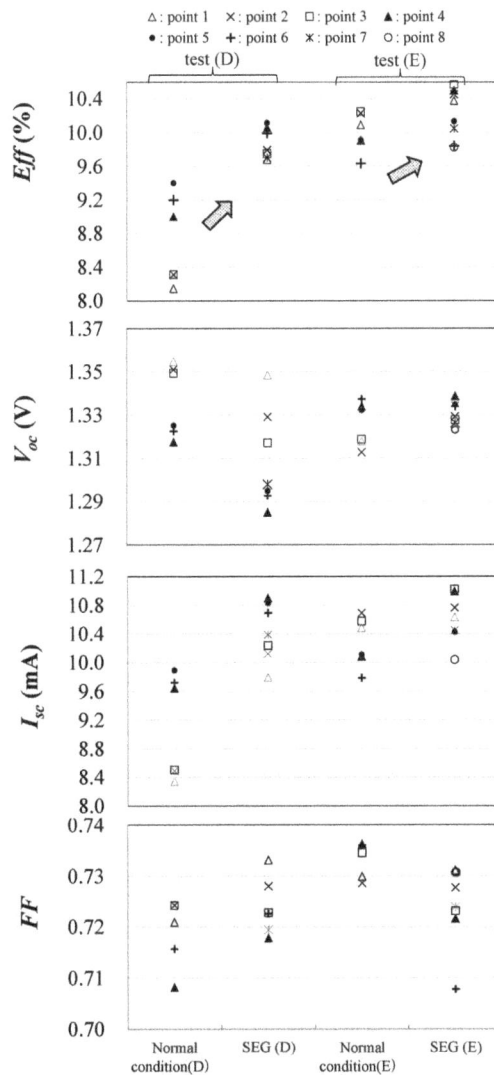

Figure 3. Results of each test cell: cell D, cell E, normal-condition cell D, and normal-condition cell E. For each cell type, eight cells were fabricated. Points 1-8 were cells location of substrate.

The results for the all I–V characteristics are relative to the bottom cell condition, because the thickness of the i-layer (a-Si:H) of the top layer is sufficient. For cell B, for which the thickness of the i-layer on top is 240 nm, the J_{sc} values of the top layer and bottom layer were 11.59 and 11.31 mA/cm^2, respectively, assuming external quantum efficiency (EQE). For cell D, for which the thickness of the top i-layer is 240 nm, the J_{sc} values of the top layer and bottom layer were 11.12 and 9.38 mA/cm^2, respectively, with external quantum efficiency. The bottom layer current has a rate-controlling effect on J_{sc}, because a top i-layer thickness of 240 nm was adequate.

3.2. Analysis Results of Actual Crystallinity Profile of µc-Si:H along with Depth

The Raman analysis results of actual crystallinity profiles of the normal-condition cell and the cell A are shown in Figure 4. The relationship between the film thickness and X_c is shown from the µc-Si:H thickness of zero. The obtained figure was plotted from the TCO side, and it also includes the thickness of the p-layer. This is because the p-layer of the bottom cell was µc-Si:H, but its thickness was very small (20 nm). The values of X_c were normalized by the maximum value of X_c among the normal-condition cell and the cell A.

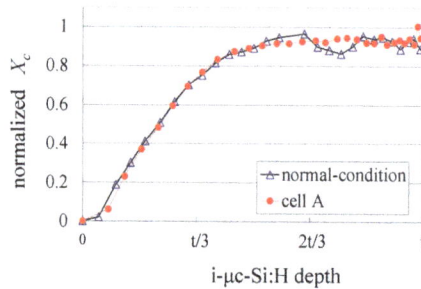

Figure 4. Results of Raman analysis of polished test cells; normal-condition cell and cell A. Values of X_c were normalized by the maximum value of X_c among normal-condition cell and cell A. X_c was expressed from the TCO side.

Both the X_c values did not show any difference between the normal-condition cell and the cell A up to the half thickness (t/2) of µc-Si:H. However, X_c of the normal-condition cell decreased at an around 2t/3. This decrease in X_c was probably caused by a defective region produced after the film thickness increased [8]. These decreases in the values of X_c were measured at four points successively. The values of X_c are connected smoothly, unlike the surrounding values. These decreases in X_c are therefore not due to measurement error. The decreasing X_c during the formation of a thick film inhibited the high X_c of µc-Si:H and effected the *Eff*.

Conversely, there was no reduction in X_c in the thickness direction with cell A. It means that *SEG* has prevented the defective region. In addition, X_c increased with the increasing film thickness, and a maximum X_c was confirmed. In the comparison of the X_c profiles between the normal-condition cell and cell A, *SEG* yielded high values of X_c.

3.3. Secondary Ion Mass Spectrometry Analyses with Cell

Nitrogen impurity levels were analyzed with the normal-condition cell and the cell A shown in Figure 5. Since it was analyzed from the back-contact side by SIMS, the results were expressed in the direction reverse to the deposition direction. In the no *SEG* region of µc-Si:H, nitrogen was detected by about 2.5×10^{18} (atom/cm^3). Nitrogen increased by approximately 10%–15% (about 2.9×10^{18} (atom/cm^3)) in the layers of the *SEG* region: at t/3 and 2t/3. In comparison, under normal conditions, it did not change significantly during the i-layer deposition. Nitrogen gas was never been used at all even after the deposition process in the process chamber. This nitrogen was brought form outside air by the substrate of the cell or film, or was released as degassing from the wall of the deposition chamber. Nitrogen gas was incorporated into the μc-Si:H [24].

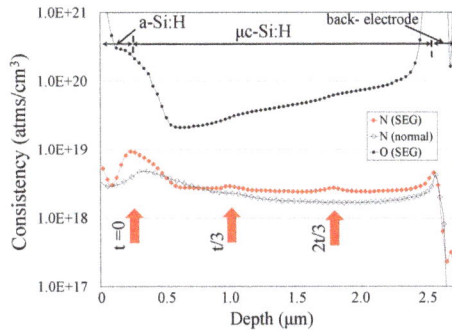

Figure 5. Profile of impurities about nitrogen and oxygen in cell A which used the *SEG* by SIMS. The nitrogen profile for normal-condition cells is also shown. Results were expressed in the direction reverse to the deposition direction.

On the one hand, between the a-Si:H layer and the μc-Si:H layer, nitrogen was detected with a threefold increase (1.0×10^{19} atom/cm^3) compared to other layers of μc-Si:H. When the processing equipment was changed from a-Si:H process to μc-Si:H process, cells were at once exposed to the atmosphere.

Oxygen impurity levels were analyzed with the cell A shown in Figure 5. Slightly more was detected in the *SEG* region of μc-Si:H in the same way as nitrogen. This was caused by the *SEG*. However, a difference level comparison between the *SEG* layer and no *SEG* layer was not as high as for nitrogen. The details are not noted here because the differences were negligible.

3.4. Film Analysis with FTIR and Raman

As a comparison of the normal-condition film and film A, the results of analysis by Raman spectrometry and FTIR are shown in Figure 6. The value of X_c' was normalized by a maximum value of film A.

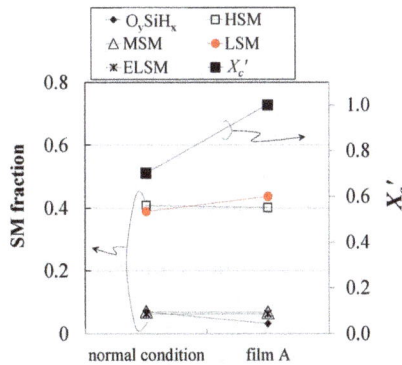

Figure 6. Analyzed results of film tests, normal-condition film and film A, by Raman spectrometry and FTIR. The values of X_c' were normalized by the value of the film A.

The film quality was different from the quality of the cell because that was deposited on the flat-glass rather than textured-TCO. The difference was spread significantly by the flat substrate. Thus, this confirmation was a relative comparison.

In film A, the LSM value was detected as stronger compared to a normal-condition film. This means that the film A contains a large amount of silicon monohydride (Si–H), so this component is expected to terminate the dangling bonds in μc-Si:H [25]. Also, O_ySiH_x which means crystalline grain boundaries was fewer than in the normal-condition film [19]. Owing to the *SEG*, the quality of μc-Si:H became better on the evaluation of the film.

However, X_c' of film A was higher than that of the normal condition film. The higher X_c', higher LSM, and lower O_ySiH_x of i-μc-Si:H such as film A can yield a high solar cell *Eff*. Thus, *SEG* successfully improved the μc-Si:H quality.

4. Discussion

To reduce the defective region correlation with the film thickness gain, the *SEG* process was carried out. Under its influence, as seen in Figure 4, X_c of μc-Si:H does not decrease the value of X_c during the process of forming a thick film. As a result, the cell achieved a high X_c and the value of X_c' increased. Also, as seen in Figure 6, the LSM value was detected as stronger than in a normal-condition film. Increasing the amorphous component such as LSM improved the quality of μc-Si:H by terminating the dangling bonds in μc-Si:H [26,27]. Eventually, the *Eff*, V_{oc}, and FF of cell A improved.

On the other hand, the difference between cell A and cell B, which was the test of discharge stop time, increased the I_{sc}. This characteristic difference was led not only by the initial discharge instability, but also by the impurity incorporation. Nitrogen and oxygen were also detected on the *SEG* cell with SIMS. On another hand, the X_c of μc-Si:H is changed by nitrogen, according to Ehara [28]. In our study of film A, the LSM which corresponds to the amorphous component (Si–H) was increased using the *SEG* layer. From this circumstantial evidence, it is reasonable to assume that nitrogen, which comes from the *SEG* process, increased the amorphous component. More nitrogen in cell B increases the N donor in the amorphous component and may increase the charged Si dangling bonds [29] and finally increased the I_{sc} more than cell A. Or oxygen was increased by *SEG* impurities in the growing film [30].

After the *SEG*, the amorphous component was deposited in the precursor defective region. The crystalline direction of μc-Si:H, which grows perpendicular to the textured substrate, was changed by the amorphous component [31]. As a result, collisions with the neighboring grains were reduced. The oxide might have influenced the X_c as well as nitrogen [32]. On the another front, a large amount of nitrogen was detected in the earlier thickness stage of μc-Si:H (around t = 0) in Figure 5. In exactly the same way, the value of X_c became low at its earlier thickness stage (around t = 0) in Figure 4. More nitrogen, which was carried from outside, might change it.

Moreover, in many of the *SEG* cell Cs, V_{oc} increased, and I_{sc} and FF decreased. Since the μc-Si:H film formation was stopped before the X_c growth increases, the continuous deposition time of μc-Si:H was shortened, and much of the amorphous phase easily formed [8]. Therefore, much of the amorphous phase increased V_{oc} [3], and decreased I_{sc} [33]. Since the shunt resistance was increased by the rising number of *SEG*, FF was decreased [34].

The other cells (D, E) showed improved *Eff*, even when certain conditions were changed. These results showed repeatability in how the *SEG* process improves efficiency. However, variations (increase or decrease) of V_{oc}, I_{sc} and FF did not show the same pattern. Thus, *Eff* changed depending on each condition. This is because the solar characteristic of i-μc-Si:H are determined by the proportion of μc-Si:H and a-Si:H. As a result, the best number for *SEG* varies depending on the i-layer condition. Double *SEG* does not always produce a high *Eff*, even though it showed improvement in this experiment. The number of *SEG*s can be determined by how high X_c becomes.

5. Conclusions

It was considered that stopping the plasma-discharge of PECVD during the deposition cannot generate a high quality film, because it is likely to contain impurities. However, in this study, stopping the plasma-discharge and applying vacuum for 30 s was done a few times very consciously during the deposition. This was aimed at increasing the X_c by avoiding collisions with the neighboring grains after the film became thicker. Then, collisions of μc-Si:H with each other were suppressed. Additionally, *SEG* improved the quality of the μc-Si:H film by increasing the amorphous component.

Finally, the X_c value of holistic µc-Si:H became high. As a result, *Eff* improved by 1.3% (0.14 points), compared with the normal-condition cell. This cell used the *SEG* process several times during the deposition of i-µc-Si:H. The optimal number of *SEG* applications depends on the conditions of the µc-Si:H film. It should be determined by how high X_c becomes. *SEG* is a process that involves simply stopping the plasma discharge of VHF PECVD. It is an easily applied process, the only difference being that the process time is a few minutes longer.

Acknowledgments: This work was in part supported by Yukiharu Uraoka and Yasuaki Ishikawa of Nara Institute of Science and Technology. The authors thank them for constructive discussion.

Author Contributions: In this paper, Takeyuki Sekimoto, Mitsuhiro Matsumoto and Akira Terakawa performed the experiments and collected and analyzed data and gave some useful suggestions for this work and helped revise the manuscript. Moreover, Yukiharu Uraoka and Yasuaki Ishikawa of Nara Institute of Science and Technology gave some useful suggestions for this work.

Conflicts of Interest: The authors declare no conflict of interest.

References

1. Cuony, P.; Marending, M.; Alexander, D.T.L.; Boccard, M.; Bugnon, G.; Despeisse, M.; Ballif, C. Mixed-phase p-type silicon oxide containing silicon nanocrystals and its role in thin film silicon solar cells. *Appl. Phys. Lett.* **2010**, *97*, 213502. [CrossRef]

2. Naruse, Y.; Matsumoto, M.; Sekimoto, T.; Hishida, M.; Aya, Y.; Shinohara, W.; Fukushima, A.; Yata, S.; Terakawa, A.; Iseki, M.; *et al.* Identification of defective regions in thin-film Si solar cells for new-generation energy devices. In Proceedings of the 38th IEEE Photovoltaic Specialists Conference, Austin, TX, USA, 3–8 June 2012; pp. 003118–003123.

3. Bailat, J.; Domine, D.; Schluchter, R.; Steinhauser, J.; Fay, S.; Freitas, F.; Bucher, C.; Feitknecht, L.; Niquille, X.; Tscharner, T.; *et al.* High-Efficiency P-I-N Microcrystalline and Micromorph Thin Film Silicon Solar Cells Deposited on LPCVD Zno Coated Glass Substrates. In Proceedings of the 4th World Conference Photovoltaic Energy Conversion, Waikoloa, HI, USA, 7–12 May 2006; Volume 2, pp. 1533–1536.

4. Boccard, M.; Despeisse, M.; Escarre, J.; Niquille, X.; Bugnon, G.; Hanni, S.; Eymard, M.B.; Meillaud, F.; Ballif, C. High-Stable-Efficiency Tandem Thin-Film Silicon Solar Cell With Low-Refractive-Index Silicon-Oxide Interlayer. *IEEE J. Photovolt.* **2014**, *4*, 1368–1373. [CrossRef]

5. Matsui, T.; Bidiville, A.; Maejima, K.; Sai, H.; Koida, T.; Suezaki, T.; Matsumoto, M.; Saito, K.; Yoshida, I.; Kondo, M. High-efficiency amorphous silicon solar cells: Impact of deposition rate on metastability. *Appl. Phys. Lett.* **2015**, *106*, 053901. [CrossRef]

6. Python, M.; Madani, O.; Dominé, D.; Meillaud, F.; Vallat-Sauvain, E.; Ballif, C. Influence of the substrate geometrical parameter son microcrystalline silicon growth for thin-film solar cells. *Sol. Energy Mater. Sol. Cells* **2009**, *93*, 1714–1720. [CrossRef]

7. Despeisse, M.; Bugnon, G.; Feltrin, A.; Stueckelberger, M.; Cuony, P.; Meillaud, F.; Billet, A.; Ballif, C. Resistive interlayer for improved performance of thin film silicon solar cells on highly textured substrate. *Appl. Phys. Lett.* **2010**, *96*, 073507. [CrossRef]

8. Hishida, M.; Ueno, H.; Sekimoto, T.; Terakawa, A. Use of microcrystallinity depth profiling in an actual tandem silicon solar cell by polishing to achieve high conversion efficiency. *Jpn. J. Appl. Phys.* **2015**, *54*, 052302. [CrossRef]

9. Das, U.; Morrison, S.; Centurioni, E.; Madan, A. Thin film silicon materials and solar cells grown by pulsed PECVD technique. *IEE Proc. Circuits Devices Syst.* **2003**, *150*, 282–286. [CrossRef]

10. Choong, G.; Vallat-Sauvain, E.; Multone, X.; Fesquet, L.; Kroll, U.; Meier, J. Measurements of Raman crystallinity profiles in thin-film microcrystalline silicon solar cells. *J. Phys.* **2013**, *46*, 235105. [CrossRef]

11. Kimura, H.; Maeda, H.; Murakami, H.; Nakahigashi, T.; Ohtani, S.; Tabata, T.; Hayashi, T.; Kobayashi, M.; Mitsuda, Y.; Nakamura, N.; *et al.* Study of Deposition Process in Modulated RF Silane Plasma. *Jpn. J. Appl. Phys.* **1994**, *33*, 4389–4394. [CrossRef]

12. Kondo, M.; Fujiwara, H.; Matsuda, A. Fundamental aspects of low-temperature growth of microcrystalline silicon. *Thin Solid Films* **2003**, *430*, 130–134. [CrossRef]

13. Matsui, T.; Kondo, M. Advanced materials processing for high-efficiency thin-film silicon solar cells. *Sol. Energy Mater. Sol. Cells* **2013**, *119*, 156–162. [CrossRef]

14. Urbain, F.; Smirnov, V.; Becker, J.-P.; Rau, U.; Finger, F.; Ziegler, J.; Kaiser, B.; Jaegermann, W. a-Si:H/μc-Si:H tandem junction based photocathodes with high open-circuit voltage for efficient hydrogen production. *J. Mater. Res.* **2014**, *29*, 2605–2614. [CrossRef]

15. Terakawa, A. Review of thin-film silicon deposition techniques for high-efficiency solar cells developed at Panasonic/Sanyo. *Sol. Energy Mater. Sol. Cells* **2013**, *119*, 204–208. [CrossRef]

16. Kilper, T.; Beyer, W.; Bräuer, G.; Bronger, T.; Carius, R.; van den Donker, M.N.; Hrunski, D.; Lambertz, A.; Merdzhanova, T.; Mück, A.; *et al.* Oxygen and nitrogen impurities in microcrystalline silicon deposited under optimized conditions: Influence on material properties and solar cell performance. *J. Appl. Phys.* **2009**, *105*, 074509. [CrossRef]

17. Droz, C.; Vallat-Sauvain, E.; Bailat, J.; Feitknecht, L.; Meier, J.; Shah, A. Relationship between Raman crystallinity and open-circuit voltage in microcrystalline silicon solar cells. *Sol. Energy Mater. Sol. Cells* **2004**, *81*, 61–71. [CrossRef]

18. Matsui, T.; Tsukiji, M.; Saika, H.; Toyama, T.; Okamoto, H. Correlation between Microstructure and Photovoltaic Performance of Polycrystalline Silicon Thin Film Solar Cells. *Jpn. J. Appl. Phys.* **2002**, *41*, 20–27. [CrossRef]

19. Mirabella, F.M., Jr. *Internal Reflection Spectroscopy: Theory and Applications*; Mirabella, F.M., Ed.; Marcel Dekker, Inc.: New York, NY, USA, 1993.

20. Smets, A.H.M.; Matsui, T.; Kondo, M. High-rate deposition of microcrystalline silicon p-i-n solar cells in the high pressure depletion regime. *J. Appl. Phys.* **2008**, *104*, 034508. [CrossRef]

21. Smets, A.H.M.; Matsui, T.; Kondo, M. Infrared analysis of the bulk silicon-hydrogen bonds as an optimization tool for high-rate deposition of microcrystalline silicon solar cells. *Appl. Phys. Lett.* **2008**, *92*, 033506. [CrossRef]

22. Smets, A.H.M.; Matsui, T.; Kondo, M.; van de Sanden, M.C.M. The hydride stretching modes of hydrogenated vacancies in amorphous and nanocrystalline silicon: A helpful tool for material characterization. In Proceedings of the 34th IEEE Photovoltaic Specialists Conference, Philadelphia, PA, USA, 7–12 June 2009; pp. 721–724.

23. Yamashita, K.; Minami, S. Evaluation of Numerical Filters for Smoothing Spectroscopic Data. *Jpn. J. Appl. Phys.* **1969**, *8*, 1505–1512. [CrossRef]

24. Quach, N.T.; Reif, R. Solid phase epitaxy of polycrystalline silicon films: Effects of ion implantation damage. *Appl. Phys. Lett.* **1984**, *45*, 910–912. [CrossRef]

25. Kondo, M.; Fukawa, M.; Guo, L.; Matsuda, A. High rate growth of microcrystalline silicon at low temperatures. *J. Non-Cryst. Solids* **2000**, *266–269*, 84–89. [CrossRef]

26. Xu, L.; Li, Z.P.; Wen, C.; Shen, W.Z. Bonded hydrogen in nanocrystalline silicon photovoltaic materials: Impact on structure and defect density. *J. Appl. Phys.* **2011**, *110*, 064315. [CrossRef]

27. Wen, C.; Xu, H.; He, W.; Li, Z.; Shen, W. Tuning oxygen impurities and microstructure of nanocrystalline silicon photovoltaic materials through hydrogen dilution. *Nanoscale Res. Lett.* **2014**, *9*. [CrossRef] [PubMed]

28. Ehara, T. The crystalline properties of nitrogen doped hydrogenated microcrystalline silicon thin films. *Thin Solid Films* **1997**, *310*, 322–326. [CrossRef]

29. Akiharu, M.; Minoru, M.; Masahiro, Y.; Minoru, K.; Tatsuo, S. Effect of Reduction in Impurity Content for a-Si:H Films. *Appl. Phys. Lett.* **1991**, *59*, 2130.

30. Keppner, H.; Meier, J.; Torres, P.; Fischer, D.; Shah, A. Microcrystalline silicon and micromorph tandem solar cells. *J. Appl. Phys.* **1999**, *69*, 169–177. [CrossRef]

31. Sakai, H.; Yoshida, T.; Hama, T.; Ichikawa, Y. Effects of Surface Morphology of Transparent Electrode on the Open-Circuit Voltage in a-Si:H Solar Cells. *Jpn. J. Appl. Phys.* **1990**, *29*, 630–635. [CrossRef]

32. Lambertz, A.; Grundler, T.; Finger, F. Hydrogenated amorphous silicon oxide containing a microcrystalline silicon phase and usage as an intermediate reflector in thin-film silicon solar cells. *J. Appl. Phys.* **2011**, *109*, 113109. [CrossRef]

33. Flikweert, A.J.; Zimmermann, T.; Merdzhanova, T.; Weigand, D.; Appenzeller, W.; Gordijn, A. Microcrystalline thin-film solar cell deposition on moving substrates using a linear VHF-PECVD reactor and a cross-flow geometry. *J. Phys. D* **2012**, *45*, 015101. [CrossRef]

34. Meillaud, F.; Shah, A.; Bailat, J.; Vallat-Sauvain, E.; Roschek, T.; Rech, B.; Dominé, D.; Söderström, T.; Python, M.; Ballif, C. Microcrystalline silicon solar cells: Theory and diagnostic tools. In Proceedings of the 4th World Conference Photovoltaic Energy Conversion, Waikoloa, HI, USA, 7–12 May 2006; pp. 1572–1575.

7

A Simple Operating Strategy of Small-Scale Battery Energy Storages for Energy Arbitrage under Dynamic Pricing Tariffs

Enrico Telaretti *, Mariano Ippolito and Luigi Dusonchet

Academic Editor: William Holderbaum

Department of Energy, Information Engineering and Mathematical Models, University of Palermo, Viale delle Scienze, 90128 Palermo, Italy; ippolito@dieet.unipa.it (M.I.); dusonchet@dieet.unipa.it (L.D.)
* Correspondence: telaretti@dieet.unipa.it

Abstract: Price arbitrage involves taking advantage of an electricity price difference, storing electricity during low-prices times, and selling it back to the grid during high-prices periods. This strategy can be exploited by customers in presence of dynamic pricing schemes, such as hourly electricity prices, where the customer electricity cost may vary at any hour of day, and power consumption can be managed in a more flexible and economical manner, taking advantage of the price differential. Instead of modifying their energy consumption, customers can install storage systems to reduce their electricity bill, shifting the energy consumption from on-peak to off-peak hours. This paper develops a detailed storage model linking together technical, economic and electricity market parameters. The proposed operating strategy aims to maximize the profit of the storage owner (electricity customer) under simplifying assumptions, by determining the optimal charge/discharge schedule. The model can be applied to several kinds of storages, although the simulations refer to three kinds of batteries: lead-acid, lithium-ion (Li-ion) and sodium-sulfur (NaS) batteries. Unlike literature reviews, often requiring an estimate of the end-user load profile, the proposed operation strategy is able to properly identify the battery-charging schedule, relying only on the hourly price profile, regardless of the specific facility's consumption, thanks to some simplifying assumptions in the sizing and the operation of the battery. This could be particularly useful when the customer load profile cannot be scheduled with sufficient reliability, because of the uncertainty inherent in load forecasting. The motivation behind this research is that storage devices can help to lower the average electricity prices, increasing flexibility and fostering the integration of renewable sources into the power system.

Keywords: price arbitrage; battery energy storage system; optimal operation; hourly electricity prices; energy management

---

1. Introduction

Electricity customers will face significant challenges in the near future due to the most recent developments in the energy market sector. These changes have been mainly driven by the increasing penetration of renewable and distributed energy sources in the power system, which can positively contribute to a reduction of CO_2 emissions. The diffusion of renewable sources has been made possible thanks to the introduction of support policies, such as those put in place for the photovoltaic (PV) and wind technology [1–4]. Clearly, the transition from the current centralized electricity market structure towards a decentralized market model will require major investments in the electricity grid infrastructure, in order to ensure an adequate level of quality and reliability of the energy supply.

In the spot markets, the electricity price varies stochastically from one day to the next and systematically between seasons. The marginal cost of producing energy has become much more volatile in the last decade, mainly due to the recent moves toward competitive liberalized markets. Indeed, the competition among actors has increased the range of variability in electricity prices, expanding the difference between on-peak and off-peak prices. Normally, electricity users are not exposed to these fluctuations but pay a constant price. In an attempt to reduce demand peaks, several utilities are moving from a conventional fixed-rate pricing scheme to new market-based models, where the electricity cost is free to fluctuate depending on the balance between supply and demand. Such dynamic pricing schemes reflect the prices of the wholesale market and are able to lower demand peaks and the volatility of the wholesale prices [5]. A first example of dynamic pricing tariff is time-of-use (TOU) pricing, which provides two or three periods of different electricity price (generally "on-peak", "mid-peak" and "off-peak" prices), depending on the hour of day. Electricity users are advised in advance about electricity prices that are not normally modified more than once or twice per year. A more flexible electricity-pricing scheme is real-time pricing (RTP), for which the retail electricity price closely reflects the wholesale energy price. In this case, customer electricity prices can vary hourly depending on the wholesale market and electricity users can manage their power consumption in a more flexible and economical manner, taking advantage of the price differential. The real-time prices can be notified to electricity customers with different timing, depending on the specific utility's RTP program. For example, with Ameren's RTP program (an Illinois' Electric Utility), hourly prices for the next day are set the night before and are communicated to customers so they can modify their power consumption in advance. Differently, with ComEd's RTP program (another Illinois' Electric Utility), hourly prices are based on the average of the twelve five-minute prices for each hour, and electricity users are notified in real-time, only when the hour has passed. Later on in this article, the RTP prices will be considered as day-ahead hourly prices, so electricity customers are advised a day before and can modify their power consumption accordingly.

The highly volatile behavior of the electricity price can be exploited by using an energy storage device in order to capture the price differential. Indeed, if an electricity customer is charged at an hourly-dependent rate, a storage system can be adopted with the aim to shift portions of consumption to different hours than those where they actually occur. The electricity is simply stored when it is inexpensive and resold back to the grid at a higher price [6,7].

The object of this article is to analyze, develop and demonstrate a charge/discharge scheduling method able to maximize the arbitrage benefit of a storage system, subject to technical constraints. The storage system is described by means of its performance parameters, such as the charge and generation capacity, the charge/discharge efficiency, the rated charge/discharge rate, the depth-of-discharge (DOD), etc., which are sufficient to evaluate the arbitrage potential of a storage system. The scheduling strategy is based on the definition of an objective function, able to maximize the arbitrage benefit of the storage owner subject to technical constraints, allowing the battery to be charged/discharged at different DOD, as further detailed in Section 4. The developed model is valid for any kind of storage, although the simulations refer to a lead-acid, a lithium-ion (Li-ion) and a sodium-sulfur (NaS) battery. Test results show that the proposed operating strategy is effective to maximize the profit for the customer. Unlike the studies reported in the literature, often requiring an estimate of the end-user load profile, the proposed operation strategy is able to properly identify, for each daily period, the charge/discharge hours relying only on the hourly spot market price profile, regardless of the specific facility's consumption. This is made possible thanks to some simplifying assumptions in the sizing and the operation of the battery energy storage system (BESS), as further details in Section 3. This could be particularly useful when the customer load profile cannot be scheduled with sufficient reliability because of the uncertainty inherent in load forecasting. In these cases, identifying a BESS operating strategy that does not depend on the user's power profile can be an important task, since the deviation of the scheduled power profile from the effective one could affect the results obtained using more complete methods. Furthermore, the proposed management strategy requires a low computational burden and can be implemented in simple and available software, for instance in a spreadsheet, representing a friendly but effective instrument to optimize the charge/discharge schedule of a storage device.

The next section summarizes existing literature on the topic of optimal operation of storage systems. In Section 3, the customer energy system used in this paper is briefly described and the basic operational assumptions are outlined. In Section 4, the problem formulation is provided, showing the objective function to be maximized and defining the constraint equations. In Section 5, a case study is presented and the technical and economic parameters for each storage device are provided. Section 6 shows the simulation results and some important remarks about the operating schedule of the storage devices. Finally, Section 7 summarizes the conclusion of the work.

2. Current Literature

Traditionally, most of the studies address the optimal operation of a storage system based on linear programming [8–11], nonlinear programming [12], dynamic programming [13–16] and multipass iteration particle swarm optimization approach [17]. Other charge/discharge strategies are described in [18–25].

2.1. Linear and Nonlinear Programming

In [8], the authors study the optimal operation of an energy storage unit installed in a small power producing facility using a conventional linear programming technique. In [9], the authors determine the optimal charge/discharge schedule by using a linear optimization model of the battery systems (based on Li-ion and lead-acid technology) for arbitrage accommodation. They found that the cost and the efficiency of the storage systems have the highest impact on simulation results. The developed model is linear and can thus be solved without much computational effort. Bradbury *et al.* [10] studied seven real-time US electricity markets and 14 different storage technologies, finding that the optimal profit-maximizing size of a storage device (*i.e.*, hours of energy storage) depends largely on its technological characteristics (round-trip charge/discharge efficiency and self-discharge), rather than the magnitude of market price volatility, which instead increases internal rate of return (IRR). The arbitrage benefit is maximized using a simple linear programming, subject to technical constraints. Graves *et al.* [11] emphasize the fact that using average peak and off-peak prices does not account for the variability in prices and thus leading to significant errors in the optimal management strategy. They also discuss the use of a linear programming for determining the optimal operation strategy.

In [12], the authors present an optimal operation strategy of BESSs to the real-time electricity price in order to achieve maximum profits of the BESS. The algorithm is based on a sequential quadratic programming method as to maximize the profits for the customer. The strategy is promising although operating and maintenance costs of the BESS are not taken into account.

2.2. Dynamic Programming

Linear programming is often considered to be too inflexible, as it typically does not capture the stochastic nature of load profiles. In order to overcome the restriction, dynamic programming methods are employed to capture the uncertainties in load profiles and electricity prices [13]. The algorithm developed in [14] is a multipass dynamic programming that ensures the minimization of the electricity bill for a given battery capacity, while reducing stress on the battery and prolonging battery life. In [15], the authors address the problem of organizing home energy storage purchases as a Markov decision process, showing that there exists a threshold-based stationary cost-minimizing policy. The battery is charged up to the threshold, when the battery level is below the threshold, and discharged when the level is above the threshold. The proposed strategy is interesting, even though the system cost is not considered. In [16], the authors propose a self-learning optimal operating control scheme based on adaptive dynamic programming for the residential energy system with batteries. The algorithm is effective in achieving minimization of the cost through neural network learning. The main feature of the proposed scheme is the ability of the continuous learning and adaptation to improve the performance during real-time operations under uncertain changes in the environment or new system configuration of the residential household.

2.3. Other BESS Management Strategies

In [17], a modified particle swarm optimization (PSO) algorithm (called multipass iteration PSO) is used to solve the optimal operating schedule of a BESS for an industrial TOU rate user with wind turbine generators. Thanks to the high computational efficiency, the algorithm can be used to evaluate the optimal operating policy of a BESS in real-time applications, based on the load condition of the user, the energy left in the BESS, and the output of wind turbines. In [18], the authors estimate the benefit of using energy storages for aggregate storage applications, such as energy price arbitrage, TOU energy cost reduction, ancillary services, and transmission upgrade deferral. The maximization of the arbitrage benefit is carried out by maximizing an objective function, under the assumption that the electricity prices are both dependent/independent on the battery operation. In [19], a simple methodology to charge/discharge a residential battery system for energy arbitrage in presence of TOU prices was described. The statistical variability of the household consumption was accounted through a Monte Carlo method. The economic feasibility of the storage system was determined in the context of the Australian retail electricity market, showing that, for various BESSs, the load shifting strategy can be profitable. In [20], the authors present an estimation of the economic feasibility of electricity storage in the west Danish power market, exploiting a simple operation strategy of the BESS in the spot market. The strategy includes two main conditions: (1) the price for buying must be less than the price for selling times the round trip efficiency (in order to ensure positive incomes) and (2) the amount of power bought in a given time period must equal the amount of power sold times the round trip efficiency (in order to ensure the balance of energy). Shcherbakova *et al.* (2014) [21] simulated the operation and resulting profits of small storage batteries (NaS and Li-ion) in South Korea using a charge/discharge strategy based on Hotelling rule. They concluded that neither technology generates a sufficient amount of arbitrage revenue to cover the battery's capital costs. Purvins and Summer [22] presented an optimal battery system management model in distribution grids for lithium-ion battery system used in stationary applications. The proposed approach is based on three management priorities, the first being the maximum utilization of renewable energy sources (RES) energy in distribution grids (preventing situations of reverse power flow at the distribution level), followed by efficient battery utilization (charging at off-peak prices and discharging at peak prices) and residual distribution grid demand smoothing. Finally, in [23,24], the authors evaluate the capacity of storage and active demand side management (DSM) to increase the self-consumed electricity in the residential sector, using a lead–acid battery. The operating strategy is based on self-consumption maximization, reducing the use of the grid and supplying the highest amount of energy from PV generation. In [25], the authors present a home energy management system model that uses a heuristic algorithm to manage and control home appliances based on a combination of energy pricing models including TOU and *RTP* tariffs. The algorithm aims to minimize overall usage and cost of energy without significantly degrading consumer comfort.

3. Energy System Description and Operational Assumptions

The customer energy system consists of a passive user (end-user), interconnected to a storage system through a bidirectional converter, as depicted in Figure 1. The bidirectional converter consists of a rectifier AC/DC (the battery charger) and an inverter DC/AC [26,27]. The battery system is handled in order to ensure an economic benefit for the customer, exploiting a load shifting strategy. Since the system marginal price (SMP) value is available one day ahead and it is defined each hour, the electricity prices are considered as hourly-dependent prices, where each hour of the day has a different electricity price. The reference period used in the study is one day, *i.e.*, the battery operation is defined starting from a vector of 24 elements as input data.

Three different operating modes are considered for the storage system: charging mode, activated when the electricity prices are low; standby mode, in which the power grid supplies directly the end-user without contribution of the storage; and discharging mode, activated when the electricity prices are high, where part of the load is supplied from the battery.

The following assumptions have been made:

- The end-user is allowed to buy the consumed energy at an hourly tariff (*RTP* tariff), defined by the utility on a daily basis. The *RTP* tariffs are assumed to be proportional to the SMP values, by applying a percentage increase to incorporate the benefit for the utility and taxes (electricity tax and value added tax (VAT)).
- The power flow is always directed from the grid to the load. The stored energy can only be used by the customer for load compensation and cannot be sold to the utility.
- The hourly electricity prices are known in advance in a finite horizon setting (daily period) and the use of the storage device does not influence the prices of electricity in the energy market (small price taking storage devices). Predictions about future electricity rates are not part of this work since the aim is to show results based upon the current electricity prices.
- Battery self-discharge is disregarded.
- Battery capacity is assumed constant throughout the battery life, without degradation.
- The common frictions during battery operation are accounted for by incorporating imperfect charging and discharging efficiency;

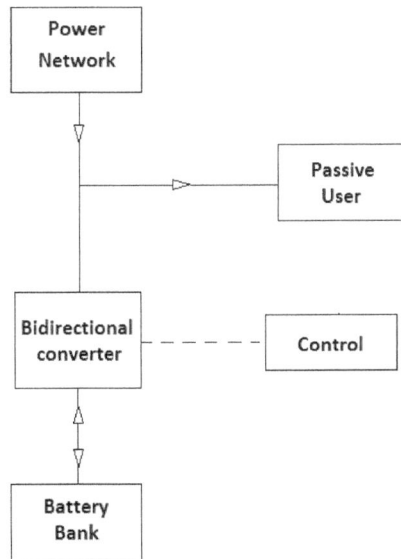

Figure 1. Grid-connected customer energy system operating in parallel with the storage system.

- The charge/discharge rate of the battery is assumed constant and equal to the rated power capacity of the device. Doing so, the storage charge/discharge constraints are automatically satisfied (*i.e.*, the energy charged/discharged into the battery at any time *t* cannot be more than the rated power capacity of the device). It is worth noting that both the battery capacity and the battery life are influenced by the charging rate. Indeed, at very high rates the capacity cell and the battery life are reduced. Fast charging may also have negative consequences on the battery efficiency [28]. Therefore, the use of a battery at constant charge/discharge rate helps

to prolong the battery life, to preserve the rated capacity and to keep the battery efficiency at appropriate values.
- The charging time is assumed equal to the discharging time, in each operating cycle. According to the last two mentioned hypotheses, the battery returns to the initial state-of-charge (SOC) at the end of each operating cycle. Such an operation means that the battery energy constraints are automatically satisfied (*i.e.*, the storage level of the battery cannot be more than the rated energy capacity of the device).
- The *DOD* of the battery can take different discrete states, depending on the value of the objective function.
- The storage capacity is assumed equal to the facilities' energy consumption during peak times (*i.e.*, the hours where electricity prices are the highest) on the day of the year of lowest consumption [29]. In other words, the battery is sized so that it can supply the entire customer load during peak price hours, on the day of the year of lowest consumption, and only a portion of the customer's load on the other days. The choice of the storage capacity is driven by a trade-off between gaining more arbitrage savings during days with relatively high peak loads and wasting idle capacity during days with low peak loads. Among all the possible solutions, the one that ensures the minimum upfront investment cost for the storage owner has been chosen. The aim of this article is to identify a battery operating strategy able to maximize the profit of the storage owner (under the considered assumptions), without attempting to identify the optimal BESS capacity. In other words, the battery has been sized according to a criterion of minimum cost, which is not necessarily the optimal one. As a consequence of this statement, the BESS can be operated regardless of the specific facility's load profile and the power flow is always directed from the grid to the load, without selling to the utility.

4. Problem Formulation

4.1. Preliminary Considerations

The optimal operating strategy of the storage device is able to uniquely determine the daily charge/discharge intervals so as to maximize the economic saving for the customer. Figure 2 shows typical daily profiles of SMP (the national single price of the Italian day-ahead market) for a reference weekly period (from 31 March to 6 April 2014) [30]. The profiles clearly show a first couple of min/max prices in the first semi-daily period and a second couple in the second half of the day. The battery thus will be charged only once a day, twice a day or it will remain idle, depending on the maximization of the objective function. Since the *RTP* tariffs are assumed to be proportional to the SMP values, hereinafter will be referred as *RTP* prices. It is worth noting that weekdays *RTP* values have a first price peak at about 8:00–10:00 a.m. and a second peak at 8:00–9:00 p.m. Differently, Sunday only retains the second peak at 9:00 p.m. As a result, we can expect that the BESS could be charged two times on weekdays (including Saturday), only one time on Sunday.

Since the battery can be charged/discharged at different *DOD*, the algorithm calculates the moving average (*MA*) of *RTP* prices (*MA RTP*) corresponding to each charge/discharge time, *d*, where *d* is a discrete variable denoting the charge/discharge time of the battery (corresponding to different *DOD* values). For example, assuming that the charge/discharge time, *d*, can take *D* different discrete values, the algorithm calculates *D* daily profiles of *MA RTP* prices, for each day of the year, *i*:

$$MA_{RTP_{i,d}}(h) = \sum_{n=h}^{h+d-1} RTP(n)/d \; h = 1, \ldots, 24 - d + 1; d = 1, \ldots, D; i = 1, \ldots, 365 \qquad (1)$$

where *d* is an index denoting the charge/discharge time of the battery, *i* is an index denoting the day of the year, *h* is an index denoting the hour of the day, *D* is the maximum charge/discharge time of the battery (corresponding to the maximum *DOD*) and $MA_{RTP_{i,d}}(h)$ is the *MA* of *RTP* prices in hour

h, corresponding to the charge/discharge time d in the day i. In the following, all equations will be referred to a generic day i, and the variability of the index over the year will be omitted.

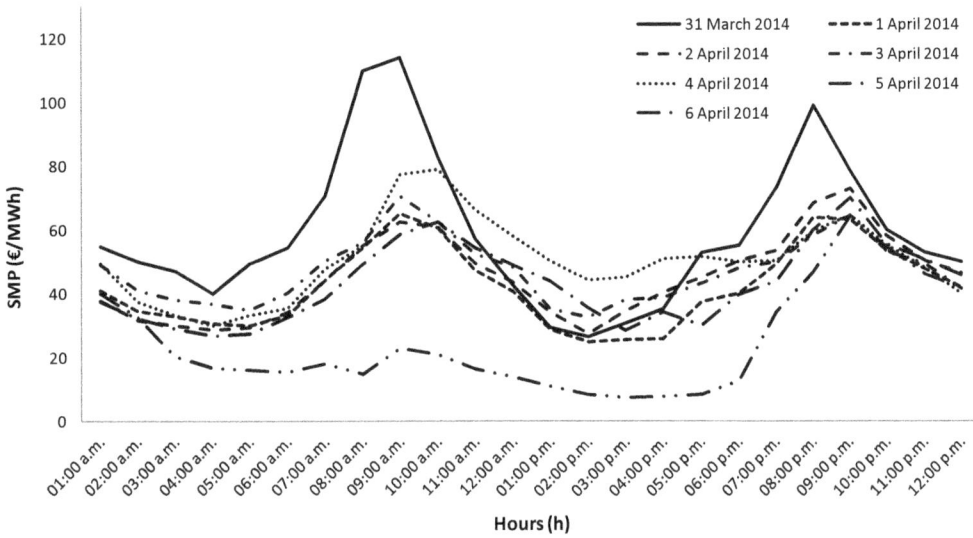

Figure 2. System marginal price (SMP) for the weekly period from 31 March to 6 April 2014.

Since the charge/discharge rate of the battery, P_{BESS}, is assumed constant, the following relation exists between the DOD and the discharged time, d:

$$DOD = \frac{E_{BESS}}{Cap} = \frac{P_{BESS} \cdot d}{P_{BESS} \cdot D_{max}} = \frac{d}{D_{max}} \qquad (2)$$

where E_{BESS} is the energy discharged from the storage device, Cap is the rated energy capacity of the BESS, and D_{max} is the maximum theoretical discharging time of the battery, corresponding to a full discharge (this is a theoretical discharging value, since the battery can never fully discharge).

Since the battery can be charged once or twice a day, depending on the maximization of the objective function, the algorithm takes into account two *MA RTP* profiles for each charge/discharge time d, the first referred to a daily period, the second to a semi-daily period. In other words, the algorithm scans both the daily and the semi-daily *MA RTP* profiles, with the aim of verifying whether the maximum of the objective function corresponds to only one cycle or to two cycles per day. Figure 3a,b shows the daily profile of *MA RTP* related to a daily period or to a semi-daily period, together with the daily/semi-daily average value, respectively:

$$Aver_{MA_{i,d}} = \sum_{h=1}^{24-d+1} MA_{RTP_{i,d}}(h) / (24 - d + 1) \; ; \; d = 1, \ldots, D \qquad (3)$$

$$\begin{cases} Aver_{MA_{i,d}^{(1)}} = \sum_{h=1}^{12} MA_{RTP_{i,d}}(h) / 12 \\ Aver_{MA_{i,d}^{(2)}} = \sum_{h=(12-d+1)}^{(24-d+1)} MA_{RTP_{i,d}}(h) / 12 \end{cases} \quad d = 1, \ldots, D \qquad (4)$$

where $Aver_{MA_{i,d}}$ is the daily average value of the *MA RTP* profile and $Aver_{MA_{i,d}^{(k)}}$ is the semi-daily average value of the *MA RTP* profile (in the semi-daily period k of the day i, with $k = 1, 2$). Figure 3a,b also shows the min/max values of *MA RTP* profiles in the daily/semi-daily period:

$$MA_{RTP_{i,d,min}}, \ MA_{RTP_{i,d,max}} \ ; d = 1, \ldots, D \tag{5}$$

$$MA_{RTP_{i,d,min}^{(k)}}, \ MA_{RTP_{i,d,max}^{(k)}} \quad d = 1, \ldots, D \ ; k = 1, \ 2 \tag{6}$$

where $\left(MA_{RTP_{i,d,min}}, \ MA_{RTP_{i,d,max}} \right)$ is the couple of min/max *MA RTP* values in a daily period and $\left(MA_{RTP_{i,d,min}^{(k)}}, \ MA_{RTP_{i,d,max}^{(k)}} \right)$ is the couple of min/max *MA RTP* values in the semi-daily period k of the day i, respectively. The average values and the min/max *MA RTP* values are calculated for each charge/discharge time d and for each day i. The daily profile in Figure 3 corresponds to the *RTP* prices when $d = 1$, to the *MA* of *RTP* prices when $d \neq 1$.

4.2. Optimization Problem Formulation

Since the battery can be charged once or twice a day, depending on the value of the objective function, the algorithm calculates the benefit for the storage owner (electricity customer) in both cases, verifying in which situation the objective function takes the maximum value. In the following sections, the objective function will be defined in both situations, by considering a daily or a semi-daily periodicity, respectively.

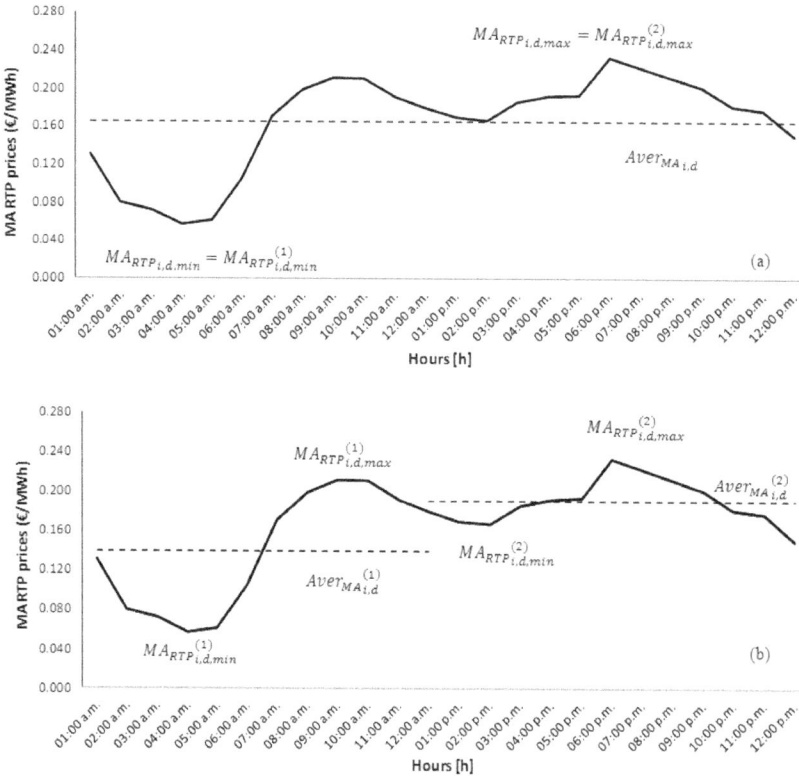

Figure 3. Daily profile of moving average of *RTP* prices (*MA RTP*) with daily average (**a**) and semi-daily average values (**b**).

4.2.1. Semi-Daily Periodicity

Under the assumption of semi-daily periodicity, the storage device will perform two charging cycles per day, according to the *MA RTP* profile shown in Figure 3b. For each battery cycle, the problem comes down to maximizing the following objective function:

$$OF_{i,d}^{(k)} = max\left(S_{BESS,i,d}^{(k)} - C_{BESS_{cycled,d}}\right) \tag{7}$$

where $S_{BESS,i,d}^{(k)}$ is the saving per kWh obtained charging/discharging the BESS over time d, in the semi-daily period k of the day i and $C_{BESS_{cycled,d}}$ is the storage cost per kWh cycled, obtained charging/discharging the BESS over time d.

The saving, $S_{BESS,i,d}^{(k)}$, can be calculated as follows:

$$S_{BESS,i,d}^{(k)} = \frac{E_{BESS,i,d}^{(k)}}{Cap} \cdot \left(MA_{RTP_{i,d,max}^{(k)}} \cdot \mu_d - \frac{MA_{RTP_{i,d,min}^{(k)}}}{\mu_c}\right) = DOD \cdot \left(MA_{RTP_{i,d,max}^{(k)}} \cdot \mu_d - \frac{MA_{RTP_{i,d,min}^{(k)}}}{\mu_c}\right) \tag{8}$$

where $E_{BESS,i,d}^{(k)}$ is the energy discharged from the storage device over time d, and μ_c and μ_d are the charge/discharge efficiencies of the battery, respectively.

The storage cost per kWh cycled can be expressed as:

$$C_{BESS_{cycled,d}} = \frac{C_{TOT_{BESS}}}{Cap \cdot N_{Full\ cycle,d}} \tag{9}$$

where $C_{TOT_{BESS}}$ is the total cost of the storage and $N_{Full\ cycle,d}$ is the number of equivalent full cycles of the battery, corresponding to a charge/discharge time d.

Denoted by $C_{BESS_{kWh}}$, the storage cost per kWh (from Equation (9)) can be expressed as:

$$C_{BESS_{cycled,d}} = \frac{C_{BESS_{kWh}}}{N_{Full\ cycle,d}} \tag{10}$$

The objective function, $OF_{i,d}^{(k)}$, can finally be expressed as:

$$OF_{i,d}^{(k)} = max\left[DOD \cdot \left(MA_{RTP_{i,d,max}^{(k)}} \cdot \mu_d - \frac{MA_{RTP_{i,d,min}^{(k)}}}{\mu_c}\right) - \frac{C_{BESS_{kWh}}}{N_{Full\ cycle,d}}\right] \tag{11}$$

The only variable that appears in the objective function is the *DOD*. Indeed, $N_{Full\ cycle,d}$ and $\left(MA_{RTP_{i,d,max}^{(k)}}, MA_{RTP_{i,d,min}^{(k)}}\right)$ are not independent variables, since they are linked to the *DOD*. The *DOD* is thus the only variable to be optimized and the search space is the set of all possible charging/discharging times, namely all integers between 1 and *D*. Ultimately, the maximization of the objective function allows one to obtain the *DOD* value that maximizes the customer's benefit, for each semi-daily charging/discharging cycle.

4.2.2. Daily Periodicity

In the same manner as was done in the previous section, in presence of a daily periodicity of the *MA RTP* profile, the objective function, $OF_{i,d}$, can be expressed as:

$$OF_{i,d} = max\left[DOD \cdot \left(MA_{RTP_{i,d,max}} \cdot \mu_d - \frac{MA_{RTP_{i,d,min}}}{\mu_c}\right) - \frac{C_{BESS_{kWh}}}{N_{Full\ cycle,d}}\right] \tag{12}$$

The maximization of the objective function allows one to obtain the DOD value that maximizes the customer's benefit, for each daily charging/discharging cycle.

4.2.3. Constraint Equations

As already stated in Section 3, the battery charging and discharging constraints are automatically satisfied, since the charge/discharge rate of the battery is assumed constant. The storage energy constraints are also satisfied, since the battery returns to the same initial SOC at the end of each charge/discharge cycle (namely the energy discharged is equal to the energy charged, in each battery cycle). Furthermore, charging/discharging periods should not overlap each other. This might happen when the battery performs two operating cycles per day. If this is the case, the charging/discharging period will be reduced accordingly.

The charge/discharge cycle of the battery would only be worth it if the difference between the maximum and minimum values of $MA\ RTP$ is higher than the cost of cycling energy plus the cost of the energy losses in the charge/discharge process. Expressed differently, Equations (11) and (12) must take positive values for the battery operation to be profitable:

$$OF_{i,d}^{(k)} > 0\ ,\ OF_{i,d} > 0 \tag{13}$$

If the constraints in Equation (13) are not satisfied, the battery will remain idle, since the arbitrage benefit is not enough to compensate for the cost of cycling energy plus the cost of the energy losses. In the following, the term "eligible" will be used to indicate an objective function whose value is greater than zero.

4.2.4. Selection of the Charging/Discharging Intervals

Once Equations (11) and (12) are calculated, the algorithm checks, for each day of the year, if the summation of the eligible objective functions corresponding to each semi-daily cycle is greater than that corresponding to the daily cycle, namely:

$$\sum_{k=1}^{2} OF_{i,d}^{(k)} \geqslant OF_{i,d}\ d = 1, \ldots, D \tag{14}$$

If Equation (14) is satisfied, the battery is charged in the first half of the day, in the second half or in both, depending on the number of the eligible objective functions, $OF_{i,d}^{(k)}$. The DOD for each battery cycle is selected according to Equation (11). If Equation (14) is not satisfied, the battery will make only one cycle per day. The corresponding DOD is selected according to Equation (12). Finally, if all the objective functions have negative value (*i.e.*, there are no eligible objective functions), the battery remains idle in the day i.

It is worth noting that the proposed operating strategy allows maximizing the customer's benefit under the assumptions described in Section 3. More complex and complete models could lead to higher benefits for the storage owner. Furthermore, the proposed method leads to an effective maximization of the objective function only if the SMP profile is assumed to have a convex form in the charging/discharging intervals, as in most spot electricity markets. If the price profile differs from a convex form, the proposed procedure could lead to suboptimal results, but it was verified that the error margin is narrow.

5. Case Study

The number of equivalent full cycles cannot be estimated directly, as it mainly depends on the energy cycled by the batteries, namely by the DOD. For most batteries, manufactures show in their datasheets the curves of number of cycles to failure, $N_{cycle,d}$ vs. the DOD (for given temperature value), as shown in Figure 4, derived for a lead-acid battery [31].

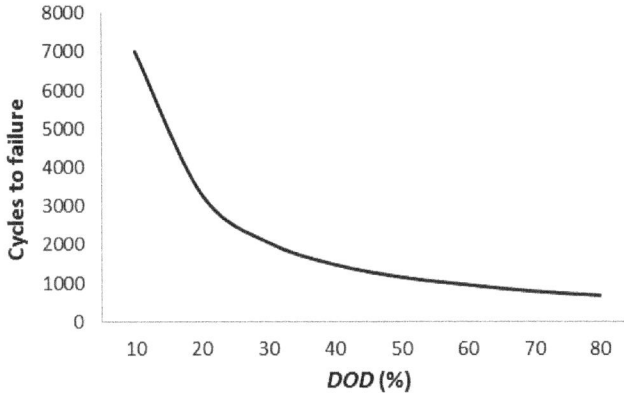

Figure 4. Typical cycles to failure *vs.* depth-of-discharge (*DOD*) curve for lead-acid-batteries.

The number of equivalent full cycles performed by the battery at a given *DOD* can be obtained as [32]:

$$N_{Full\ cycle,d} = DOD \cdot N_{cycle,d} \tag{15}$$

where $N_{cycle,d}$ is the number of cycles to failure, as derived from Figure 4.

For most of electrochemical batteries, the number of equivalent full cycles remains constant (for given operating temperature) and does not depend on the *DOD*. Expressed differently, the total Ah a battery can deliver over its life is approximately constant. However, the relationship deviates for some electrochemistries, especially at low *DOD*. With a view to highlight the changes, Figure 5 shows a comparison of cycles to failure *vs.* *DOD* curves for three different BESS technologies (lead-acid, Li-ion and NaS battery).

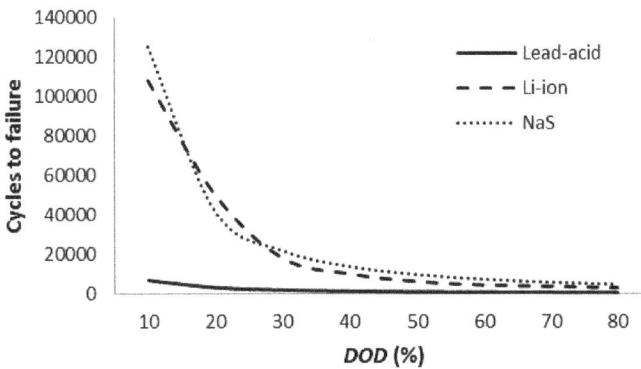

Figure 5. Cycles to failure *vs.* depth-of-discharge (*DOD*) curve for three different battery technologies.

Let us assume $D_{max} = 5\ h$, which corresponds to a discharging time $D = 4\ h$ at a $DOD = 80\%$. The number of equivalent full cycles, for each selected *DOD* (ranging from 1 to 4 h), is reported in Table 1, for each of the selected battery technologies. The values were calculated using Equation (11). The number of cycles to failure, N_{cycles_d}, was deduced from the typical cycles to failure *vs.* *DOD* curve, for each battery option [31,33,34]. Table 1 also shows the percentage increment, $\Delta N_{Full}(\%)$, with respect to the value corresponding to a $DOD = 80\%$. It is worth noting that the percentage increment is minimum for lead-acid, maximum for Li-ion battery.

Table 1. Number of equivalent full cycles for each selected *DOD*, for the three battery technologies.

DOD(%)	Lead-Acid Battery		Li-Ion Battery		NaS Battery	
	$N_{Full\ cycles,d}$	ΔN_{Full} (%)	$N_{Full\ cycles,d}$	ΔN_{Full} (%)	$N_{Full\ cycles,d}$	ΔN_{Full} (%)
80%	540		2400		3592	
60%	570	5.56	2640	10	4269	18.85
40%	590	9.26	4040	68.3	5445	51.58
20%	660	22.22	10000	316.7	8253	129.76

The analysis has been carried out by referring to a typical medium-scale public facility (Department of Energy, Information engineering and Mathematical models (DEIM), University of Palermo). For the selected facility, a reference weekly period has been considered, from 31 March to 6 April 2014. The SMP for the reference weekly period have already been reported in Figure 2.

The proposed strategy can be applied to several kinds of storages, but the test results refer to three kind of batteries, lead-acid, Li-ion and NaS, that are, nowadays, the most suitable to be used in residential, commercial or industrial buildings, for load shifting applications. Among the three technologies, Li-ion batteries are the most promising in terms of cost reduction and cycling performance [35]. The technical and economic parameters are reported in Table 2 for each of the selected battery technologies.

Table 2. Technical and economic parameters selected for the three battery technologies.

Components	Specifications		
Technology	Lead-Acid Battery	Li-Ion Battery	NaS Battery
Energy capacity (kWh)	20	20	20
Power rating (kW)	5	5	5
Roundtrip efficiency (%)	82	90	81
Operating temperature (°C)	(−20)–(+50)	(−20)–(+45/+60)	300–350
Healthy *DOD* (%)	80	80	NA
Cycles to failure (80% *DOD*)	1100	3000	4500
BESS cost (€/kWh)	171	844	256
PCS cost (€/kW)	172	125	171
BOP cost (€/kW)	70	0	53

The storage cost and the charge/discharge roundtrip efficiency have been selected calculating the arithmetic mean between low and high literature values [36]. In Table 2, the total storage cost has been decomposed as the sum of the power conversion system (PCS) cost, the BESS cost and the balance-of plant (BOP) cost [37]. The operating temperatures and the healthy *DOD* were derived from [29]. The rated energy capacity (equal to 20 kWh for each battery) was selected referring to the facility's energy consumption during peak price hours, on the day of the year of lowest consumption, as already specified in Section 3.

The storage costs per kWh cycled are on average higher than the difference between maximum and minimum electricity prices. Indeed, the average storage costs per kWh cycled are equal to 0.171 €/kWh cycled for lead-acid, 0.103 €/kWh cycled for Li-ion and 0.096 €/kWh cycled for NaS batteries, as against a maximum value of 0.1 €/kWh for the difference between maximum and minimum electricity price. For this reason, a grant equal to 75% of the upfront investment cost is considered in this analysis. The storage costs per kWh cycled have been obtained considering average values of $C_{BESS_{kWh}}$ and $N_{Full\ cycle}$, according to [36].

6. Simulation Results

For each day of the reference period, the algorithm handles the *MA RTP* prices, corresponding to each *DOD*, calculating the value of the objective functions and verifying the fulfillment of condition in Equation (14).

The values of the objective functions together with the charge/discharge time, for the three battery technologies, are reported in Table 3. If Equation (14) is satisfied, Table 3 reports the value of $\sum_{k=1}^{2} OF_{i,d}^{(k)}$ and the column *d* shows a couple of values, (x,y), denoting the charging/discharging time of the first and the second half day period, respectively. If Equation (14) is not satisfied, the value of the daily objective function, $OF_{i,d}$, is reported and the column *d* shows a single value denoting the charging/discharging time in the daily period. Finally, if all the objective functions have negative value (*i.e.*, there are no eligible objective functions) the battery remains idle and the corresponding values of the objective function and the charging/discharging times are missing in Table 3.

Table 3. Values of the objective functions in the reference weekly period.

	Lead Acid		Li-ion		NaS	
	OF	d	OF	d	OF	d
31/03/2014	0.038	4,4	0.036	2,1	0.122	4,4
01/04/2014	-	-	-	-	0.049	4,4
02/04/2014	-	-	0.002	-,1	0.047	4,3
03/04/2014	-	-	-	-	0.018	3,2
04/04/2014	-	-	0.004	1,-	0.042	4,-
05/04/2014	-	-	0.001	1	0.043	4,4
06/04/2014	0.028	-,4	0.01	-,4	0.071	-,4
Weekly OF	0.066		0.053		0.392	

The values reported in Table 3 lead to the following fundamental results (valid under the assumption that a subsidy equal to 75% of the upfront investment cost is granted to the storage owner):

- Among the three considered storage options, the use of NaS batteries leads to the maximum benefit for the storage owner (the value of the weekly objective function is around six times the one observed for the lead-acid battery); indeed, although NaS batteries have an acquisition cost higher than lead-acid, the number of cycles to failure is more than three times higher than that of lead-acid battery (see Table 2).
- The lead-acid technology appears to be the least convenient for arbitrage applications, despite its lower cost. This is essentially due to the low number of equivalent full cycles compared to the other battery technologies. The Li-ion technology also has a low profitability for arbitrage applications, essentially because of the high upfront investment cost. However, the situation could rapidly change since Li-ion batteries are the most promising in terms of cost reduction and cycling performance [31].
- Lead-acid battery remains idle during most of the days, since the gap between maximum and minimum electricity price is not enough to compensate for the low number of equivalent full cycles.
- As previously stated in Section 4.1, NaS battery is charged two times per day on weekdays (except on Friday), and only one time on Sunday. This is because weekdays have two price peeks, and the gap between max/min electricity price is high enough to compensate for the cost of cycling energy plus the cost of the energy losses in the charge/discharge process.
- The NaS battery often performs two operating cycles, whereas the Li-ion battery performs two operating cycles only on Monday. This is essentially due to the high upfront investment cost of Li-ion battery compared with NaS technology, and to the lower number of equivalent full cycles.

- On Sunday, the batteries perform only one cycle in the second half of the day, lasting four hours (as previously stated in Section 4.1).

It is worth noting that the battery cycle lasts four hours when the objective function takes a high value, *i.e.*, when the gap between high and low electricity prices is large. Indeed, in this case the first term of the objective function prevails over the second term and the higher *DOD* resulting from the greater discharge duration offsets the number of equivalent full cycles.

Finally, it is possible to assert that, at the current price of storage technologies, the use of batteries for arbitrage applications is not profitable for the storage owner. The battery is charged once a day or twice a day depending on the shape of *RTP* profiles, being the BESS operating cycle dependent on the specific battery technology.

In order to highlight the advantages of the proposed approach compared to other simple methods, a comparison is made with respect to a simple strategy (base case) where the battery is operated in the hours where the gap between the lowest and the highest prices is maximized. The base case differs from the proposed operating strategy since the battery can be operated at different hours, not necessarily uninterrupted, but always regardless of the facility's load profile. Besides, in the base case, the battery is operated always at its maximum *DOD* (4 h), if the discharge duration is compatible with the objective function values, under the fulfillment of constraint conditions.

The values of the objective functions together with the charge/discharge time, in the base case, are shown in Table 4. When the objective functions have negative value, the corresponding values and the charging/discharging times are missing in Table 4.

Table 4. Values of the objective functions for the base case.

	Lead acid		Li-ion		NaS	
	OF	*d*	*OF*	*d*	*OF*	*d*
31/03/2014	0.038	4,4	0.014	4,4	0.122	4,4
01/04/2014	-	-	-	-	0.049	4,4
02/04/2014	-	-	-	-	0.046	4,4
03/04/2014	-	-	-	-	0.013	4,4
04/04/2014	-	-	-	-	0.042	4,-
05/04/2014	-	-	-	-	0.043	4,4
06/04/2014	0.028	-,4	0.009	-,4	0.071	-,4
Weekly *OF*	0.066		0.023		0.386	
% weekly increase	-		130%		1.5%	

It was found that the percentage increase of the weekly objective function, compared to the base case, is 130% for Li-ion and 1.5% for NaS batteries, as reported in Table 4.

According to the values reported in Table 4, the comparison between the proposed operating strategy and the base case leads to the following considerations:

- For lead acid battery, the values of the objective function are the same (the weekly percentage increase is zero). Indeed, this kind of battery performs the same charging/discharging cycles both in the proposed operating strategy and in the base case.
- For Li-ion battery, the weekly percentage increase of the objective function is large (130%). Indeed, in the base case the Li-ion battery remains idle for most of the days and the value of the objective function on Monday is more than halved compared with the corresponding value reported in Table 3.
- For NaS battery, the weekly percentage increase of the objective function is 1.5%, as a result of an increase of the objective functions on Wednesday and Thursday.

The last conclusion is particularly meaningful since it confirms that operating the battery at low *DOD* can be advantageous for the storage owner when the gap between high and low electricity prices is limited (e.g., when the objective function takes a small value).

Figure 6a,b show the graphic comparison between the objective function values of the two approaches, for NaS and Li-ion battery, respectively.

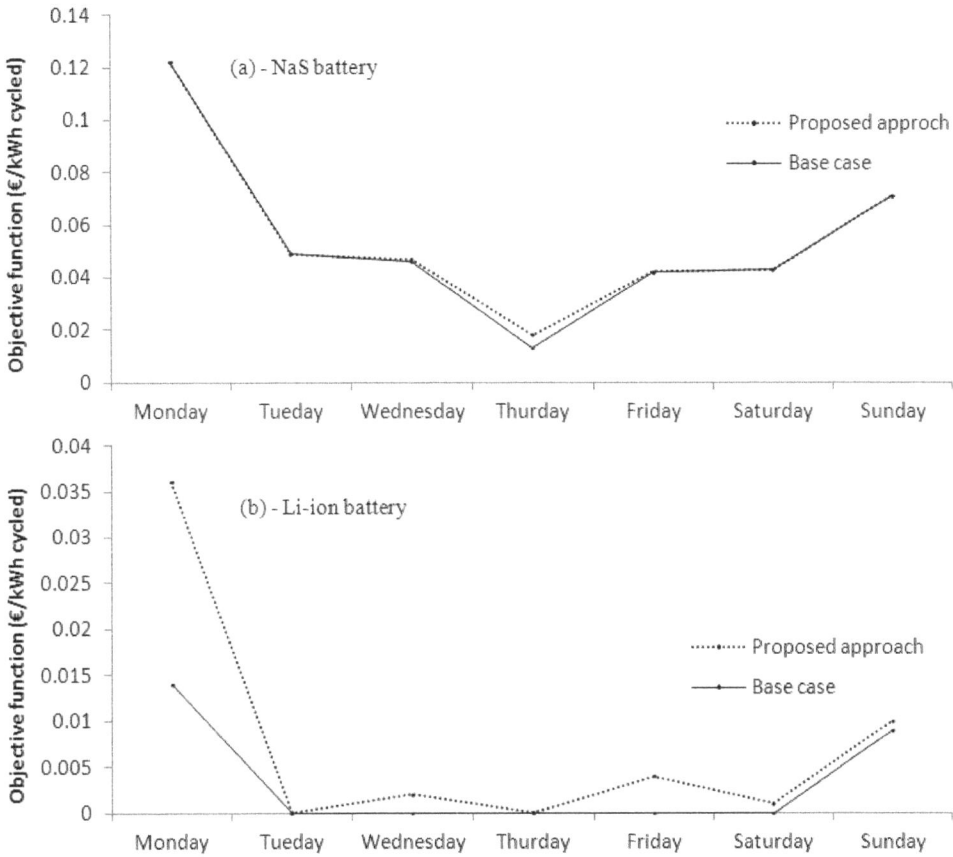

Figure 6. Graphic comparison between the objective function values of the two approaches: **(a)** NaS battery and **(b)** Li-ion battery.

The results obtained from the proposed approach show the effectiveness of the proposed operating strategy compared to the base case.

Finally, the effect of the proposed operating strategy on the daily curve of the energy extracted from the main grid is evaluated. To this aim, the power consumption of the department was registered over a reference period of one week (from 31 March to 6 April 2014).

Figure 7 shows the DEIM power diagram for the reference period, without (Figure 7a) and with (Figure 7b) storage contribution.

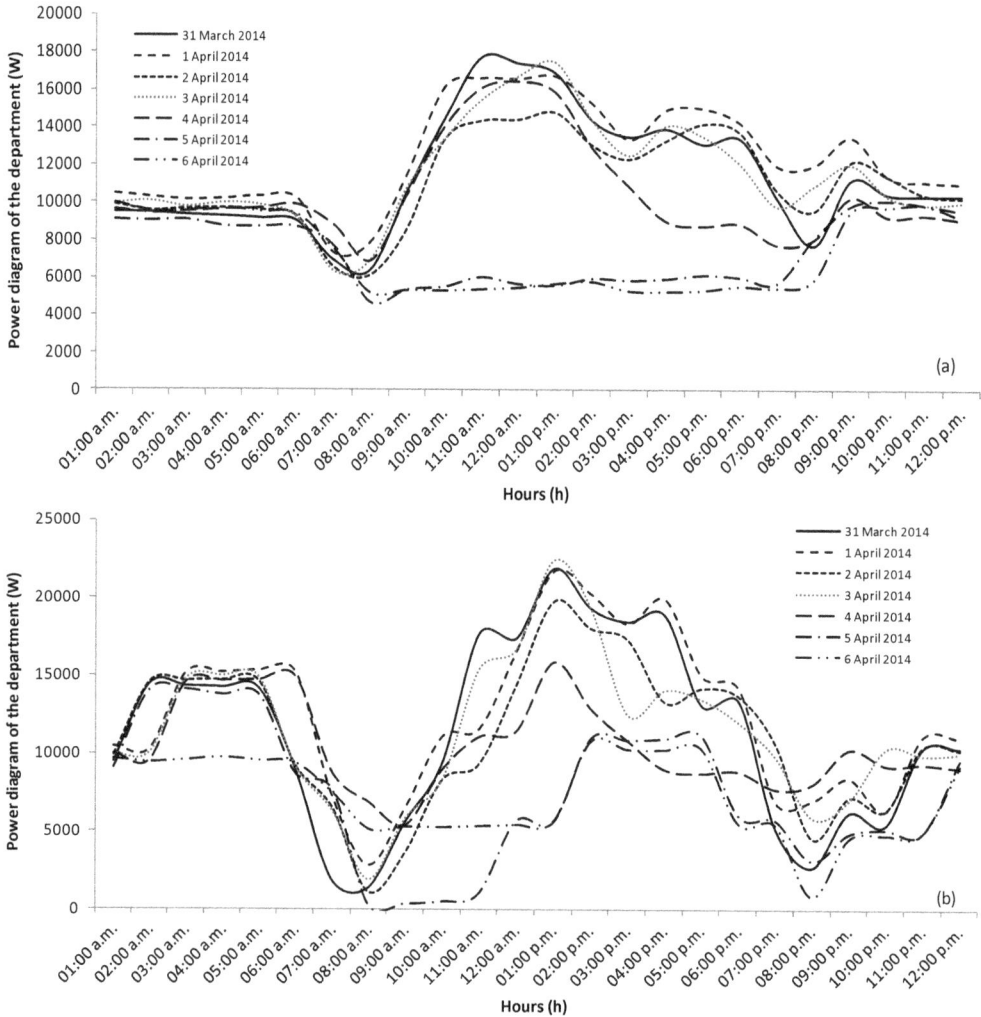

Figure 7. Power diagram of the department without (**a**) and with storage contribution (**b**).

Figure 7b shows power spikes due to the BESS charging/discharging. The maximum weekly peak load is increased at 23 kW (against a value of 18 kW without storage) when the proposed operating strategy is applied. Conversely, the minimum weekly peak load is reduced to zero when the storage is operated (against a value of 5 kW without storage). Therefore, the implementation of the proposed strategy does not lead to a flattening of the power profile but to an increase in the gap between peak and off-peak loads.

7. Conclusions and Future Work

This paper develops a detailed storage model linking together technical, economic and electricity market parameters. The storage system is described by means of its performance parameters, such as the charge and generate capacity, the charge/discharge efficiency, the rated charge/discharge rate, the *DOD*, *etc.*, which are sufficient to evaluate the arbitrage potential of the storage device. The proposed operating strategy aims to maximize the profit of the storage owner (electricity customer) by determining the optimal charge/discharge schedule. Unlike the studies reported in the literature, often requiring an estimate of the end-user load profile, the proposed operating strategy is able to identify the proper charging schedule of the device regardless of the specific facility's consumption. This is made possible since the battery is sized referring to the facilities' energy consumption during peak price hours, on the day of the year of lowest consumption. Under this assumption, the storage will be able to supply the entire customer load during the day of the year of lowest consumption, but only a portion of the customer's load on the other days. This could be particularly useful when the customer load profile cannot be scheduled with sufficient reliability, because of the uncertainty inherent in load forecasting. In these cases, identifying a BESS operating strategy that does not depend on the user's power profile can be an important task, since the deviation of the scheduled power profile from the effective one could affect the results obtained using more complete methods. In order to highlight the advantages of the proposed approach compared to other methods, a comparison is made with respect to a simple strategy (base case) where the battery is charged only one time per day at its maximum DOD (equal to four hours). The results obtained from the proposed approach show the effectiveness of the proposed operating strategy. The proposed model can be applied to several kinds of storages but the test results refer to three electrochemical technologies: lead-acid, Li-ion and NaS battery. The simulation results show that the operating schedule of the storage device differs in the various days of the week and it depends on the specific battery used (the most critical parameters being the acquisition cost of the battery bank and the number of cycles to failure). The operating cycle lasts four hours (*i.e.*, the maximum available charge/discharge time) when the objective function takes high values. However, in the days when the objective function has a lower value, the storage device is operated at a lower discharging time. This is because the higher gap between high and low electricity prices and the higher value of equivalent full cycles fully offset the less benefit due to the lower DOD (which results in a lower energy discharged). Simulation results show that, at current prices, no BESS technology is cost effective, due to the high upfront investment costs. However, if a subsidy is granted to reduce the initial investment cost, the use of NaS batteries leads to the maximum benefit among the three considered storage options. This is essentially due to the high number of equivalent full cycles (four times higher than that of lead-acid batteries). Conversely, the lead-acid technology appears to be the least convenient for arbitrage applications, despite its lower cost. This is essentially due to the low number of equivalent full cycles compared to the other battery technologies. In addition, the Li-ion technology has a low profitability for arbitrage applications, essentially because of the high upfront investment cost. However, the situation could rapidly change since Li-ion batteries are the most promising in terms of cost reduction and cycling performance.

In a future work, the authors will evaluate the effect of load forecasting uncertainty on the accuracy of storage operating strategies, in order to demonstrate that often the deviation of the scheduled power profile from the effective one could affect the results of more complete methods.

Acknowledgments: This work was supported by the project i-NEXT (Innovation for greeN Energy and eXchange in Transportation), identification code: PON04a2_Hi-NEXT (CUP B71H12000700005).

Author Contributions: This work was conceived by Enrico Telaretti. Preparation of the manuscript has been performed by Enrico Telaretti. Simulation and analysis of the results have been perfomed by Enrico Telaretti. Luigi Dusonchet and Mariano Ippolito supervised the work, giving a final review of the paper. All authors read and agreed to the final article.

Conflicts of Interest: The authors declare no conflict of interest.

Abbreviations

BESS	Battery Energy Storage System
BOP	Balance-of Plant
DEIM	Department of Energy, Information Engineering and Mathematical Models
DOD	Depth-of-Discharge
DSM	Demand Side Management
IRR	Internal Rate of Return
Li-ion	Lithium-Ion
MA	Moving Average
MA RTP	Moving Average of *RTP* Prices
NaS	Sodium-Sulphur
PCS	Power Conversion System
PSO	Particle Swarm Optimization
PV	Photovoltaic
RES	Renewable Energy Sources
RTP	Real-Time Pricing
SMP	System Marginal Price
SOC	State-of-Charge
TOU	Time-of-Use
VAT	Value Added Tax

References

1. Campoccia, A.; Dusonchet, L.; Telaretti, E.; Zizzo, G. Feed-in tariffs for grid-connected PV systems: The situation in the European community. In Proceedings of IEEE Power Tech Conference, Lausanne, Switzerland, 1–5 July 2007; pp. 1981–1986.

2. Campoccia, A.; Dusonchet, L.; Telaretti, E.; Zizzo, Z. Financial Measures for Supporting Wind Power Systems in Europe: A Comparison between Green Tags and Feed'in Tariffs. In Proceedings of IEEE Power Electronics, Electrical Drives, Automation and Motion, Ischia, Italy, 11–13 June 2008; pp. 1149–1154.

3. Sgroi, F.; Tudisca, S.; Di Trapani, A.M.; Testa, R.; Squatrito, R. Efficacy and Efficiency of Italian Energy Policy: The Case of PV Systems in Greenhouse Farms. *Energies* **2014**, *7*, 3985–4001. [CrossRef]

4. Giannini, E.; Moropoulou, A.; Maroulis, Z.; Siouti, G. Penetration of Photovoltaics in Greece. *Energies* **2015**, *8*, 6497–6508. [CrossRef]

5. Borenstein, S. The long-run efficiency of real-time electricity pricing. *Energy J.* **2005**, *26*, 93–116. [CrossRef]

6. Dusonchet, L.; Ippolito, M.G.; Telaretti, E.; Graditi, G. Economic impact of medium-scale battery storage systems in presence of flexible electricity tariffs for end-user applications. In Proceedings of IEEE International Conference on the European Energy Market, Florence, Italy, 10–12 May 2012; pp. 1–5.

7. Dusonchet, L.; Ippolito, M.G.; Telaretti, E.; Zizzo, G.; Graditi, G. An optimal operating strategy for combined RES–based Generators and Electric Storage Systems for load shifting applications. In Proceedings of IEEE International Conference on Power Engineering, Energy and Electrical Drives, Instanbul, Turkey, 13–17 May 2013; pp. 552–557.

8. Youn, L.T.; Cho, S. Optimal operation of energy storage using linear programming technique. In Proceedings of the World Congress on Engineering and Computer Science, San Francisco, CA, USA, 20–22 October 2009; pp. 480–485.

9. Ahlert, K.; Van Dinther, C. Sensitivity analysis of the economic benefits from electricity storage at the end consumer level. In Proceedings of IEEE Power Tech Conference, Bucharest, Romania, 28 June–2 July 2009; pp. 1–8.

10. Bradbury, K.; Pratson, L.; Patino-Echeverri, D. Economic viability of energy storage systems based on price arbitrage potential in real-time U.S. electricity markets. *Appl. Energy* **2014**, *114*, 512–519. [CrossRef]

11. Graves, F.; Jenkin, T.; Murphy, D. Opportunities for Electricity Storage in Deregulating Markets. *Electr. J.* **1999**, *12*, 46–56. [CrossRef]

12. Hu, W.; Chen, Z.; Bak-Jensen, B. Optimal operation strategy of battery energy storage system to real-time electricity price in Denmark. In Proceedings of the IEEE Power and Energy Society General Meeting, Minneapolis, MN, USA, 25–29 July 2010; pp. 1–7.

13. Mokrian, P.; Stephen, M. A stochastic programming framework for the valuation of electricity storage. In Proceedings of 26th USAEE/IAEE North American Conference, Ann Arbor, MI, USA, 24–27 September 2006; pp. 1–34.

14. Maly, D.K.; Kwan, K.S. Optimal battery energy storage system (BESS) charge scheduling with dynamic programming. *IEE Proc. Sci. Meas. Technol.* **1995**, *142*, 454–458. [CrossRef]

15. Van de Ven, P.M.; Hegde, N.; Massoulié, L.; Salonidis, T. Optimal control of residential energy storage under price fluctuations. In Proceedings of International Conference on Smart Grids, Green Communications and IT Energy-aware Technologies, Venice, Italy, 22–27 May 2011; pp. 159–162.

16. Huang, T.; Liu, D. Residential energy system control and management using adaptive dynamic programming. In Proceedings of the International Joint Conference on Neural Networks, San Jose, CA, USA, 31 July–5 August 2011; pp. 119–124.

17. Lee, T.Y. Operating Schedule of Battery Energy Storage System in a Time-of-Use Rate Industrial User With Wind Turbine Generators: A Multipass Iteration Particle Swarm Optimization Approach. *IEEE Trans. Energy Conv.* **2007**, *22*, 774–782. [CrossRef]

18. Abeygunawardana, A.; Ledwich, G. Estimating benefits of energy storage for aggregate storage applications in electricity distribution networks in Queensland. In IEEE Power and Energy Society General Meeting, Vancouver, BC, Canada, 21–25 July 2013; pp. 1–5.

19. Byrne, C.; Verbic, G. Feasibility of Residential Battery Storage for Energy Arbitrage. In Proceedings of Power Engineering Conference (AUPEC), 2013 Australasian Universities, Hobart, TAS, Australia, 29 September–3 October 2013; pp. 1–7.

20. Ekman, C.K.; Jensen, S.H. Prospects for large scale electricity storage in Denmark. *Energy Convers. Manag.* **2010**, *51*, 1140–1147. [CrossRef]

21. Shcherbakova, A.; Kleit, A.; Cho, J. The value of energy storage in South Korea's electricity market: a Hotelling approach. *Appl. Energy* **2014**, *125*, 93–102. [CrossRef]

22. Purvins, A.; Sumner, M. Optimal management of stationary lithium-ion battery system in electricity distribution grids. *J. Power Sources* **2013**, *242*, 742–755. [CrossRef]

23. Castillo-Cagigal, M.; Caamaño-Martín, E.; Matallanas, E.; Masa-Bote, D.; Gutiérrez, A.; Monasterio-Huelin, F.; Jiménez-Leube, J. PV self-consumption optimization with storage and Active DSM for the residential sector. *Sol. Energy* **2011**, *85*, 2338–2348. [CrossRef]

24. Matallanas, E.; Castillo-Cagigal, M.; Gutiérrez, A.; Monasterio-Huelin, F.; Caamaño-Martín, E.; Masa, D.; Jiménez-Leube, J. Neural network controller for Active Demand-Side Management with PV energy in the residential sector. *Appl. Energy* **2012**, *91*, 90–97. [CrossRef]

25. Abushnaf, J.; Rassau, A.; Górnisiewicz, W. Impact of dynamic energy pricing schemes on a novel multi-user home energy management system. *Electr. Power Syst. Res.* **2015**, *125*, 124–132. [CrossRef]

26. Ippolito, M.G.; Telaretti, E.; Zizzo, G.; Graditi, G.; Fiorino, M. A Bidirectional Converter for the Integration of LiFePO$_4$ Batteries with RES-based Generators. Part I: Revising and finalizing design. In Proceedings of 3rd Renewable Power Generation Conference, Naples, Italy, 24–25 September 2014; pp. 1–6.

27. Ippolito, M.G.; Telaretti, E.; Zizzo, G.; Graditi, G.; Fiorino, M. A Bidirectional Converter for the Integration of LiFePO$_4$ Batteries with RES-based Generators. Part II: Laboratory and Field Tests. In Proceedings of 3rd Renew. Power Generation Conference, Naples, Italy, 24–25 September 2014; pp. 1–6.

28. Viera, J.C.; Gonzalez, M.; Liaw, B.Y.; Ferrero, F.J.; Alvarez, J.C.; Campo, J.C.; Blanco, C. Characterization of 109 Ah Ni–MH batteries charging with hydrogen sensing termination. *J. Power Sources* **2007**, *171*, 1040–1045. [CrossRef]

29. Zheng, M.; Meinrenken, C.J.; Lackner, K.S. Agent-based model for electricity consumption and storage to evaluate economic viability of tariff arbitrage for residential sector demand response. *Appl. Energy* **2014**, *126*, 297–306. [CrossRef]

30. GME home page. Available online: http://www.mercatoelettrico.org/it/Default.aspx (accessed on 26 May 2015).

31. Dufo-López, R. Optimisation of size and control of grid-connected storage under real time electricity pricing conditions. *Appl. Energy* **2015**, *140*, 395–408. [CrossRef]

32. Dufo-López, R.; Bernal-Agustin, J.L. Techno-economic analysis of grid-connected battery storage. *Energy Conv. Manag.* **2015**, *91*, 394–404. [CrossRef]

33. The Lithium-Ion Battery. Service Life Parameters. Available online: https://www2.unece.org/wiki/download/attachments/8126481/EVE-06-05e.pdf?api=v2 (accessed on 26 May 2015).

34. Lu, N.; Weimar, M.R.; Makarov, Y.V.; Ma, J.; Viswanathan, V.V. The Wide-Area Energy Storage and Management System–Battery Storage Evaluation. Available online: http://www.pnl.gov/main/publications/external/technical_reports/PNNL-18679.pdf (accessed on 26 May 2015).

35. Divya, K.C.; Østergaard, J. Battery energy storage technology for power systems—An Overview. *Electr. Power Syst. Res.* **2009**, *79*, 511–520. [CrossRef]

36. Battke, B.; Schmidt, T.S.; Grosspietsch, D.; Hoffmann, V.H. A review and probabilistic model of life cycle costs of stationary batteries in multiple applications. *Renew. Sustain. Energy Rev.* **2013**, *25*, 240–250. [CrossRef]

37. Telaretti, E.; Sanseverino, E.R.; Ippolito, M.; Favuzza, S.; Zizzo, G. A novel operating strategy for customer-side energy storages in presence of dynamic electricity prices. *Intell. Ind. Syst.* **2015**, *1*, 233–244. [CrossRef]

8

Assessing the Environmental Sustainability of Electricity Generation in Turkey on a Life Cycle Basis

Burcin Atilgan and Adisa Azapagic *

Academic Editor: Vasilis Fthenakis

School of Chemical Engineering and Analytical Science, The University of Manchester, The Mill, Room C16, Sackville Street, Manchester M13 9PL, UK; burcin.atilgan@manchester.ac.uk
* Correspondence: adisa.azapagic@manchester.ac.uk

Abstract: Turkey's electricity mix is dominated by fossil fuels, but the country has ambitious future targets for renewable and nuclear energy. At present, environmental impacts of electricity generation in Turkey are unknown so this paper represents a first attempt to fill this knowledge gap. Taking a life cycle approach, the study considers eleven impacts from electricity generation over the period 1990–2014. All 516 power plants currently operational in Turkey are assessed: lignite, hard coal, natural gas, hydro, onshore wind and geothermal. The results show that the annual impacts from electricity have been going up steadily over the period, increasing by 2–9 times, with the global warming potential being higher by a factor of five. This is due to a four-fold increase in electricity demand and a growing share of fossil fuels. The impact trends per unit of electricity generated differ from those for the annual impacts, with only four impacts being higher today than in 1990, including the global warming potential. Most other impacts are lower from 35% to two times. These findings demonstrate the need for diversifying the electricity mix by increasing the share of domestically-abundant renewable resources, such as geothermal, wind, and solar energy.

Keywords: electricity generation; environmental impacts; life cycle assessment; Turkey

1. Introduction

With its young population, fast growing economy and industrialisation, Turkey has become one of the fastest growing energy markets in the world. Its energy demand has been increasing rapidly over the past few decades, rising almost six-fold in the period between 1970 and 2010 [1,2]. The demand for electricity has also followed this trend. In 2010, the total installed capacity of 49,524 MW generated 211,208 GWh of electricity, four times more than in 1990 [3]. However, only 43% of the demand is met by domestic energy resources (lignite, hydropower, geothermal and wind) with the rest of electricity generated by imported fuels (hard coal and natural gas) [4]. As a result, Turkey has become dependent on other countries for the supplies of fuels, particularly natural gas, which provides 46.5% of electricity (see Figure 1). In 2010, around 98% of the country's natural gas demand was met through imports, mainly from Russia [5]. At current production levels (730 million m^3 in 2010), Turkey has the equivalent of only nine years of domestic gas reserves remaining [5]. There are currently 187 gas power plants with 18,213 MW of installed capacity, which, in 2010, generated 98,144 GWh [4]. The majority of the installations are combined cycle power plants [6].

The next most significant fuel is coal which supplies 26% of electricity [4]. Turkey has significant coal reserves with lignite being much more abundant than hard coal: 10.8 billion tonnes *vs.* 515 million tonnes of proven reserves, respectively [7]. However, most of Turkish lignite is of low quality with poor calorific value and high sulphur content. Hard coal is also of low grade, but of cokeable or semi-cokeable quality. In 2010, total coal production reached 73.4 Mt of which 69.7 Mt was lignite,

2.5 Mt hard coal and 1.2 Mt asphaltite [7]. By comparison, 24.3 Mt were imported, of which 60% from Russia and Colombia and 40% from the USA and South Africa [7]. There are 16 lignite and eight hard coal plants in Turkey, with the total installed capacity of 11,891 MW, which generated 55,046 GWh of electricity in 2010 [4]; the majority (65%) of this was from the lignite plants. Around 85% of the capacity is based on pulverised coal and the rest on circulating fluidised bed installations.

In contrast to gas and coal, the contribution from oil power plants has been declining over the years, and today almost no oil power plants remain in Turkey, as most have been converted to natural gas combined cycle power plants [6].

The contribution of hydropower is close to that of coal, providing around a quarter of electricity (Figure 1). The theoretically-viable hydroelectric potential of Turkey is estimated at 433 TWh/year [8]. Almost half of this is technically and nearly 30% economically exploitable [8]. Currently, there are 55 reservoir and 205 run-of-river plants with the total installed capacity of 15,831 MW, generating on average 51,795 GWh per year.

The only other sources of renewable energy available present in the electricity mix in Turkey are wind and geothermal, providing 1.4% and 0.3% of the total, respectively. Turkey has a good wind potential, estimated at 48,000 MW with an annual production capacity of 130 TWh/year [9]. There are currently 39 onshore wind power plants with an installed capacity of 1320 MW that produced 2916 GWh in 2010. The capacity of the individual turbines varies from 0.85 to 3 MW. At present, there are no offshore wind turbines in Turkey.

Located on the Alpine–Himalayan tectonic belt, Turkey is one of the richest countries in the world in terms of geothermal energy resources, with an estimated capacity of 31,500 MW [9]. However, the majority of this (88%) has the temperature below 200 °C, which is less suitable for electricity generation (unless an organic Rankine cycle system is used) but is still useful for direct use as heat [10]. Thus, there are only six geothermal power plants in Turkey with the installed capacity of 94.2 MW which generated 668 GWh in 2010 [3].

As a result of the high share of fossil fuels, the direct greenhouse gas (GHG) emissions from electricity (largely from burning fossil fuels) reached 99 Mt CO_2-eq. in 2010, a quarter of the total national emissions in that year [11]. Since Turkey is a signatory to the Kyoto Protocol (Annex I), it is important that it identifies sustainable energy technologies suitable for the country to reduce the GHG emissions and other environmental impacts from the electricity sector. However, environmental impacts from the electricity sector in Turkey are currently unknown and there is no baseline to help the country identify the best way forward.

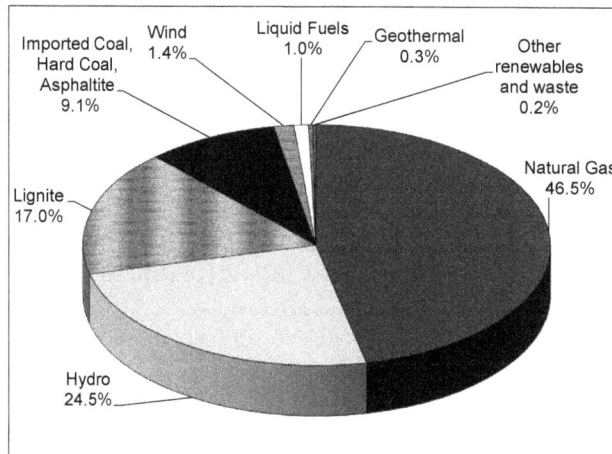

Figure 1. Share of different fuels in electricity generation in Turkey in 2010 [4].

Therefore, this paper sets out to estimate the environmental impacts of electricity in Turkey on a life cycle basis over the last 25 years (1990–2014). Although many life cycle assessment (LCA) studies of different electricity technologies have been carried out elsewhere, as far as the authors are aware, there are no such studies for Turkey. The only exception is the previous work by Atilgan and Azapagic [12,13], but that only considered electricity from individual technologies rather than the whole electricity supply mix which is the focus of this work. Therefore, this is the first study of its kind for Turkey and the findings will provide an insight into the past trends and current environmental impacts of electricity generation, helping to identify opportunities for improving the environmental sustainability of the sector. A further feature is that each of the 516 plants currently present in the Turkish electricity system have been considered individually. This has seldom been the case in other studies, which typically use average values for different technologies.

Some of the existing LCA studies relevant to the technologies present in Turkey's electricity mix are summarised in Table 1. The results from these studies are compared to the current study later in the paper (Section 3.1). As can be seen in the table, the scope varied across the studies, including the type of technologies and environmental impacts considered. All studies, however, considered GHG emissions and the related global warming potential. Most also considered eutrophication, acidification and photochemical smog, except Kannan, *et al.* [14] who considered the global warming potential and energy and Weisser [15] who focused on the global warming potential alone. Most studies used the Institute of Environmental Sciences (CML) 2001 method [16] to estimate the impacts. The same method has been applied in the current study, which considers 11 environmental categories, as discussed in the next sections.

2. Methods

The study has been carried out according to the LCA methodological guidelines in the International Organization for Standardization (ISO) 14040 and 14044 standards [17,18]. The LCA modelling has been carried out in GaBi v.6 [19].

Table 1. Some life cycle assessment (LCA) studies of electricity generation.

Aim and Scope of the Study	Technologies	Country	Environmental Impacts	Reference
Life cycle energy use, GWP and cost assessment of gas fired combined cycle plant	Natural gas	Singapore	Global warming, energy use	[14]
Life cycle greenhouse gas emissions from electric supply technologies	Lignite, hard coal, oil, natural gas, nuclear, CCS, hydro, wind, solar PV, biomass, energy storage	Not specified	Global warming	[15]
Setting up life cycle models for the environmental analysis of hydropower generation, considering technical and climatic boundary conditions	Run-of-river, storage and pumped storage hydro	Germany	Global warming, acidification, eutrophication, photochemical smog, energy demand	[20]
LCA of carbon dioxide capture and storage from lignite power plants	Pulverised coal (PC), PC with CO_2 capture, Integrated gasification combined cycle (IGCC), IGCC with CO_2 capture, oxyfuel plant with CO_2 capture	Germany	Global warming, energy demand, photochemical smog, eutrophication, acidification	[21]
LCA of a 2 MW rated power wind turbine	Onshore wind	Spain	Global warming, resource depletion, ecotoxicity, ozone layer depletion, acidification, eutrophication, photochemical smog, human toxicity	[22]
LCA of mini-hydropower plants	Run-of-river hydro	Thailand	Global warming, resource depletion, acidification, human toxicity, photochemical smog, water ecotoxicity	[23]

Table 1. *Cont.*

Aim and Scope of the Study	Technologies	Country	Environmental Impacts	Reference
LCA of electricity generation	Nuclear, coal, natural gas, oil, renewables	Mexico	Global warming, ecotoxicity, ozone layer depletion, acidification, eutrophication, photochemical smog, human toxicity, resource depletion	[24]
LCA of the wind turbines	Onshore wind	Spain	Global warming, ecotoxicity, ozone layer depletion, acidification, eutrophication, photochemical smog, human toxicity, resource depletion	[25]
LCA of a hydroelectric power	Run-of-river hydro	Thailand	Global warming, ecotoxicity, ozone layer depletion, acidification, eutrophication, photochemical smog, resource depletion	[26]
Life cycle sustainability assessment of electricity generation	Nuclear, coal, natural gas, offshore wind, solar PV	UK	Global warming, ozone layer depletion, acidification, eutrophication, photochemical smog, land use, ecotoxicity, human toxicity, resource depletion	[27]
LCA of 1 kWh generated by a Gamesa onshore wind farm	Onshore wind	Europe	Cumulative energy demand, global warming, summer smog, ecotoxicity, eutrophication, acidification, human toxicity, land use	[28]
Life cycle data for hydroelectric generation at Embretsfoss 4 power station	Run-of-river hydro	Norway	Global warming, acidification, eutrophication, photochemical smog, ozone layer depletion, waste	[29]
Life cycle data for hydroelectricity from Trollheim power station	Reservoir hydro	Norway	Global warming, acidification, eutrophication, photochemical smog, ozone layer depletion, waste	[30]
Life cycle assessment of electricity from an onshore V90-3.0 MW wind plant	Onshore wind	Not specified	Global warming, ecotoxicity, ozone layer depletion, acidification, eutrophication, photochemical smog, human toxicity, resource depletion	[31]
Life cycle assessment of wind power	Onshore wind	Not specified	Global warming, ecotoxicity, ozone layer depletion, acidification, eutrophication, photochemical smog, human toxicity, resource depletion	[32]
LCA of a wind plant	Onshore wind	France	Resource depletion, acidification, eutrophication, global warming, photochemical smog	[33]

The CML 2001 method [16], November 2010 update, has been used to estimate the following impacts: abiotic depletion potential (ADP elements and fossil), acidification potential (AP), eutrophication potential (EP), fresh water aquatic ecotoxicity potential (FAETP), global warming potential (GWP), human toxicity potential (HTP), marine aquatic ecotoxicity potential (MAETP), ozone layer depletion potential (ODP), photochemical oxidants creation potential (POCP) and terrestrial ecotoxicity potential (TETP).

The goal of the study, key assumptions and data sources are detailed in the following sections.

2.1. Goal and Scope Definition

The goal of the study has been to estimate life cycle environmental impacts of electricity generation in Turkey and to identify improvement opportunities for the future. To enable that, the impacts of the individual power plants had to be estimated first. For these purposes, 2010 has been chosen as the base year because most complete data were available for the individual plants for that year. In addition to this, the impacts have been estimated for electricity generation over the past 25 years, covering the period from 1990 to 2014, to identify the impact trends and find out whether the electricity sector is becoming more or less environmentally sustainable.

The study is based on two functional units. The first is defined as "generation of 1 kWh of electricity" to enable comparisons of the impacts for individual electricity technologies as well as for different electricity mixes for over the period. The second functional unit is defined as the "total annual electricity generation" to estimate the total annual impacts from electricity over the last 25 years.

As indicated in Figure 2, the study considers all the options present in the Turkish electricity mix: coal, gas, hydro (large- and small-scale reservoir, run-of-river), wind and geothermal. The scope is from "cradle to grave", comprising the extraction, processing and transport of fuels (where relevant) and raw materials as well as the construction, operation and decommissioning of power plants. Since the focus of the work is on electricity generation, its transmission, distribution and consumption are outside the system boundary.

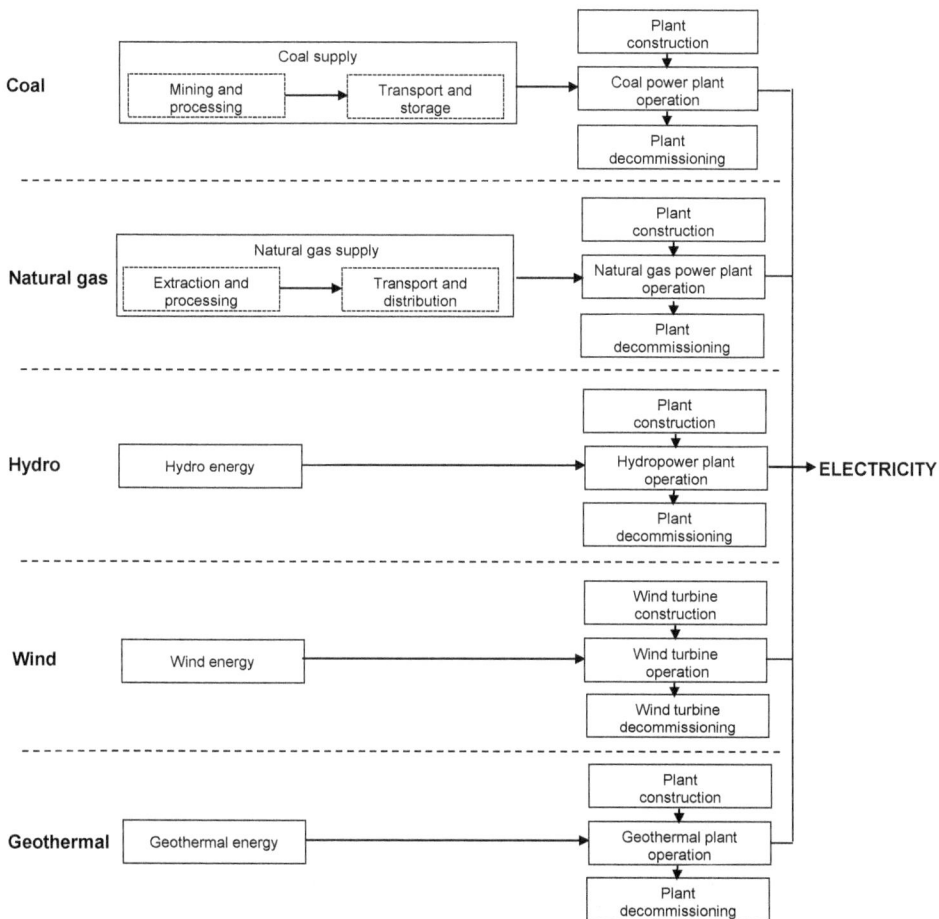

Figure 2. The life cycles of electricity from coal, natural gas, hydro, wind and geothermal power.

2.2. Inventory Data

In 2010, 211,208 GWh of electricity were generated by a total of 516 power plants, specifically, by 16 lignite, eight hard coal, 187 gas, 55 reservoir and 205 run-of-river hydropower, 39 wind and six geothermal installations. All of these plants are considered in this study (see Table 2 and Supplementary information). The key assumptions are summarised in Tables 3 and 4 with further details provided below.

Table 2. Power plants in Turkey (2010).

Type of Power Plant	Number of Plants	Installed Capacity (MW)	Annual Generation (GWh/yr)
Lignite	16	8140	35,942
Hard coal [a]	8	3751	19,104
Natural gas	187	18,213	98,144
Large-reservoir hydropower (capacity > 500 MW)	8	8459	30,583
Small-reservoir hydropower (capacity < 500 MW)	47	4608	13,885
Run-of-river hydropower	205	2764	7327
Onshore wind	39	1320	2916
Geothermal	6	94	668
Total	516	47,349 (49,524) [b]	208,569 (211,208) [c]

[a] Hard coal type power plant includes hard coal, imported coal and asphaltite power plants in Turkey. [b] The total installed capacity in 2010 was 49,524 MW. The difference from the installed capacity shown in the table is due to multi-fuel, liquid fuel and other renewable-waste plants not included in the table. [c] The total generation was 211,208 GWh. The difference from the generation shown in the table is due to liquid fuel and other renewable-waste plants not included in the table. However, the total actual electricity generation has been used to estimate the impacts from electricity mix.

Table 3. The amount of fuels used for electricity generation in 2010 [12].

	Natural Gas (million m³)	Hard Coal (million tonnes)	Lignite (million tonnes)	Transport Distances (km) Gas [a]	Hard Coal [b]
Domestic fuel	-	0.20	55.89	-	-
Imported fuel					
Russia	9921	4.45 [c]	-	5750	5000
Iran	4383	-	-	2700	-
Azerbaijan	2551	-	-	1150	-
Algeria	2205	-	-	4000	-
Nigeria	671	-	-	4500	-
USA	-	1.48	-	-	10,500
South Africa	-	1.48	-	-	13,000
Other	1738	-	-	1750	-
Total	21,469	7.61	55.89	19,850	28,500

[a] Transport by pipeline. Total weighted average distance of 4000 km used for LCA modelling, taking into account the amounts of gas imported from each country as listed in the table. [b] Russia: 4500 km by rail, 500 km by shipping; USA: 1000 km by rail, 9500 km by shipping; South Africa: 500 km by rail, 12,500 km by shipping. [c] This includes the amount of hard coal imported from Colombia but as there are no LCA data for the Colombian coal, the LCA impacts from the Russian coal have been used instead.

2.2.1. Electricity from Fossil Fuels

Detailed data have been available for each lignite and hard coal power plants so that each has been modelled separately based on the data in Tables S1 and S2 in Supplementary information. For the natural gas plants, data availability has been more limited. Therefore, an average efficiency of 55% has been assumed for all the gas plants; this matches the average efficiency of the combined cycle gas turbine (CCGT) plants for which the data have been available but also the efficiency of the plants in Turkey reported by IEA and NEA [34] as well as some other sources [35]. Specific data have not been available for the auto-producer plants; as mentioned earlier, they are not connected to the grid but generate electricity for own consumption. However, the total generation from these plants has been considered.

The data for the coal and natural gas power installations (Tables S1–S3 in Supplementary information) have been used together with the fuel composition data (Table 4) to estimate the emissions from each plant using GEMIS 4.8 software [36]. These have then been combined with the other inventory data in Tables 3 and 4 to estimate the life cycle impacts from the power plants in GaBi. The background life cycle inventory data have been sourced from Ecoinvent v2.2 [37] but have been adapted as far as possible to Turkey's conditions.

2.2.2. Electricity from Renewables

The data for the hydropower plants and onshore wind turbines are summarised in Tables 2 and 4; for details for the individual plants see Tables S4–S7 in Supplementary information. The background inventory data for hydropower have been sourced from the Ecoinvent v2.2 [37,38] and ESU [39] databases. Since the data for construction materials for reservoir plants in the databases correspond to a different size to the ones considered here, it has been necessary to apply some scaling of the impacts to reflect the difference in the size of the installations. Typically in LCA, this is carried out by assuming a linear relationship between the size of the plant and its impacts. However, in reality, the relationship is non-linear. To account for this, the "six-tenths" approach has been applied [40]. This method is typically used to estimate costs of different plant sizes accounting for the "economies of scale", *i.e.*, larger plants cost less to build per unit of capacity than smaller installations. The same analogy has been applied to the impacts, which would also be lower per unit of installed capacity for bigger than smaller plants. Thus, the impacts have been scaled according to the following relationship:

$$E_2 = E_1 \times \left(\frac{C_2}{C_1}\right)^{0.6} \tag{1}$$

where:
E_1: environmental impacts of the larger plant
E_2: environmental impacts of the smaller plant
C_1: capacity of the larger plant
C_2: capacity of the smaller plant
0.6: the "six-tenths" scaling factor.

The construction materials for the hydro plants are assumed to be transported for 200 km by rail and 100 km by lorry (see Table 4). At the end of its service life, the plants are dismantled and components are recycled using the assumed recycling rates given in Table 4.

The inventory data for onshore wind turbines have been sourced from [41] based on the Vestas V80 2 MW turbine. The size of the turbine has been scaled down from 2 MW using Equation (1) to match the average turbine size in Turkey of 1.94 MW (see Table 4). The construction materials are assumed to be transported 100 km by lorry and 100 km by rail. The turbines are imported into Turkey from Europe, assuming an average distance of 2000 km by rail and a further 150 km by lorry to the installation locations in Turkey. At the end of life, the turbine construction materials are recycled as specified in Table 4.

Detailed data for the geothermal plants have not been available so that the LCA data have been taken directly from the GaBi database [19]. The only dataset available is for a 30 MW flash-steam plant with an annual electricity generation of 250 GWh. The data are aggregated so that it has not been possible to adapt them to Turkey-specific conditions. However, as geothermal power contributes only 0.3% of the total electricity (see Figure 1), this limitation is not deemed significant. Moreover, the database model is representative of the standard, widely-used type of geothermal plant.

The total annual environmental impacts for the base year have been calculated based on the impacts of all the options present in the Turkish electricity mix, their share in the electricity mix and the total electricity generation in 2010. The same approach has been applied across all the other years considered in this work.

To model the electricity mix in Turkey over the period, the following assumptions have been made:

- the contribution of the liquid-fuel power plants to the total generation of electricity is small and for simplicity has been substituted with the equivalent amount of electricity generated by the gas power plants;
- the data on the specific technologies for other renewables and waste have not been available. As their contribution to the total electricity generation is small, they have been substituted by small-reservoir hydropower.

It has also been assumed that there were no technology changes over the period considered, using the characteristics of technologies in 2010 to model the impacts for the previous years. This is arguably a reasonable assumption as fossil and hydro-technologies are well established and have long lifetimes so that no technological changes would be expected over the period. Wind and geothermal are less well established, but their contribution to the total generation in the earlier years was negligible (<0.03% and <0.1%, respectively).

3. Results and Discussion

The results are given in Figures 3–9 and are discussed in the following sections, first for different electricity technologies, then for electricity in the base year (2010) and finally for electricity generation from 1990 to the present.

3.1. Environmental Impacts of Different Electricity Technologies

The life cycle environmental impacts of different electricity generation options in Turkey are compared in Figure 3 and in Tables S8–S18 in Supplementary information. The results show that coal is the worst performer for all the impact categories, except for the ODP, which is lower than for gas power because of leakages of halon 1211 and halon 1301 used as fire suppressants and coolants in the life cycle of gas. Wind power has the highest impacts among the renewable options, with nine out of 11 impacts higher than for hydropower and geothermal power. This is due to the impacts from the extraction and processing of the construction materials. Wind power also has the second highest ADP elements, after hard coal. On the other hand, its GWP is 88% lower than for geothermal power and 11% smaller than for large-reservoir hydropower. The AP of geothermal electricity is lower than lignite power and higher than any other power options considered in this study: almost all of the impact is due to the hydrogen sulphide emissions from geothermal steam to air.

Comparison of Results with the Literature

As far as the authors are aware, there are no other LCA studies of electricity technologies in Turkey, except the previously mentioned studies [12,13]. However, as discussed in the introduction, there are many studies of electricity technologies in other countries, which are compared to the results from the current study in Figures 4 and 5. As can be seen, a wide range of values has been reported in the literature across different impacts. This is primarily due to different assumptions, such as capacities and efficiencies, pollution control measures, fuel origin, plant lifetimes and, in the case of hydropower and wind, local water and wind characteristics and plant design.

Table 4. Assumptions and summary of inventory data.

	Coal	Gas	Hydropower	Wind
Mining and Processing	Lignite: • Domestic lignite, open pit and underground mining • Composition (% w/w) [a]: sulphur: 0.8%–4.5%; ash: 19%–40%; water: 20%–50% • Net heating value (NHV): 7.2–13.9 MJ/kg Hard coal: • Domestic and imported, open pit and underground mining • Composition (% w/w) [a]: sulphur: 0.5%–0.9%; ash: 7%–11%; water: 4%–7% • NHV: 27–27.5 MJ/kg	• Composition (% vol.) [a]: C_1: 94.7%–97.3%; C_2: 1–3.4%; C_3: 0.3%–0.6%; C_4: 0.1%–0.4%; C_{5+}: 0.02%–0.1%; CO_2: 0.06%–0.6%; N_2: 0.1%–4.6% • NHV: 36.5–40.4 MJ/kg • Leakage during extraction: 0.38%		
Transport [b]	Lignite: • Power plants adjacent to the mine Hard coal: • Russia: Freight train (4500 km); freight ship (500 km) • USA: Freight train (1000 km); freight ship (9500 km) • South Africa: Freight train (500 km); freight ship (12,500 km)	• Pipeline: Russia: 5750 km; Iran: 2700 km; Azerbaijan: 1150 km; Nigeria: 4000 km; Other: 4500 km	• Construction materials [c]: Freight train: 200 km; Lorry > 16 tonne: 100 km	• Construction materials: Freight train: 100 km; Lorry > 16 tonne: 100 km • Turbine: Freight train: 2000 km; Lorry > 16 tonne: 150 km • Maintenance: Passenger car: 100 p.km/year

Table 4. Cont.

	Coal	Gas	Hydropower	Wind
Plant Construction	• Lifetime: 30 years [d] Lignite: • Data from Ecoinvent [e] based on average size of the plant of 380 MW (a mix of 500 MW and 100 MW plants in a 70:30 ratio) Hard coal: • Data from Ecoinvent [e] based on average size of the plant of 460 MW (a mix of 500 MW and 100 MW plants at the 90:10 ratio)	• Lifetime: 25 years [d] • Data from Ecoinvent assuming 400 MW plant	Large reservoir: • Life time: 150 years [e,f] • Large reservoir • Data based on Ecoinvent [e,g] with average size of 175.6 MW plant and scaled up to 1057 MW plant Small reservoir: • Data based on ESU [f] with average size of 95 MW plant and scaled up to 98 MW plant Run-of-river: • Life time: 80 years [e,f] Data based on ESU f for the average size of 8.6 MW plant and scaled up to 13.5 MW plant	• Lifetime: 40 years for fixed parts and 20 years for moving parts [e] • Number of turbines: 682 Data based on the average size of 2 MW turbine [h] and scaled down to the average size of 1.94 MW
Plant Operation	• See Table 3 for details Lignite: • Average water use: 37.3 kg/kWh Hard coal: • Average water use: 32.7 kg/kWh	• All plants assumed to be CCGT with efficiency of 55% • Average water use: 3.4 kg/kWh	Large reservoir: • Lubricating oil: 7 mg/kWh Small reservoir: • Lubricating oil: 0.03 mg/kWh Run-of-river: • Lubricating oil: 0.12 mg/kWh	• Lubricating oil: 43.1 mg/kWh
Plant Decommissioning [i]	• Metals: 50% recycled, 50% landfilled • Concrete: 50% recycled, 50% landfilled • Plastics: 20% recycled, 80% landfilled	• Metals: 50% recycled, 50% landfilled • Concrete: 50% recycled, 50% landfilled • Plastics: 20% recycled, 80% landfilled	• Metals: 50% recycled, 50% landfilled • Concrete: 50% recycled, 50% landfilled • Plastics: 20% recycled, 80% landfilled	• Metals: 50% recycled, 50% landfilled • Concrete: 50% recycled, 50% landfilled • Plastics: 20% recycled, 80% landfilled

[a] Based on data from different mines and countries. [b] Estimated by using online mapping. [c] It is assumed that gravel is extracted at the construction site. [d] Source: [42]. [e] Source: [37]. [f] Source: [39]. [g] Source: [38]. [h] Source: [41]. [i] The system has been credited for recycling. The recycling rates are assumed due to a lack of data.

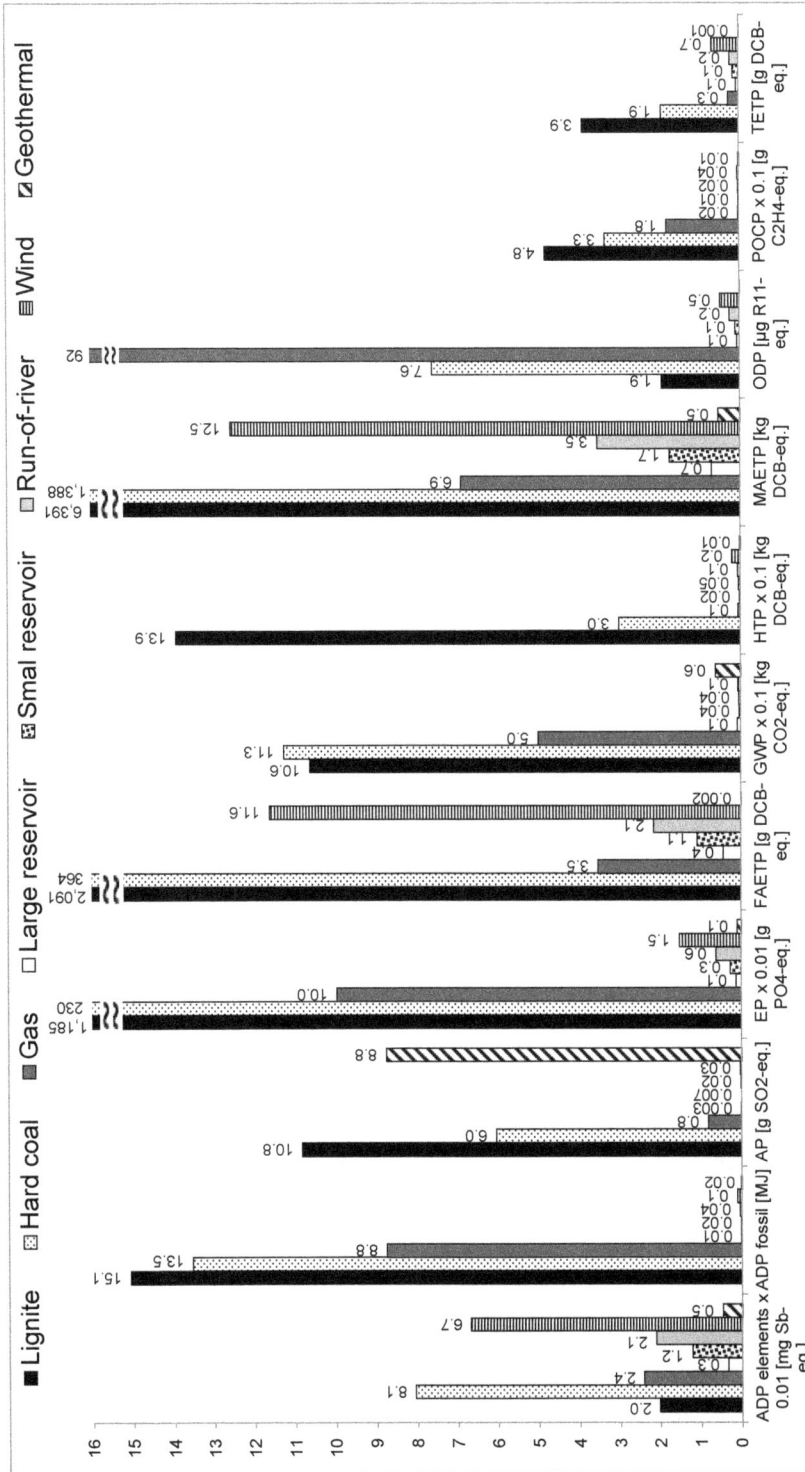

Figure 3. Environmental impacts for different electricity options in Turkey. [All impacts expressed per kWh of electricity generated. The values shown on top of each bar represent the total impact after the recycling credits for the plant construction materials have been taken into account. Some values have been rounded off and may not correspond exactly to those quoted in the text. ADP: Abiotic depletion of elements; ADP fossil: Abiotic depletion of fossil; AP: Acidification potential; EP: Eutrophication potential; FAETP: Fresh water aquatic ecotoxicity potential; GWP: Global warming potential; HTP: Human toxicity potential; MAETP: Marine aquatic ecotoxicity potential; ODP: Ozone layer depletion potential; POCP: Photochemical oxidants creation potential; TETP: Terrestrial ecotoxicity potential.].

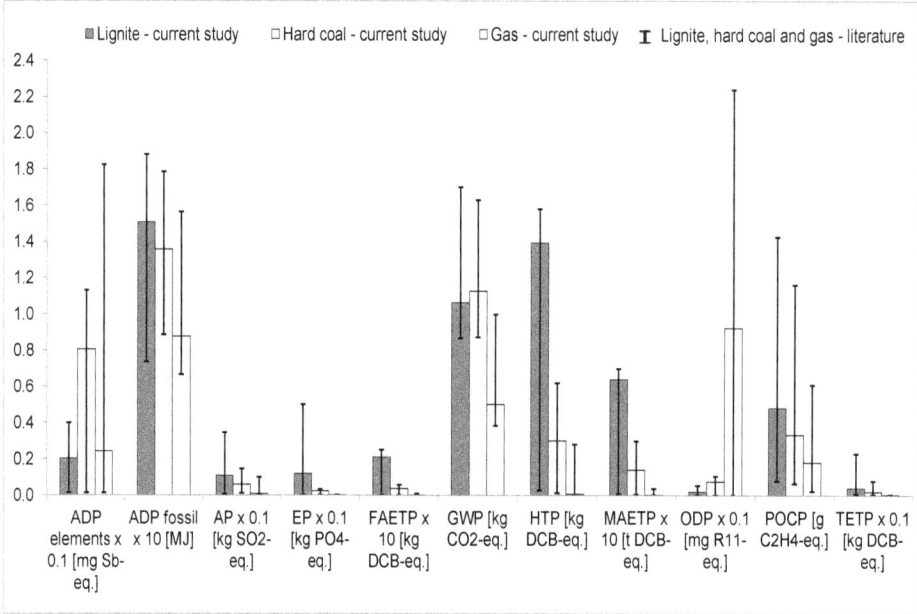

Figure 4. Comparison of the results from current study with literature for coal and gas. [All impacts expressed per kWh of electricity generated. Literature data for lignite from [15,19,21,43]; for hard coal from [19,27,43]; for natural gas from [14,15,19,24,27,43]. All impacts estimated using the CML 2001 method. For impacts nomenclature, see Figure 3.].

Nevertheless, all the impacts estimated in this study for lignite and hard coal power are well within the ranges reported in the literature (Figure 4). For example, for lignite electricity, the GWP falls between 866 and 1700 g CO_2-eq./kWh, which compares well with the estimate in this study of 1062 g CO_2-eq./kWh. For hard coal, the GWP in the literature ranges between 872 and 1628 g CO_2-eq./kWh so that the value of 1126 g CO_2-eq./kWh obtained in the current study is around the middle of the range. A good agreement of the results is also found for gas electricity. For instance, the GWP reported in the literature is between 383 and 996 g CO_2-eq./kWh, and, in this work, it is estimated at 499 g CO_2-eq./kWh.

There are only four LCA studies of reservoir hydropower plants so that comparison with the literature is limited. As shown in Figure 5, the results for the GWP, AP, EP, ODP, POCP and TETP are comparable to the lower values reported in the literature. For example, the GWP falls between 2.7 and 11.6 g CO_2-eq./kWh in the literature, which compares well to this study's estimate of 6.9 g CO_2-eq./kWh for reservoir hydropower. On the other hand, the results for the ADP elements and fossil, FAETP, HTP and MAETP are below the range of values in the literature. One of the reasons for this could be the very different size of large and small scale plants which are not distinguished in the literature.

All the impacts estimated in this study for run-of-river hydropower and wind electricity are well within the ranges reported by other authors (Figure 5). For example, the GWP for run-of-river hydropower is between 0.3 and 5.2 g CO_2-eq./kWh, which compares well to the value of 4.1 g CO_2-eq./kWh obtained in the current study. For onshore wind, the GWP ranges from 6.2 to 31 g CO_2-eq./kWh, and the estimate in the present work is 7.3 g CO_2-eq./kWh.

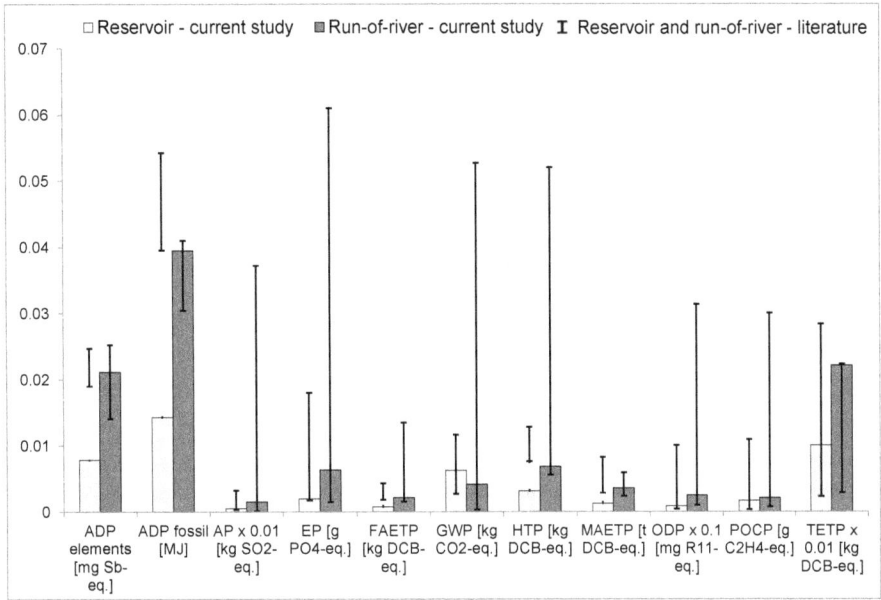

(a)

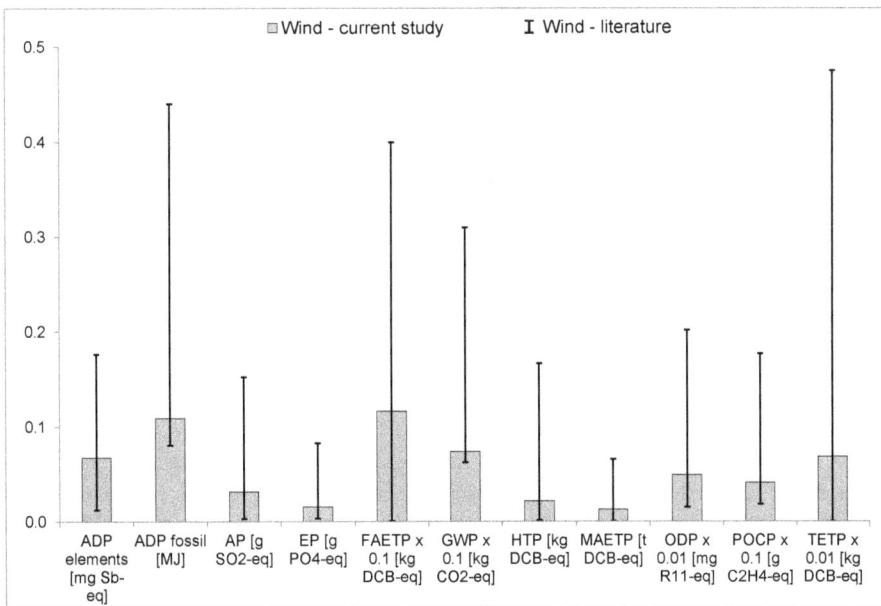

(b)

Figure 5. Comparison of the results from current study with literature for hydropower and wind power. (**a**) Reservoir and run-of-river hydropower. (**b**) Onshore wind. [All impacts expressed per kWh of electricity generated. The current study results for the reservoir hydropower present the average value for large and small reservoir. Literature data for reservoir from: [30,43,44]; for run-of-river from: [20,23,26,29,43–45]; for onshore wind from [22,25,28,31,33,43]. All impacts estimated using the CML 2001 method. For impacts nomenclature, see Figure 3.].

3.2. Environmental Impacts of Electricity Generated in the Base Year

Since the base year is used as a basis for the estimates of impacts in the other years, these results are discussed in more detail than for the rest of the period. Thus, the impacts for the base year are first presented for the functional unit of "1 kWh of electricity", followed by the "total annual generation of electricity".

3.2.1. Impacts per kWh

The impacts per kWh of electricity generated in 2010 are given in Figure 6. The results suggest that fossil fuels cause the majority of the impacts, with coal contributing 43%–54% to the depletion of elements and fossil resources as well as the GWP; it also causes 84%–99% of the toxicity related impacts. The exception is ozone layer depletion, 98% of which arises from gas power.

These results are discussed for each impact in more detail below. Note that all the results incorporate the credits for material recycling after decommissioning of the plants.

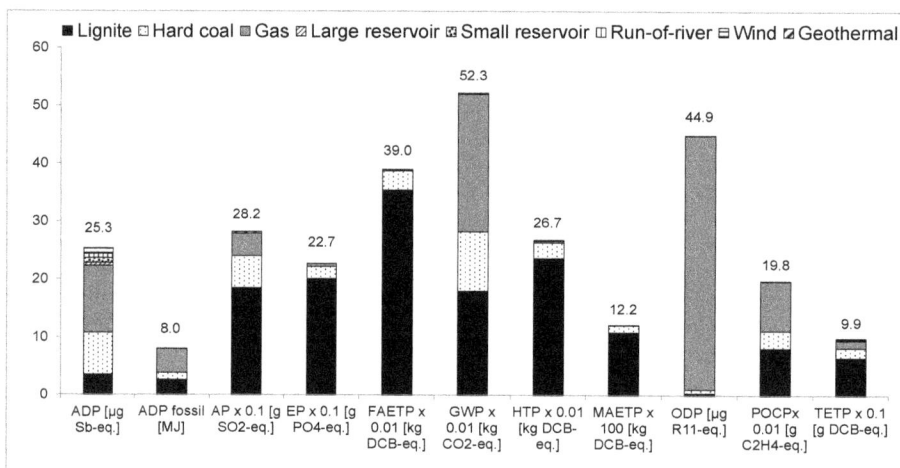

Figure 6. Environmental impacts per kWh of electricity for the base year (2010). [For impacts nomenclature, see Figure 3.].

Abiotic Depletion Potential

The depletion of elements from electricity generation is estimated at 25.3 μg Sb-eq./kWh (Figure 6). Lignite, hard coal and natural gas power contribute 14%, 29% and 46% to the total, respectively. The high contribution of hard coal, despite its small share in the electricity mix (9.1%) is due to its long-range transport, which contributes 63% to the ADP of hard coal. By contrast, lignite is sourced domestically, so there are no impacts from transport.

The remaining contributions are from hydropower (8%) and wind power (4%); the impact from geothermal power is negligible (0.1%). This impact is primarily due to the use of chromium, copper, molybdenum and nickel during the construction of the plants and fuel supply infrastructure. Coal and gas also account for the majority (99.9%) of fossil resource depletion, equivalent to 8 MJ/kWh (Figure 6).

Acidification and Eutrophication Potential

The AP of 2.8 g SO_2-eq./kWh is mainly due to SO_2 (82%) and N_2O (14%) emissions. The biggest contributor is the electricity from lignite (66%) because of the high sulphur content and a lack of desulphurisation systems at some of the plants. The second biggest contributor is hard coal power (20%).

Lignite is also the main cause of the EP, estimated at 2.3 g PO_4-eq./kWh, with the majority (87%) related to the emissions of phosphates during mining. As for the AP, the next biggest contributor is hard coal (9%).

Ecotoxicity Potential

All ecotoxicity related impacts (freshwater, marine and terrestrial) are also caused mainly by lignite power. For the FAETP, equivalent to 0.39 kg DCB-eq./kWh, 91% is due to the emissions of nickel, beryllium, cobalt and vanadium during mining. In the case of the MAETP, estimated at 1.2 t DCB-eq./kWh, the emissions to water of beryllium (40%) and hydrogen fluoride to air (33%) are the main burdens contributing to this impact. Finally, lignite contributes 66% to the TETP of 0.1 g DCB-eq./kWh, mainly because of the emissions of mercury (~75%).

Global Warming Potential

The global warming potential for the electricity mix in Turkey is estimated at 523 g CO_2-eq./kWh. Emissions of CO_2 account for 92% of the total, with CH_4 contributing 7% and N_2O 1%. The main source of the GHG emissions is the combustion of coal and natural gas in power plants. The renewables contribute only 0.4% to the total GWP (Figure 6).

Human Toxicity Potential

Like ecotoxicity, this impact is also mainly due to lignite mining (88%). The main burdens contributing to the HTP of 267 g DCB-eq./kWh include emissions of selenium (36%) and hydrogen fluoride (11%).

Ozone Layer Depletion Potential

The ODP is estimated at 45 μg R11-eq./kWh. The single largest contributor (98%) is electricity from gas. As mentioned earlier, it is mainly due to leakages of halon 1211 and halon 1301 used as fire suppressants and coolants in the life cycle of gas.

Photochemical Oxidants Creation Potential

This impact is equal to 198 mg C_2H_4-eq./kWh with 41%, 15% and 43% arising from lignite, hard coal and gas power plants, respectively. The main contributing burdens include SO_2 (47%), non-methane volatile organic compounds (32%) and N_2O (11%).

3.2.2. Total Annual Impacts

As can be seen from Figure 7, lignite and hard coal power contribute together more than 40% to most of the impacts in the base year, despite providing only a quarter of the total electricity. The exception is ozone layer depletion which is almost entirely (98%) from gas power, which generates 46.5% of electricity. The renewables, which supply around 27% of the demand, add 0.04%–12% to the impacts, mainly related to hydropower because of its high share in the mix (24.5%).

The total GWP is estimated at 111 Mt CO_2-eq., 54% of which is due to coal and 46% to gas power. The estimated direct CO_2 emissions (emitted from the power plants as opposed to the life cycle emissions) are equal to 92 Mt CO_2-eq. This is in good agreement with the total direct emissions of 99 Mt CO_2-eq. from the whole Turkish electricity sector in 2010 [11]. The slight difference between the two estimates stems from different assumptions, including plant efficiencies and the amount of fuel used in different power plants.

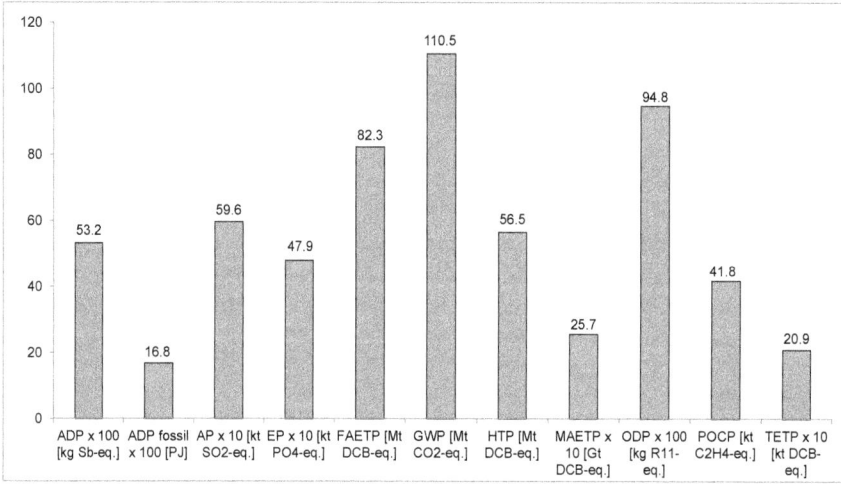

Figure 7. Total annual environmental impacts for the base year (2010). [For impacts nomenclature, see Figure 3.].

To put these results into context, they have been normalised to the annual impacts in the European Union comprising 28 countries (EU28). As indicated in Figure 8, Turkish electricity sector contributes almost 5% to the depletion of fossil fuels and 2% to the GWP relative to these impacts in the EU28 countries. The contribution to the AP, EP and POCP is also significant, ranging from 3.5% for the former to 2.4% for the latter. The contribution of the ADP elements and ODP is small (0.1%). However, these results should be interpreted with care for at least two reasons: first, the base year for the EU28 impacts is 2000 (the latest available in the CML 2001 method) while the base year here is 2010; secondly, all impacts but the GWP for the EU28 are associated with high uncertainty and are likely underestimated, particularly the toxicity-related categories [46,47], which is the reason why they have not been considered in the normalised results.

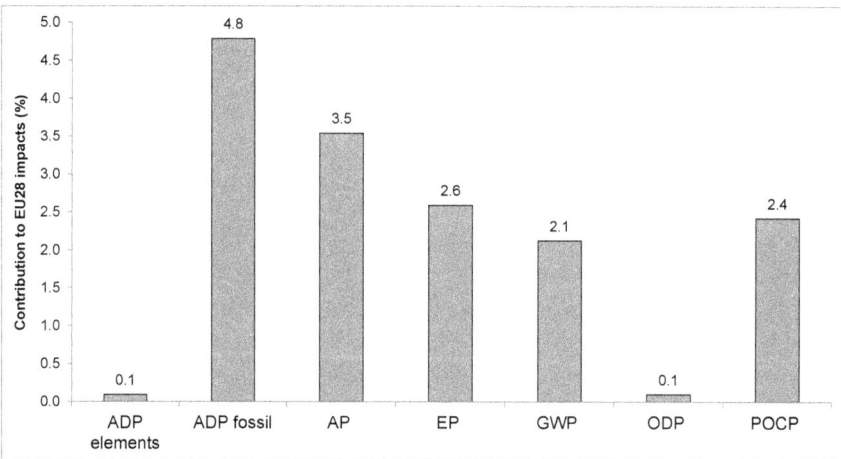

Figure 8. Annual impacts from electricity generation in Turkey in the base year (2010) normalised to the annual EU28 impacts. [EU28: the European Union with 28 countries. All data for the EU28 impacts are for the year 2000, the latest available data in the CML 2001 method. For impacts nomenclature, see Figure 3.].

3.3. Environmental Impacts from Electricity Generation from 1990–2014

The total annual environmental impacts from the electricity mix in Turkey over the period have been calculated based on the impacts of different technologies presented in Section 3.1, their contribution to electricity generation in a particular year and the total electricity generation in that year. The results are summarised in Figure 9; for brevity, they only show the impacts for selective years in five-year intervals: 1990, 1995, 2000, 2005, 2010 and 2014 (data for 2015 were not available at the time of writing).

Figure 9 also shows the generation mix for each of the years considered. In 1990, total generation was 57 TWh and in 2014 it reached 250 TWh, representing a four-fold increase over the period. During this time, the electricity mix has also changed. In the past, it was mainly based on lignite, gas and reservoir hydropower but, over time, the contribution of other sources grew. For example, the share of hard coal (mainly imported) increased from 0.6 TWh in 1990 to 36.7 TWh in 2014. Since 2005, there has also been an increase in the share of renewables, mainly run-of-river hydropower (to 5% in 2014) and onshore wind (to 3%). The reservoir hydropower generation was reduced by almost a half in 2000 compared to 1995 and by another half in 2014 compared to 2010 because of severe drought; the shortfall in electricity generation was made up by gas power.

It can be noticed from Figure 9 that the annual impacts have been going up steadily over the period, increasing from two times (EP, HTP, FAETP and MAETP) to nine-fold (ODP). The GWP increased by five-fold, from 28.7 Mt CO_2-eq. in 1990 to 143 Mt CO_2-eq. in 2014. This is largely due to a growing electricity demand, although for some impacts the rate of increase is higher than the demand growth, including the GWP, because of the higher rate of growth of fossil-fuel electricity. The only exception to the trend are the reductions in the EP, FAETP, HTP and MAETP in 2005 on the previous period as a result of a decrease in the share of lignite power. The latter, together with hard coal and gas dominate all the impacts because of their dominance in the generation mix, with coal providing 26%–35% and gas 18%–48% of electricity across the years.

The impact trends per kWh are less uniform than for the total annual impacts because they depend on the electricity mix only. On the other hand, the annual impacts also depend on the amount of electricity generated and increase proportionally to the demand. Although, in general, most impacts are higher for the fossil fuel options, some impacts are also high for the renewables (e.g., ADP for wind power; see Figure 3) so that their contribution to the total impacts from the electricity mix may be disproportionate to their share in the mix. The upward trend is found for four impacts: ADP elements and ODP, which are two times higher in 2014 than in 1990, and ADP fossil and GWP, which increased by 13% in the same period. The increase in the depletion of elements is due to the growth in renewable electricity generation because wind and hydropower have high resource requirements per kWh of electricity generated. The other three impacts have increased because of the increasing share of fossil fuels; for example, the share of gas increased from 25% in 1990 to 50% in 2014, leading to the significant increase in the ODP. Some impacts were also affected by the low hydropower generation during the drought in 2000 and 2014, with the GWP, ADP fossil, ODP and POCP having the highest values per kWh in these two years. For example, the GWP in 2000 was 550 g CO_2-eq./kWh, 13% higher than in 1995. In 2014, it was 570 g CO_2-eq./kWh or 8% higher than in 2010, despite an overall 4% increase in the share of other renewables.

However, the opposite trend is found for the remaining impacts, with the EP, FAETP, HTP and MAETP being two times lower, the TETP 40% and AP 34% smaller at the end of the period than in 1990. This is a result of the increase in natural gas and renewables in the mix. The only exception is the POCP, which remained more or less unchanged over the period. This is because of the cancelling effects of the change over time of the share of coal and gas in the mix and the difference in their respective POCP: lignite and hard coal have 1.8–2.7 times higher impact than gas but their contribution to power generation reduced by 20% from 1990 to 2014 while the contribution of gas doubled.

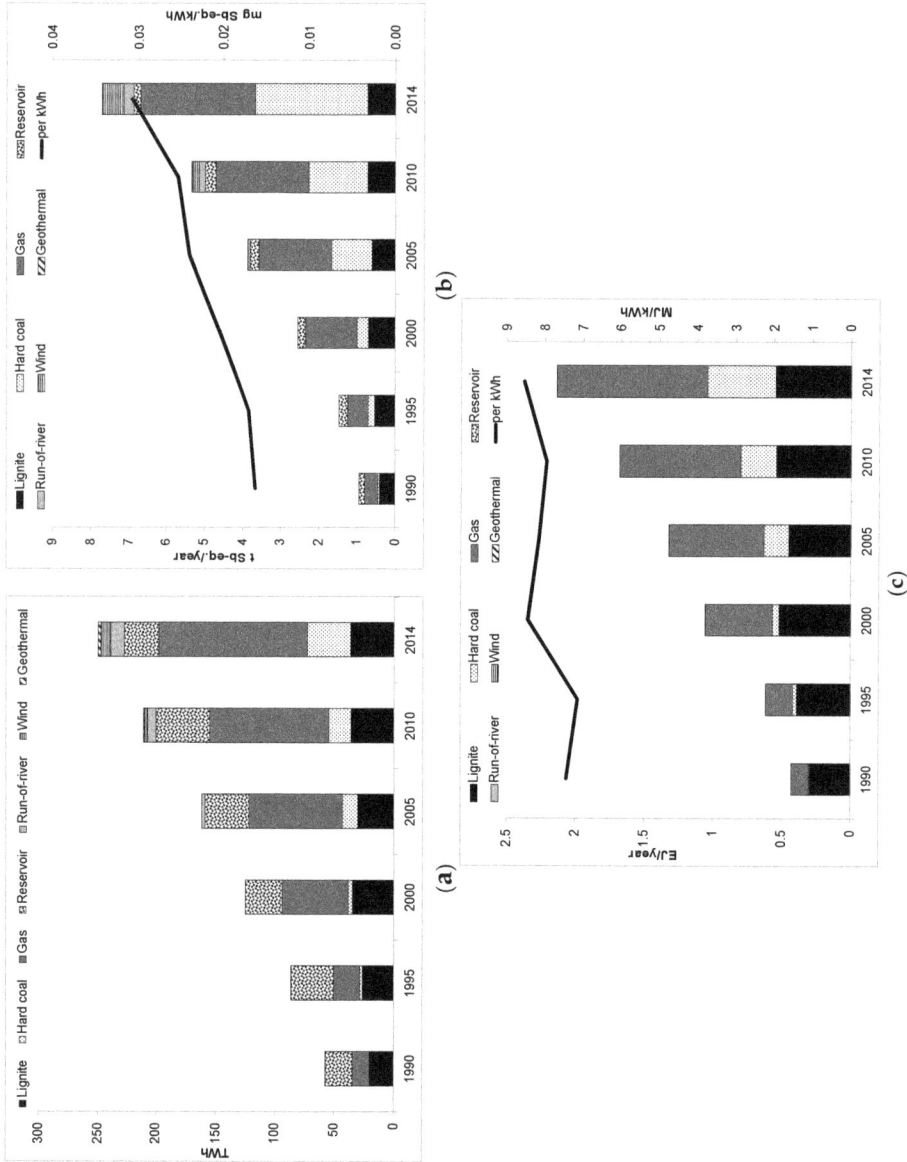

Figure 9. *Cont.*

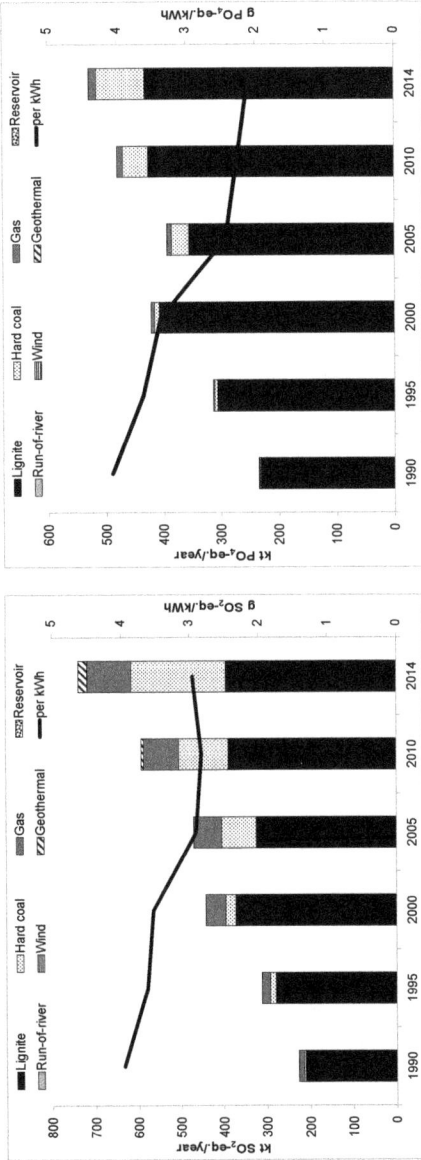

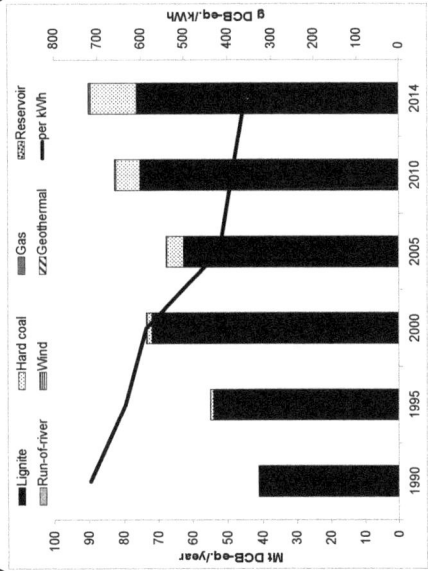

Figure 9. *Cont.*

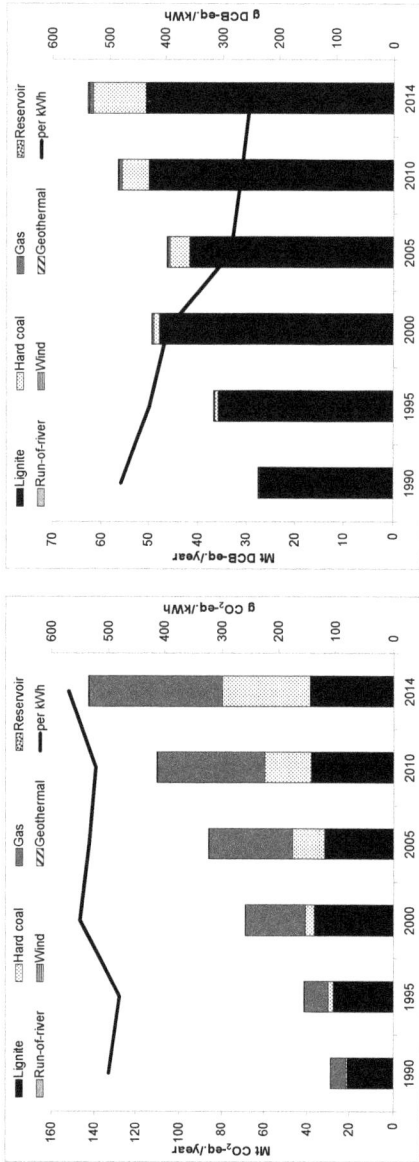

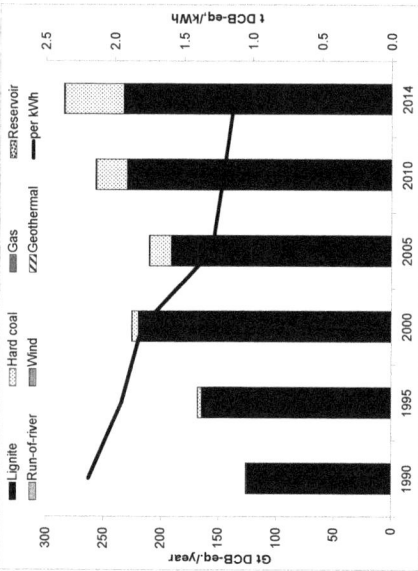

Figure 9. *Cont.*

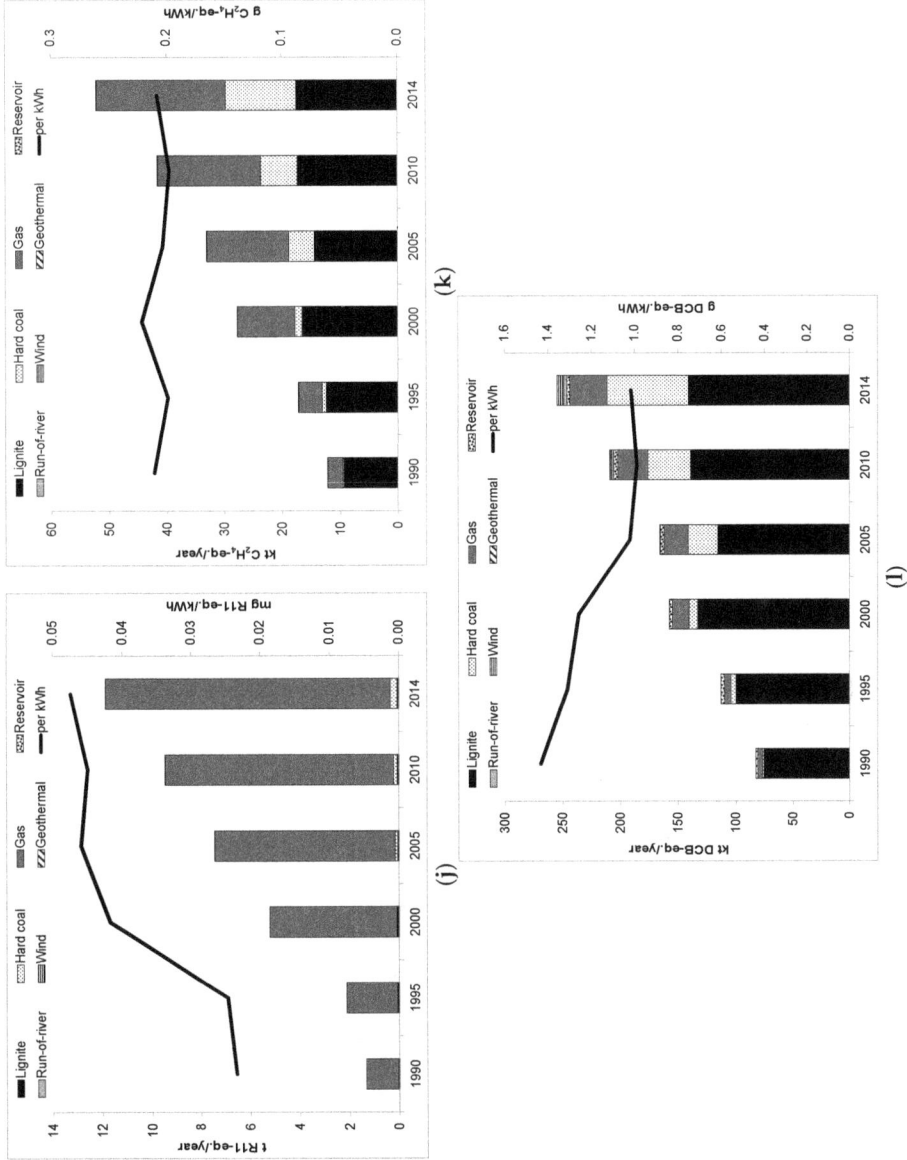

Figure 9. Environmental impacts of electricity in Turkey in the period 1990–2014 for total annual generation and per kWh. (**a**) Electricity generation mix in Turkey; (**b**) ADP elements; (**c**) ADP fossil; (**d**) AP; (**e**) EP; (**f**) FAETP; (**g**) GWP; (**h**) HTP; (**i**) MAETP; (**j**) ODP; (**k**) POCP; (**l**) TETP. [For impacts nomenclature, see Figure 3.].

4. Conclusions

This study has estimated for the first time the life cycle environmental impacts of electricity generation in Turkey for the period 1990–2014. Eleven impacts have been considered for 516 power plants using lignite, hard coal, natural gas, hydro (large and small scale reservoir hydro and run-of-river), onshore wind and geothermal energy. The impacts have been estimated per kWh and for the total amount of electricity generated annually over the period.

The results suggest that, in comparison to the other options, lignite power has the highest impacts for eight out of eleven categories. Gas power is the worst option for ozone layer depletion. Geothermal electricity scores the best for six impacts (eutrophication, ozone layer depletion and all the toxicity categories). Large-reservoir hydropower has the lowest depletion of elements and fossil resources as well as acidification. Run-of-river is the best option for the global warming potential.

The annual impacts have been going up steadily over the period, increasing from two times (EP, HTP, FAETP and MAETP) to nine-fold (ODP); the GWP increased by a factor of five. This is due to meeting the growing demand by fossil fuels. The trends per unit of electricity generated are less uniform than for the annual impacts. The upward trend is found for four impacts only: ADP elements and ODP, which are two times higher in 2014 than in 1990, and ADP fossil and GWP, which increased by 13% in the same period. However, the opposite trend is found for the other impacts, with the EP, FAETP, HTP and MAETP being two times lower, the TETP 40% and AP 34% smaller at the end of the period than in 1990. The only exception is the POCP, which remained more or less unchanged.

As a signatory to the Kyoto Protocol, it is important that Turkey identified more sustainable pathways for future electricity supply. Being a developing country, it is likely that the demand will continue to grow. Therefore, the key question is how to decouple the expected demand growth with the environmental impacts. The results of this work show that reducing the share of coal in the electricity mix would lead to significant reductions in GHG emissions as well as the other impacts such as acidification, eutrophication and all the toxicity categories. This could be achieved in the short term by expanding the natural gas capacity; however, both ozone layer depletion and energy dependency on other countries would increase. Given the great potential of renewable energies in Turkey, the expansion of renewables should be pursued more aggressively in the medium to long term. Among the renewable energy sources, hydro, onshore wind and geothermal power are well established in Turkey and have a large potential for development. However, increasing the proportion of onshore wind power in the electricity mix would increase depletion of elements while increasing the share of geothermal power would increase acidification. As the results show, hydropower options would lead to a reduction in the environmental impacts; however, that would be at the expense of some other impacts not included in this study, such as biodiversity and land use. Other renewables such as solar power could play a role in the future given that solar energy is abundant in Turkey. The country is also trying to introduce nuclear power, which would help to reduce GHG emissions, but it raises various other concerns, such as human health impacts from radiation, risk of accidents and long-term storage of nuclear waste.

The findings from this work have provided an insight into the previously unknown environmental impacts of electricity generation in Turkey, helping to identify opportunities for future improvements. However, in addition to the environmental sustainability evaluated here, it is also important to understand various economic and social aspects to help the industry and policy makers in Turkey identify and implement most sustainable electricity options for the future. This is the subject of a forthcoming paper by the authors.

Acknowledgments: This work was funded by the Republic of Turkey Ministry of National Education and the UK Engineering and Physical Sciences Research Council, EPSRC (Grant no. EP/K011820/1). This funding is gratefully acknowledged.

Author Contributions: Adisa Azapagic conceived and supervised the research. Burcin Atilgan collected the data and carried out the LCA study. Adisa Azapagic and Burcin Atilgan co-wrote the paper.

Conflicts of Interest: The authors declare no conflict of interest.

References

1. MENR. *Mavi Kitap (Blue Book)*; Ministry of Energy and Natural Resources: Ankara, Turkey, 2012; pp. 20–69.
2. WEC. Turkey Energy Balance Table (1970–2004). Available online: http://www.dektmk.org.tr/incele.php?id=MTQ3 (accessed on 13 April 2014).
3. TEIAS. Electricity Generation & Transmission Statistics of Turkey. Available online: http://www.teias.gov.tr/TurkiyeElektrikIstatistikleri.aspx (accessed on 2 June 2014).
4. EUAS. *Annual Report*; Turkish Electricity Generation Company (EUAS): Ankara, Turkey, 2011.
5. TPAO. *The Oil and Gas Sector Report of Turkey*; Turkish Petroleum Corporation (TPAO): Ankara, Turkey, 2011; pp. 1–26.
6. MMO. *Turkiye'de Termik Santraller*; Makina Muhendisleri Odasi: Ankara, Turkey, 2010; pp. 1–81.
7. TKI. *Lignite Sector Report of Turkey*; General Directorate of Turkish Coal Enterprises (TKI): Ankara, Turkey, 2012; pp. 1–61.
8. DSI, Turkey Water Report 2009. Available online: http://www2.dsi.gov.tr/english/pdf_files/TurkeyWaterReport.pdf (accessed on 20 March 2012).
9. EMRA. *Data on Energy Potential of Turkey*; Personel communication, 15.06.2014; Republic of Turkey Energy Market Regulatory Authority: Ankara, Turkey, 2014.
10. Komurcu, M.I.; Akpinar, A. Importance of geothermal energy and its environmental effects in Turkey. *Renew. Energy* **2009**, *34*, 1611–1615. [CrossRef]
11. FutureCamp. *Baseline Emission Calculations, Verified Carbon Standard (VCS)*; Version 3; FutureCamp: Ankara, Turkey, 2011.
12. Atilgan, B.; Azapagic, A. Life cycle environmental impacts of electricity from fossil fuels in Turkey. *J. Clean. Prod.* **2015**, *106*, 555–564. [CrossRef]
13. Atilgan, B.; Azapagic, A. Renewable electricity in Turkey: Life cycle environmental impacts. *Renew. Energy* **2015**. in press. [CrossRef]
14. Kannan, R.; Leong, K.C.; Osman, R.; Ho, H.K.; Tso, C.P. Gas fired combined cycle plant in Singapore: Energy use, GWP and cost and Life cycle approach. *Energy Convers. Manag.* **2005**, *46*, 2145–2157. [CrossRef]
15. Weisser, D. A guide to life-cycle greenhouse gas (GHG) emissions from electric supply technologies. *Energy* **2007**, *32*, 1543–1559. [CrossRef]
16. Guinée, J.B.; Gorrée, M.; Heijungs, R.; Huppes, G.; Kleijn, R.; Koning, A. *Life Cycle Assessment: An Operational Guide to the ISO Standards: Part 2a*; Ministry of Housing, Spatial Planning and Environment (VROM) and Centre of Environmental Science (CML): Den Haag and Leiden, The Netherlands, 2001.
17. International Standard Organization (ISO). *Life Cycle Assessment—Requirements and Guidelines*; ISO: Geneva, Switzerland, 2006.
18. International Standard Organization (ISO). *Life Cycle Assessment—Principles and Framework*; ISO: Geneva, Switzerland, 2006.
19. PE International. *GaBi version 6*; PE International: Stuttgart, Echterdingen, Germany, 2013.
20. Schuller, O.; Albrecht, S. *Setting up Life Cycle Models for the Environmental Analysis of Hydropower Generation, Considering Technical and Climatic Boundary Conditions*; Life Cycle Assessment VIII: Seattle, WA, USA, 2008.
21. Pehnt, M.; Henkel, J. Life cycle assessment of carbondioxide capture and storage from lignite power plants. *Int. J. Greenh. Gas Control* **2009**, *3*, 49–66. [CrossRef]
22. Martínez, E.; Sanz, F.; Pellegrini, S.; Jiménez, E.; Blanco, J. Life-cycle assessment of a 2 MW rated power wind turbine: CML method. *Int. J. Life Cycle Assess.* **2009**, *14*, 52–63. [CrossRef]
23. Suwanit, W.; Gheewala, S. Life cycle assessment of mini-hydropower plants in Thailand. *Int. J. Life Cycle Assess.* **2011**, *16*, 849–858. [CrossRef]
24. Santoyo-Castelazo, E.; Gujba, H.; Azapagic, A. Life cycle assessment of electricity generation in Mexico. *Energy* **2011**, *36*, 1488–1499. [CrossRef]
25. Lahuerta, F.; Saenz, E. Life cycle assessment of the wind turbines installed in Spain until 2008. In Proceedings of the Europe's Premier Wind Energy Conference and Exhibition, Brussels, Belgium, 14–17 March 2011.
26. Pascale, A.; Urmee, T.; Moore, A. Life cycle assessment of a community hydroelectric power system in rural Thailand. *Renew. Energy* **2011**, *36*, 2799–2808. [CrossRef]

27. Stamford, L.; Azapagic, A. Life cycle sustainability assessment of electricity options for the UK. *Int. J. Energy Res.* **2012**, *36*, 1263–1290. [CrossRef]
28. Pereg, J.R.M.; Hoz, J.F. *Life Cycle Assessment of 1 kWh Generated by a Gamesa Onshore Windfarm G90 2.0 MW*; Gamesa: Zamudio (Vizcaya), Spain, 2013.
29. Arnøy, S.; Modahl, I.S. *Life Cycle data for Hydroelectric Generation at Embretsfoss 4 (E4) Power Station: Background Data for Life Cycle Assessment (LCA) and Environmental Product Declaration (EPD)*; Ostfold Research: Kråkerøy, Norway, 2013.
30. Arnøy, S. *Hydroelectricity from Trollheim Power Station*; Østfoldforskning AS: Oslo, Norway, 2013.
31. Garrett, P.; Rønde, K. *Life Cycle Assessment of Electricity Production from an Onshore V90–3.0 MW Wind Plant*; Vestas Wind Systems A/S: Aarhus, Denmark, 2013.
32. Garrett, P.; Rønde, K. Life cycle assessment of wind power: Comprehensive results from a state-of-the-art approach. *Int. J. Life Cycle Assess.* **2013**, *18*, 37–48. [CrossRef]
33. Palomo, B.; Gaillardon, B. *Life Cycle Assessment of a French Wind Plant*; Europe's Premier Wind Energy Event: Barcelona, Spain, 2014.
34. IEA/NEA. *Projected Costs of Generating Electricity, 2005 Update*; International Energy Agency and Nuclear Energy Agency: Paris, France, 2005.
35. Aslanoglu, S.Y.; Koksal, M.A. Elektrik uretimine bagli karbondioksit emisyonunun bolgesel olarak belirlenmesi ve uzun donem tahmini. *Hava Kirliligi Arastirma Derg. (HKAD)* **2012**, *1*, 19–29.
36. Öko Institute Global Emission Model for Integrated Systems (GEMIS) v.4.8. Available online: http://www.oeko.de/service/gemis/en/index.htm (accessed on 10 September 2012).
37. Dones, R.; Bauer, C.; Bolliger, R.; Burger, B.; Faist Emmenegger, M.; Frischknecht, R.; Heck, T.; Jungbluth, N.; Röder, A.; Tuchschmid, M.; *et al. Ecoinvent Report: Life Cycle Inventories of Energy Systems: Results for Current Systems in Switzerland and Other UCTE Countries*; Swiss Centre for Life Cycle Inventories: Dübendorf, Switzerland, 2007.
38. Bauer, C.; Bolliger, R. *Ecoinvent Report: Wasserkraft*; Swiss Centre for Life Cycle Inventories: Dübendorf, Switzerland, 2007.
39. Flury, K.; Frischknecht, R. *Life Cycle Inventories of Hydroelectric Power Generation, ESU Database*; Öko-Institute e.V.: Uster, Switzerland, 2012.
40. Coulson, J.M.; Sinnott, R.K.; Richardson, J.F. *Coulson & Richardson's Chemical Engineering*; Butterworth Heinemann Ltd.: Oxford, UK; Boston, MA, USA, 1993.
41. Kouloumpis, V.; Stamford, L.; Azapagic, A. Decarbonising electricity supply: Is climate change mitigation going to be carried out at the expense of other environmental impacts? *Sustain. Prod. Consum.* **2015**, *1*, 1–21. [CrossRef]
42. TEIAS. *Turkiye Elektrik Enerjisi Uretim Planlama Calismasi (2012–2030)*; Turkish Electricity Transmission Corporation (TEIAS), Research Planning and Coordination Department: Ankara, Turkey, 2013.
43. Swiss Centre for Life Cycle Inventories. *Ecoinvent, Ecoinvent Database v2.2*; Swiss Centre for Life Cycle Inventories: St. Gallen, Switzerland, 2010.
44. ESU. *ESU Database*; Öko-Institute e.V., ESU-services Ltd.: Uster, Switzerland, 2012.
45. Raadal, H.L.; Gagnon, L.; Modahl, I.S.; Hanssen, O.J. Life cycle greenhouse gas (GHG) emissions from the generation of wind and hydro power. *Renew. Sustain. Energy Rev.* **2011**, *15*, 3417–3422. [CrossRef]
46. Benini, L.; Mancini, L.; Sala, S.; Manfredi, S.; Schau, E.M.; Pant, R. *Normalisation Method and Data for Environmental Footprints*; European Commission, Joint Research Centre, Institute for Environment and Sustainability: Ispra, Italy, 2014.
47. Heijungs, R.; de Koning, A.; Lightart, T.; Korenromp, R. *Improvement of LCA Characterisation Factors and LCA Practice for Metals*; TNO Environment, Energy and Process Innovation, Netherlands Organisation for Applied Scientific Research: Apeldoorn, The Nertherland, 2004.

9

A Power Prediction Method for Photovoltaic Power Plant Based on Wavelet Decomposition and Artificial Neural Networks

Honglu Zhu [1], Xu Li [1,*], Qiao Sun [2], Ling Nie [2], Jianxi Yao [1] and Gang Zhao [3]

Academic Editor: Tapas Mallick

[1]   School of Renewable Energy, North China Electric Power University, Beijing 102206, China;
     hongluzhu@126.com (H.Z.); jianxiyao@ncepu.edu.cn (J.Y.)
[2]   Beijing Guodiantong Network Technology Co., Ltd., Beijing 100070, China;
     sunqiao@sgepri.sgcc.com.cn (Q.S.); nieling@sgepri.sgcc.com.cn (L.N.)
[3]   School of Electronic Engineering, Xidian University, Xian 710071, China; gangzhao@mail.xidian.edu.cn
*   Correspondence: lixu_ncepu@126.com

Abstract: The power prediction for photovoltaic (PV) power plants has significant importance for their grid connection. Due to PV power's periodicity and non-stationary characteristics, traditional power prediction methods based on linear or time series models are no longer applicable. This paper presents a method combining the advantages of the wavelet decomposition (WD) and artificial neural network (ANN) to solve this problem. With the ability of ANN to address nonlinear relationships, theoretical solar irradiance and meteorological variables are chosen as the input of the hybrid model based on WD and ANN. The output power of the PV plant is decomposed using WD to separated useful information from disturbances. The ANNs are used to build the models of the decomposed PV output power. Finally, the outputs of the ANN models are reconstructed into the forecasted PV plant power. The presented method is compared with the traditional forecasting method based on ANN. The results shows that the method described in this paper needs less calculation time and has better forecasting precision.

Keywords: photovoltaic power prediction; wavelet decomposition; artificial neural network; theoretical solar irradiance; signal reconstruction

1. Introduction

As energy supply and the environmental situation becomes increasingly tight and critical around the world, the contradiction between electricity power supply and demand stands out. The development and utilization of conventional energy sources suffer from increasing limitations. Solar energy is recognized as an ideal renewable energy power generation sourcef. PV power generation is an important solar energy utilization pattern, but the output of PV power plants is highly random, with a fluctuating and intermittent nature. The grid-connection of large-scale PV power generation will bring about severe challenges to the safe and stable operation of grid and guaranteed electric power quality. Forecasting the output power of PV plant is a significant problem for electric power departments to adjust dispatch planning in time, boost the reliability of electric system operation and the connection level of PV power plants and reduce spinning the reserve capacity of generation systems [1,2].

In the past few years, PV power forecasting has been widely studied. Short-term power prediction methods for PV power plants mainly include two categories: physical methods and statistical methods. Physical methods mean that a physical equation is established for forecasting according to the PV power generation process and system characteristics and in combination with forecast weather data [3,4]. Statistical methods aim to summarize inherent laws to forecast the output power of PV power plants based on historical power data [1,2,5–9]. The above methods have their respective advantages, but the non-stationary characteristics of PV power output has an important influence on the convergence and properties of the above methods.

Since solar irradiance received at a site on the Earth's surface shows periodicity and non-stationary characteristics due to the influence of Earth's rotation and revolution, output power data of PV plants shows one day periodicity. In other words, the power output presents a rising trend before noon, and presents a declining trend after noon. If an effective method to reduce the non-stationary characteristics of PV output power is not adopted, conventional power prediction methods cannot guarantee the precision of forecasting results, or even the convergence of the method [10]. Wavelet analysis methods can effectively extract the nonlinear and non-stationary features from original signals [1,5,11]. They are applicable to analyzing the fluctuations of renewable energy sources with obvious intermittent and non-stationary characteristics. Relevant research has achieved good results in wind power prediction [12].

Thus, in order to deal with the periodic and non-stationary problems of PV output power, a hybrid modeling method based on WD and ANN is proposed in the paper, to achieve both good algorithm convergence and prediction results. This paper is organized as follows: Section 2 analyzes the periodicity and non-stationary characteristics of PV output. Section 3 describes the PV power output decomposition process. Section 4 demonstrates the hybrid prediction method which combines wavelet decomposition and neural networks. Section 5 verifies and compares the effectiveness of the proposed method. Section 6 offers the conclusions of this study.

2. Analysis of Photovoltaic (PV) Output Characteristics

2.1. Periodicity and Non-Stationary Characteristics of Photovoltaic (PV) Output

The data used in the paper are from the PV testing platform of the State Key Laboratory of Alternate Electrical Power System with Renewable Energy Sources at North China Electric Power University in Changping District (Beijing, China), recorded with a sampling interval of 1 min. Table 1 shows the detailed information of the PV testing platform.

Table 1. PV power plant information.

Item	Data	Item	Data
Longitude	116.3059°E	Mounting disposition	Flat roof
Latitude	40.08914°N	Field type	fixed tilted plane
Altitude	80m	Installed capacity	10 kWp
Azimuth	0°	Technology	polycrystalline silicon
Tilt	37°	PV module	JKM245P

Figure 1 shows the output of the PV power plant for 5 days. It is known from the figure that the power output curve is influenced by the Earth's rotation with a period of one day, the output is 0 during the non-illuminated time at night and rises in the forenoon, and then declines in the afternoon. Influenced by the surrounding meteorological factors, the output power of the PV power plant shows different characteristics every day. The output characteristics of the PV power plant under different weather conditions are analyzed. Figure 2 shows the output curves of the PV power plant under four typical weather conditions: clear day, cloudy day, overcast day and rainy day. The solid line is the measured output of the PV power plant, while the dotted line is the theoretical output of the PV power plant, while the four colors correspond to the four seasons. Figure 2a, corresponding to a clear day, shows the measured output of the PV power plant is similar to the theoretical output; In Figure 2b, on a cloudy day, the measured output curve of the PV power plant fluctuates frequently due to the movement of clouds in the sky; In Figure 2c, on an overcast day, the measured PV power plant output declines significantly, and the fluctuation is smaller than in cloudy weather; In Figure 2d, on a rainy day, the measured output of the PV power plant may show a huge decline. Influenced by seasonal factors, the global solar irradiance curve has obvious amount differences. It is highest in summer, followed by spring, autumn and winter. In brief, the daily output power of the PV power plant shows nonlinear characteristics from day to night; while the output of the PV power plant has obvious differences in the total amount during different seasons in a 1-year period. It can be seen that conventional power prediction methods based on time series are not applicable to the output power of a PV power plant.

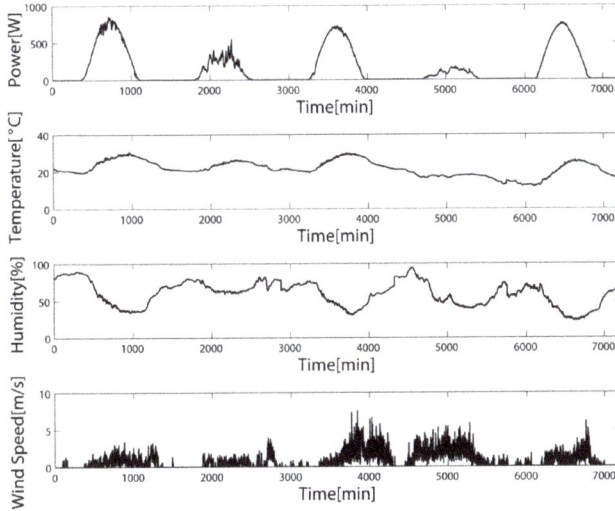

Figure 1. Output and meteorological factors of a PV power plant.

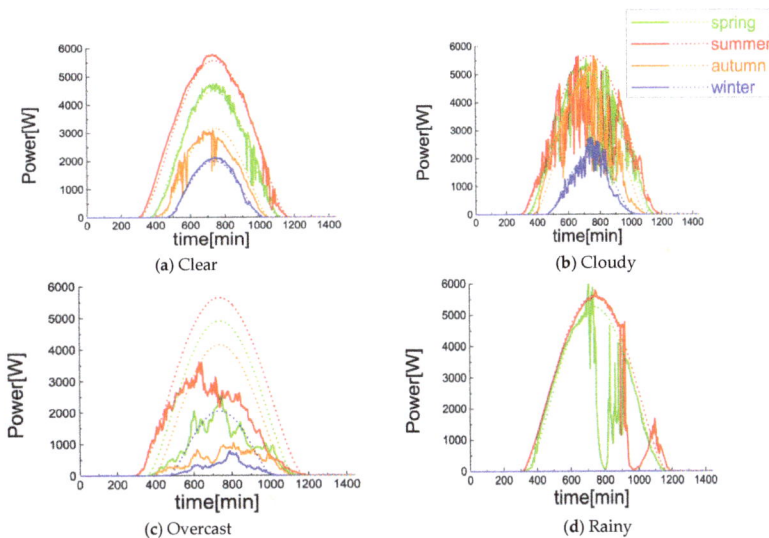

Figure 2. Output of a PV power plant during typical seasons and under typical weather conditions. (**a**) Daily PV output on clear day during typical seasons, (**b**) Daily PV output on cloudy day during typical seasons, (**c**) Daily PV output on overcast day during typical seasons, (**d**) Daily PV output on rainy day during typical seasons.

2.2. Influence of Meteorological Factors on Photovoltaic (PV) Output and Model Input Selection

To establish an accurate and reliable output power prediction model for a PV power plant, it is necessary to analyze the effect factors for the PV power plant output. In the physical sense, global solar irradiance received on the ground is a direct influencing factor on the voltage effect of PV cells. Meanwhile, PV output is also influenced by various kinds of meteorological conditions (Figure 1). Table 1 illustrates the correlation of output power with irradiance, temperature and humidity for each typical weather condition. The Pearson product-moment correlation coefficient, also known as r, can measure the direction and strength of the linear relationship between two variables, which is a method to quantify non-deterministic relationship. We use r to select the inputs for the forecasting model. The term r can be defined as follows:

$$r = \frac{\sum\limits_{i=1}^{N} \left(X(i) - \overline{X} \right) \times \left(Y(i) - \overline{Y} \right)}{\sqrt{\sum\limits_{i=1}^{N} \left(X(i) - \overline{X} \right)^2} \times \sqrt{\sum\limits_{i=1}^{N} \left(Y(i) - \overline{Y} \right)^2}} \tag{1}$$

where, N is the length of time series; i is the series number; X is the output power of the PV power plant; Y is one of the meteorological conditions, namely irradiance, temperature, humidity or wind speed.

It can be seen from Table 2 that the correlation coefficient between the power output of a PV power generator and solar irradiance is greater than 0.8, which means they are highly correlated, while the correlation coefficient between PV power generation output and temperature is greater than 0.3, which means these factors are positively and low-level correlated. The correlation coefficient of humidity indicates a low but negative correlation. The correlation between PV power generation output and wind speed is small.

Table 2. Pearson product-moment correlation coefficient between PV output and environmental factors under typical weather conditions.

Weather Condition	Pearson Product-Moment Correlation Coefficient			
	Irradiance	Temperature	Humidity	Wind Speed
Clear	0.966	0.322	−0.527	−0.229
Cloudy	0.891	0.441	−0.511	−0.025
Overcast	0.987	0.409	−0.478	0.125
Rainy	0.923	0.410	0.039	−0.178

3. Wavelet Decomposition (WD) for Photovoltaic (PV) Power Output

3.1. Wavelet Decomposition (WD) Fundamentals

The wavelet transform method [1,13] is a mathematical tool, much like a Fourier transform, that in analyzing a time series signal, can be used to analyze nonlinear and non-stationary signals. The wavelet transform method decomposes a signal into different scale layers with different levels of resolution. The decomposition into different scales is made possible by the fact that the wavelet transform method is based on a square-integrable function and group theory representation. The wavelet transform provides a local representation (both in time and frequency) of a given signal, so it is suitable for analyzing a signal with varying time-frequency resolution, such as the output power of a PV power plant. The mathematical definition for a wavelet transform is as follows.

Let $x(t)$ is a finite energy signal, which satisfies:

$$\int_{-\infty}^{+\infty} |x(t)|^2 \, dt < \infty \tag{2}$$

The wavelet transform for $x(t)$ is:

$$CWT_\psi x(a,b) = W_x(a,b) = \int_{-\infty}^{+\infty} x(t)\psi_{a,b}^*(t)dt \tag{3}$$

where:

$$\psi_{a,b}(t) = |a|^{-\frac{1}{2}} \psi(\frac{t-b}{a}) \tag{4}$$

$\psi(t)$ is the base function of the mother wavelet, the asterisk denotes the complex conjugate, and a, b are the transform parameters. By the selection of a and b, the wavelet transform provides an elegant algorithm which is known as the WD technique, to decompose a signal into different layers with different time and frequency resolution.

The WD decomposes the given signal $x(n)$ into its detailed smoothed layers. The signal for the power output of a PV power plant contains sharp edges and jumps caused by the fluctuation of the solar radiation, and it has nonlinear characteristics and periodicity. By using the WD method, the output power of the PV power plant is decomposed into two parts; one is the smoothed version of the signal, and the other part contains the detailed version of the signal. Therefore, the WD method discriminates disturbances from the original signal, and can analyze them separately. The idea of the WD is presented in Figure 3.

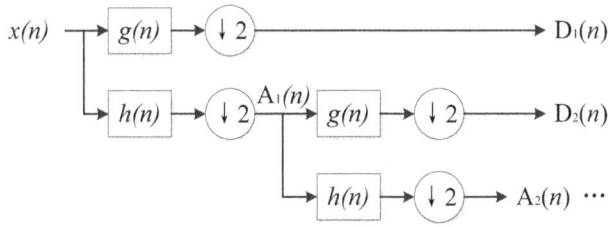

Figure 3. Wavelet decomposition algorithm.

Let $x(n)$ be the discrete-time signal for the output power of a PV power plant. $x(n)$ is to be decomposed into detailed layers and one smoothed layer. From the WD method, the decomposed signals at scale 1 are $A_1(n)$ and $D_1(n)$, where $A_1(n)$ is the smoothed version of the input signal, and $D_1(n)$ is the detailed version of the input signal $x(n)$ in the form of the wavelet transform coefficients. They are defined as:

$$A_1(n) = \sum_k h(k-2n)x(k) \tag{5}$$

$$D_1(n) = \sum_k g(k-2n)x(k) \tag{6}$$

where $h(n)$ and $g(n)$ are the associated filter coefficients that decompose $x(n)$ into $A_1(n)$ and $D_1(n)$. The higher scale decomposition is based on the $A_1(n)$. The decomposed signal at higher scale is given by:

$$A_j(n) = \sum_k h(k-2n)A_{j-1}(k) \tag{7}$$

3.2. Wavelet Decomposition (WD) of Power Signals of a Photovoltaic (PV) Power Plant

WD is conducted for output power signals collected for 5 days from a PV power plant, by a Dmeyer wavelet function with five decomposition layers, shown in Table 3. The decomposition results of the PV power plant output in time-domain signals are shown in Figure 4.

Table 3. Layers Definitions of WD.

Reconstructed Sequence	Definition	Meaning
A5	smoothed signal at 5th layer	reflects change trend of output power of PV power plant, close to theoretically calculated solar irradiance
D5	detailed signal at 5th layer	reflect composition and change rules of high frequency part of signal
D4	detailed signal at 4th layer	
D3	detailed signal at 3rd layer	
D2	detailed signal at 2nd layer	
D1	detailed signal at 1st layer	

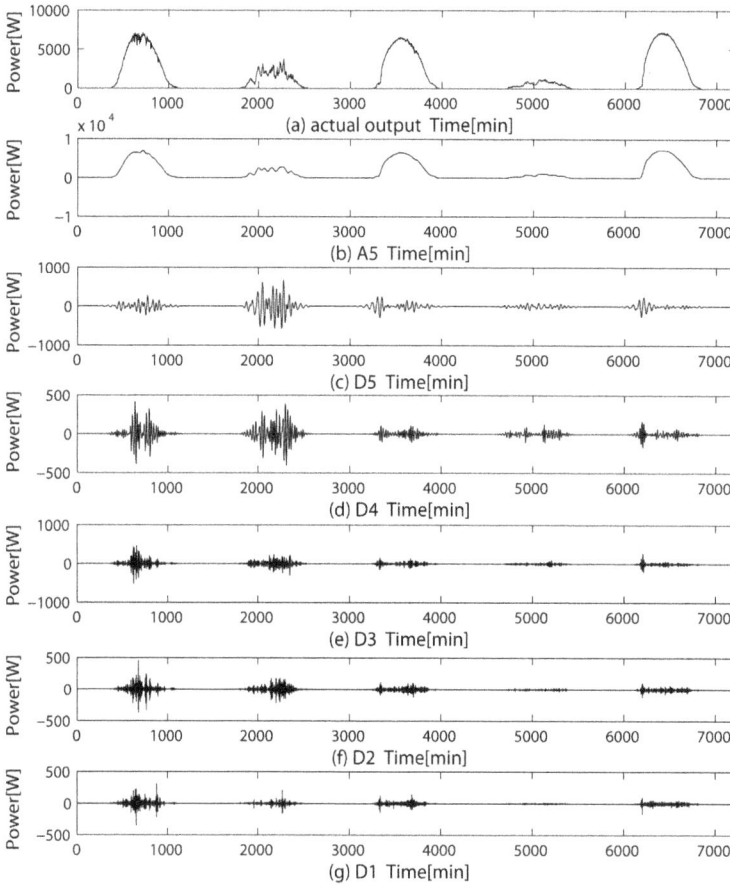

Figure 4. WD of PV output power. (a) Original signal of actual PV output, (b) Smoothed signal of actual PV output at 5th layer, (c) Detailed signal of actual PV output at 5th layer, (d) Detailed signal of actual PV output at 4th layer, (e) Detailed signal of actual PV output at 3rd layer, (f) Detailed signal of actual PV output at 2nd layer, (g) Detailed signal of actual PV output at 1st layer.

It is known from Figure 5 that after 5-layer decomposition layers A5, D5 and D4 can represent the major information of the PV output. Detail signals at other layers represent high-frequency disturbances of meteorological environmental factors on the PV output, which convey less information. Hence, chosen low frequency layers are used in the forecasting model after wavelet decomposition.

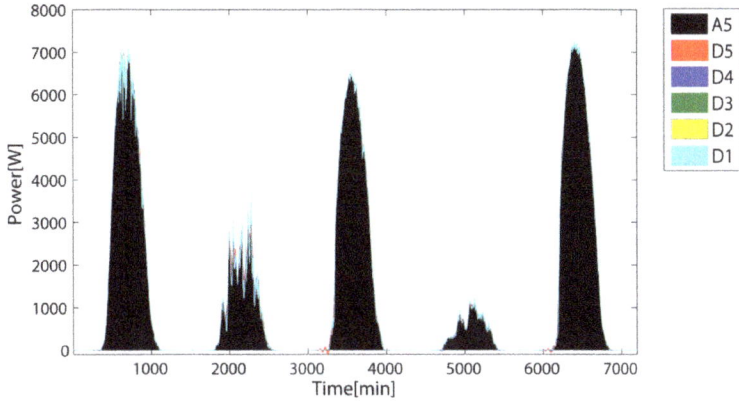

Figure 5. Area graph at each layer.

4. Intelligent Forecasting Model Based on Wavelet Decomposition (WD) and Artificial Neural Network (ANN)

4.1. Artificial Neural Network (ANN) Fundamentals

ANN [6,14–16] is the most popular and widely used tool for artificial intelligence modeling. It is formed by a large numbers of highly related processing elements through an adaptive learning process. ANN has self-adaptiveness, fault tolerance, robustness, and strong inference ability, so it has been successfully applied for power forecasting of wind farms and PV power plant. The ANN is able to learn the complex relationships between the output and the input after training. A basic neural network consists of three layers. They are input layer, hidden layer and output layer as shown in Figure 6.

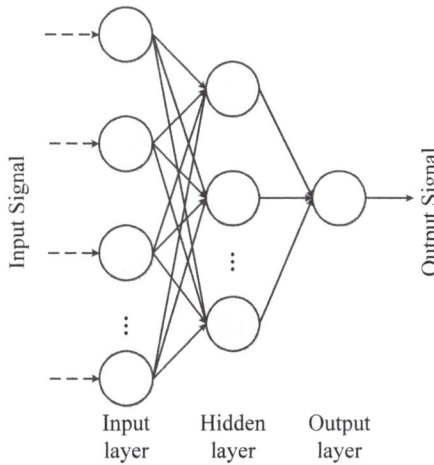

Figure 6. A three-layers artificial neural network.

We use ANN to build the intelligent forecasting model for the output power of the PV power plant. The training of ANN starts with random weights and then neurons works to make sure the error is minimal. The values of the hidden layer neurons is defined by:

$$h_j = f_1(\sum_{i=1}^{n} v_{ij}x_i + \theta_j) \tag{8}$$

where x_i and h_j are the values of the different layer's neuron; $f_1(\cdot)$ is the sigmoid transfer function; v_{ij} is the adjustable weight between the input and hidden layers and θ_j is the bias of the hidden layer neuron. The hidden layer is the input for the output layer, and the output layer neurons is defined by:

$$o_j = f_2(\sum_{i=1}^{n} w_{ij}h_i + \gamma_j) \qquad (9)$$

where o_j is the output value of the output layer neuron; $f_2(\cdot)$ is a linear transfer function; w_{ij} is the adjustable weight between the hidden and output layers; γ_j is the bias of the output layer neuron. The three-layer feed-forward neural network is chosen to build the forecasting model in the paper.

4.2. Forecasting Process

Figure 7 shows the working framework for the method presented in the paper. The theoretical solar irradiance [17,18] is the input of the A5 neural network model. Detailed calculation Equations (A1)–(A13) of theoretical solar irradiance are given in Appendix and Figure A1 is the curves of daily theoretical solar irradiance in 15th day of every month.

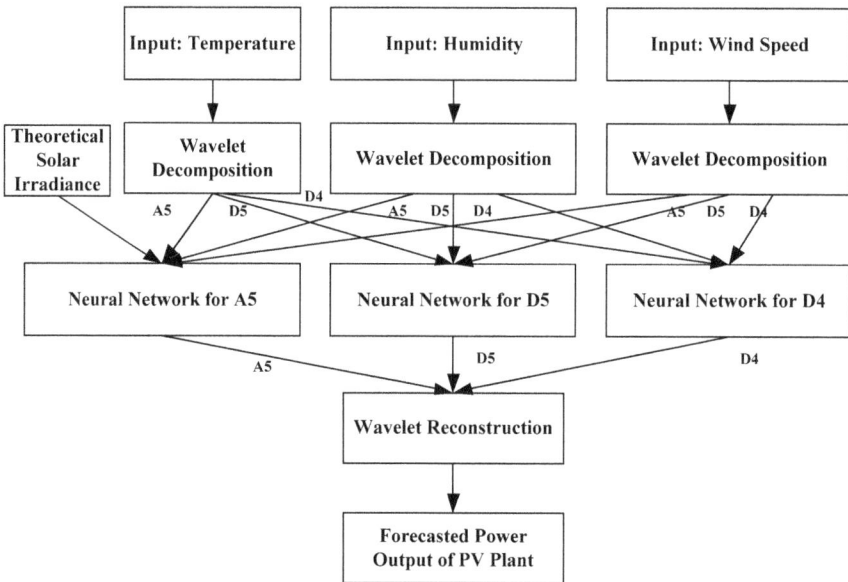

Figure 7. Frame diagram of algorithm.

The forecasting process includes three steps:

Step 1: WD of the output for PV power plant

5-layer WD is carried out for the output power of the PV plant, temperature signal, humidity signal and wind speed signal, followed by the comparison between the approximation layer and detailed layer. Figures 8 and 9 show the smoothed layer A5 and detailed layer D5 after WD. It is known through comparison that the detailed signal part of output power series of PV power plant after multi-scale WD has stationary properties. It is known through Figures 4, 8 and 9 that, through 5-layer WD, the smoothed signal layer can represent the major low frequency information of the original signal. The smoothed signal layers of different signals have good proximity. The periodicity of different signals extracted from the A5 layer is more obvious. Meanwhile, the high frequency interfering noise is filtered well.

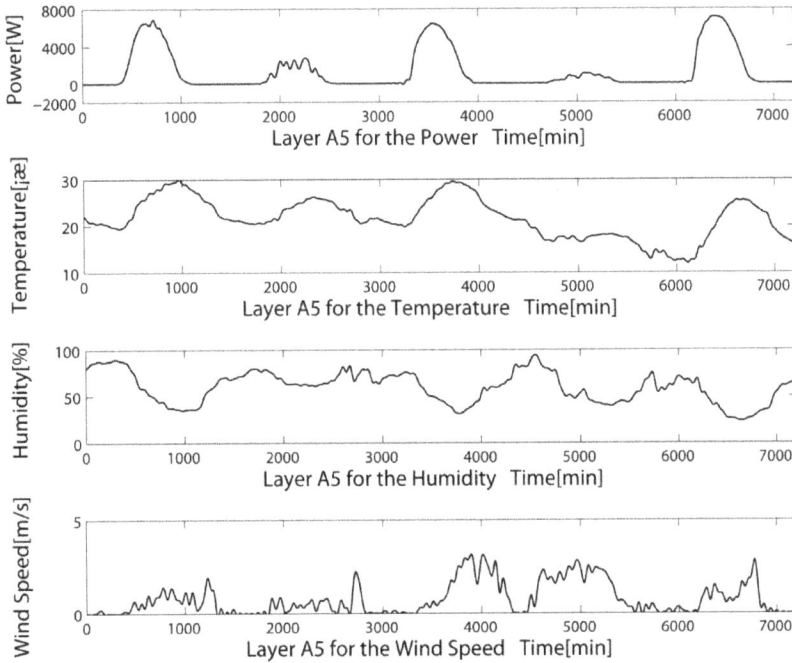

Figure 8. A5 layer comparison of different signals.

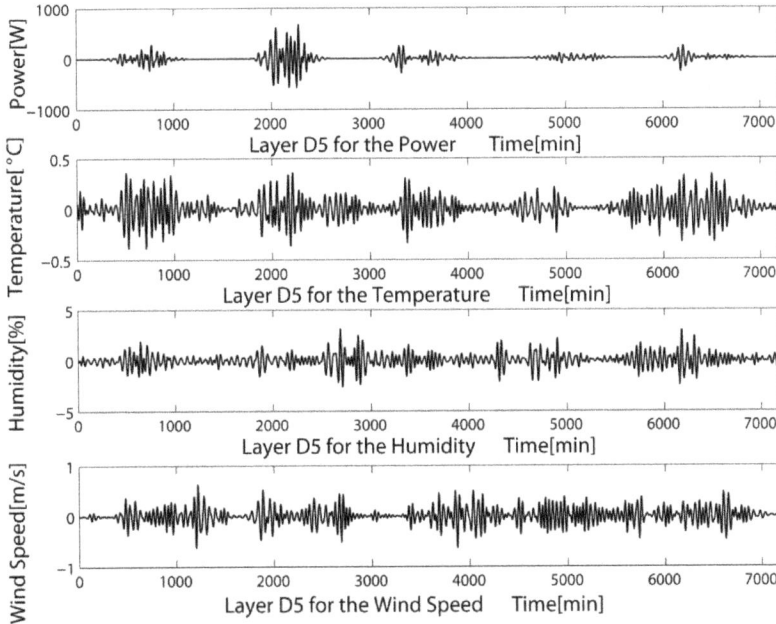

Figure 9. D5 layer comparison of different signals.

Step 2: build ANNs forecasting model

In the physical sense, solar radiation is a direct influencing factor of the voltage effect of PV cells. The irradiance directly influences the output of a photovoltaic cell. Solar irradiance received by a

PV array is influenced by the array installation angle, cloud quantity in the sky and solar position. Meanwhile, the output of a PV power plant is also influenced by meteorological conditions and its own characteristics, but for a known PV power plant, the output time series data of the PV system has a certain autocorrelation. This is because power plant information has been contained in the power output data of the given PV power plant, and complex analysis of the influence of random installation site and operation time on PV system performance degeneration can be avoided. Thus, this paper chooses to use ANNs to establish output power forecasting modeling for PV power plants.

The structure of the forecasting model is shown in Figure 6. An A5 layer neural network is established by selecting A5 smoothed signal layers after WD, including theoretical solar irradiance, temperature, humidity and wind speed signal. The D5 and D4 detailed signal layer after wavelet decomposition of temperature, humidity and wind speed signal are chosen to establish D5 and D4 ANNs. High frequency information at other layers are treated as disturbances for establishing the model, so they are not used.

Step 3: Reconstruction of signals

After different signals are trained for ANNs at each layer, the forecasting results of different signal layers of PV output are obtained. According to Equation (7)—the computational formula of coefficients at different signal layers—signal reconstruction is carried out for the PV power plant power output.

5. Example Analysis and Verification

This paper proposes a hybrid method to forecast the power output of a PV power plant in combination of WD and ANN. To evaluate the effectiveness of this method, it is compared with the forecasting method based on ANN. Figure 10 gives the forecasting performance contrast between the ANN method and WD + ANN method. The solid line shows the measured power output of the PV power plant, while the dotted line shows the output forecast by the two methods. Similarly, the theoretical solar irradiance, temperature, humidity and wind speed are chosen as inputs of the neural network model. The PV power plant power output serves as the target value of the neural network. The WD + ANN method is set up according to Figure 6. It is known from Figure 9 that both the ANN method and WD + AN method can effectively forecast the power output of the PV power plant, but the WD + ANN method can selectively get rid of high-frequency invalid information, and the forecasted power value is closer to the actual measured output.

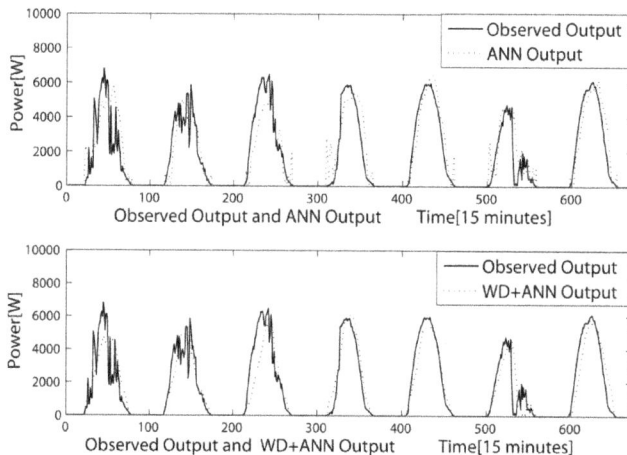

Figure 10. Comparison of forecasted results.

Through comparing forecasting results, on a clear day, WD + ANN has good forecasting precision. On cloudy or rainy days, WD + ANN can reflect different total amounts of output power. However, on cloudy or rainy days, due to abrupt changes in the cloud layer and other environmental factors, the measured power fluctuates greatly, and the forecasting results have a certain deviation. Since it is not possible to effectively forecast the ultra-short term output power variation of the PV power plant on the basis of current weather forecasting technology, it is necessary to consider the application of ground-based observation cloud pictures to correct the forecasting results.

We use root mean square error (RMSE), mean absolute error (MAE) and mean absolute percentage error (MAPE) indexes [1,7] for the assessment of the forecast results based on WD + ANN and ANN. Their definitions are as follows:

$$RMSE = \frac{\sqrt{\sum\limits_{i=1}^{n}[P_M(i) - P_P(i)]^2}}{Cap \times \sqrt{n}} \times 100\% \tag{10}$$

$$MAE = \frac{\sum\limits_{i=1}^{n}|P_M(i) - P_P(i)|}{Cap \times n} \times 100\% \tag{11}$$

$$MAPE = \frac{1}{n}\sum\limits_{i=1}^{n}\left|\frac{P_M(i) - P_f(i)}{P_M(i)}\right| \times 100\% \tag{12}$$

where, $P_M(i)$ is measured power at time i; $P_f(i)$ is the forecast power at time i; n is the number of samples; Cap is the mean running capacity.

Calculation of mean running capacity in PV power forecasting is decided by the initial power of the photovoltaic inverter, installed capacity of the PV system and operation time. Factor $Run(i)$ is defined to describe the working state of the PV plant at time i. When the measured power $P_M(i)$ is higher than P_S, the initial power of photovoltaic inverter, then $Run(i)$ equal to 1, meaning that the PV plant is running, else, $Run(i)$ is equal to 0, meaning that the PV plant is not running, as shown in Equation (13):

$$Run(i) = \begin{cases} 1 & , if\ P_M(i) > P_S \\ 0 & , if\ P_M(i) \leqslant P_S \end{cases} \tag{13}$$

The definition of Cap is shown in Equation (14):

$$Cap = \frac{\sum\limits_{i=1}^{n} Run(i)}{n} \times P_r \tag{14}$$

where, P_S is initial power of the photovoltaic inverter; P_r is installed capacity of the PV system. Equations (13) and (14) can eliminate the accumulated calculation during nighttime when the PV plant stops working.

Table 4 shows the assessment results of different methods to to better understand the effect of using WD + ANN and ANN under various different weather conditions. It is known from Table 4 that the WD + ANN method has better forecasting precision and convergence speed than the ANN method under different weather conditions.

Table 4. Comparison of forecasting performance between ANN and WD + ANN model on clear, cloudy, overcast, rainy days.

Model	Weather	Error			Convergence Epochs
		RMSE(%)	MAE(%)	MAPE(%)	
ANN	clear	9.313	4.978	13.858	
	cloudy	18.472	10.259	21.550	4521
	overcast	18.511	10.220	35.226	
	rainy	22.948	13.062	30.926	
WD + ANN	clear	7.193	3.639	9.240	
	cloudy	16.817	9.578	21.294	2677
	overcast	17.607	10.544	26.767	
	rainy	19.663	10.349	25.373	

6. Conclusions

This paper puts forward a method to forecast the power output of PV power plants based on WD and ANN. Due to the periodic and non-stationary characteristics of the power output series of a PV power plant, the wavelet analysis method is adopted to carry out multi-scale decomposition of the PV output. A smoothed signal and detailed signal of the PV output are obtained. Forecasting models at different signal layers are established through ANN. Finally, through construction of the forecasting results of different signal layers, the forecasting results of the PV power plant are obtained. Through comparison to the ANN method, it is shown that the forecasting method proposed in this paper has better forecasting precision, and less algorithm convergence time.

Acknowledgments: The authors would like to acknowledge the financial support of the Beijing Higher Education Young Elite Teacher Project (YETP0714), the Program of the Co-Construction with Beijing Municipal Commission of Education, State Key Laboratory of Alternate Electrical Power System with Renewable Energy Sources (LAPS14006).

Author Contributions: All authors contributed to this work by collaboration. Honglu Zhu is the first author in this manuscript. All authors revised and approved for the publication.

Conflicts of Interest: The authors declare no conflict of interest.

Appendix

This Appendix gives details on the equation for calculating the theoretical solar irradiance (Figure 6):

$$E_g = E_{dir} + E_{dif} \tag{A1}$$

where, E_g is the global horizontal irradiance; E_{dir} is the direct horizontal irradiance; E_{dif} is the diffuse horizontal irradiance. The direct and diffuse horizontal irradiance can be calculated according to the following steps [15,16]: according to the Cooper formula, the declination angle is calculated by the following equation:

$$\delta = 23.45 \times \sin(360° \times \frac{284 + n}{365}) \tag{A2}$$

where, n is the ordinal number of the date.

Calculation of the solar hour angle starts from 0° at 12:00 am, and every hour represents 15°. For instance, 14:00 pm and 10:00 am represent 30° and −30°. The azimuth angle takes south direction as 0°, and west as positive. We account for the influence of the time difference on hour angle, since the longitude of Beijing is 120°E, so the computational equation of hour angle ω in China is:

$$\omega = (12 - t) \times 15° + (120° - \psi) \tag{A3}$$

where, t is the Beijing time; Ψ is the longitude.

Calculation of the solar altitude angle can be conducted as follows:

$$\sin h = \sin\phi \times \sin\delta + \cos\phi \times \cos\delta \times \cos\omega \tag{A4}$$

where, ϕ is the latitude.

The computational formula of the extraterrestrial normal irradiance is:

$$E_0 = E_{SC} \times [1 + 0.034 \times \cos\left(\frac{2\pi n}{365}\right)] \tag{A5}$$

where, E_{SC} is solar constant, equal to 1367 W/m^2.

The degree of influence of the the atmosphere on sunlight received by the Earth surface is defined as air mass (AM). Air mass is a dimensionless quantity. It is the ratio of the path of solar rays passing through the atmosphere and the path of solar rays passing through the atmosphere in the zenith angle direction. Assuming that standard atmospheric pressure is 101,325 Pa and air temperature is 0 °C, the vertical incidence path of solar rays at sea level is defined as 1. According to the value of the solar altitude angle h, air mass m should be calculated in two situations. When h $\geqslant 30°$, air mass m is:

$$m(h) = 1/\sin h \tag{A6}$$

when h $< 30°$:

$$m(h) = [1229 + (614 \times \sinh)^2]^{1/2} - 614 \times \sin h \tag{A7}$$

Generally, the influence of temperature on m is neglected, but in high altitude regions, the atmospheric pressure should be corrected, i.e.,

$$m(z, h) = m(h)\frac{P(z)}{P_0} \tag{A8}$$

$$P(z)/P_0 = [(288 - 0.0065 \times z)/288]^{5.256} \tag{A9}$$

where, z is the elevation.

The atmospheric transparency coefficient is used to represent the degree of attenuation of the the solar radiation by the atmosphere. On a clear day, the solar direct radiation is influenced by air mass. Meanwhile, molecular scattering, ozone absorption and selective absorption of some gases (CO_2, O_2, N_2) influence the penetrability of solar radiation. Hence, the atmospheric transparency coefficient is related to local atmospheric conditions:

$$\tau_{dir} = 0.56 \times (e^{-0.56m(z,h)} + e^{-0.096m(z,h)}) \tag{A10}$$

The atmospheric transparency coefficient of diffuse irradiance is:

$$\tau_{dif} = 0.2710 - 0.2939 \times \tau_{dir} \tag{A11}$$

On a clear day, the direct solar radiation accounts for the main part of global solar radiation. In view of the atmospheric attenuation and solar altitude, direct horizontal irradiance at any time can be figured out as follows:

$$E_{dir} = E_0 \times \tau_{dir} \times \sinh \times k_1 \tag{A12}$$

where, k_1 is an empirical coefficient of atmospheric turbidity. In direct solar radiation calculations, it ranges from 0.8 to 0.9.

Diffuse horizontal radiation is complex. It is related to scattering distribution (cloud shape, cloud quantity and atmosphere). Diffuse irradiance arriving at horizontal level is E_{dif}, and the calculation formula is as follows:

$$E_{dif} = \frac{1}{2} \times E_0 \times \sinh \times \frac{1 - \tau_{dif}}{1 - 1.4\ln[\tau_{dif}/m(z, h)]} \times k_2 \tag{A13}$$

where, k_2 is an empirical coefficient of atmospheric turbidity. In diffuse solar radiation calculations, it ranges from 0.60 to 0.90. When the atmosphere is turbid, $0.60 \leqslant k_2 \leqslant 0.70$. When it is normal, $0.710 \leqslant k_2 \leqslant 0.80$. When it is good, $0.810 \leqslant k_2 \leqslant 0.90$.

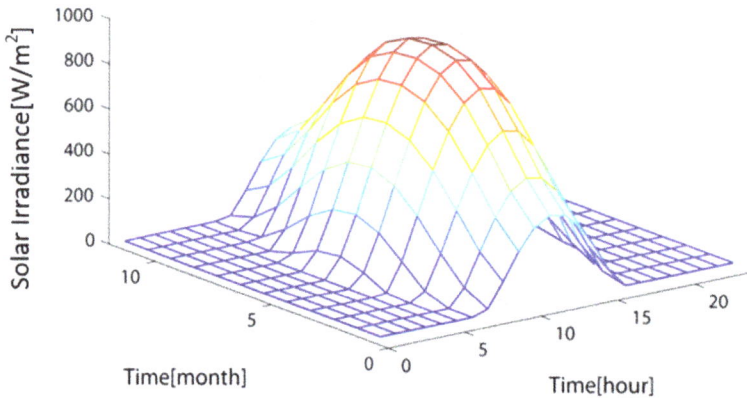

Figure A1. Theoretical solar irradiance.

References

1. Mandal, P.; Madhira, S.T.S.; haque, A.U.; Meng, J.; Pineda, R.L. Forecasting power output of solar photovoltaic system using wavelet transform and artificial intelligence techniques. *Procedia Comput. Sci.* **2012**, *12*, 332–337. [CrossRef]

2. Ogliari, E.; Grimaccia, F.; Leva, S.; Mussetta, M. Hybrid predictive models for accurate forecasting in PV systems. *Energies* **2013**, *6*, 1918–1929. [CrossRef]

3. Lorenz, E.; Scheidsteger, T.; Hurka, J.; Heinemann, D.; Kurz, C. Regional PV power prediction for improved grid integration. *Prog. Photovolt.* **2010**, *19*, 757–771. [CrossRef]

4. Lorenz, E.; Heinemann, D.; Kurz, C. Local and regional photovoltaic power prediction for large scale grid integration: Assessment of a new algorithm for snow detection. *Prog. Photovolt.: Res. Appl.* **2012**, *20*, 760–769. [CrossRef]

5. Karthikeyan, L.; Nagesh Kumar, D. Predictability of nonstationary time series using wavelet and EMD based ARMA models. *J. Hydrol.* **2013**, *502*, 103–119. [CrossRef]

6. Mellit, A.; Kalogirou, S.A. Artificial intelligence techniques for photovoltaic applications: A review. *Prog. Energy Combust. Sci.* **2008**, *34*, 574–632. [CrossRef]

7. Pedro, H.T.C.; Coimbra, C.F.M. Assessment of forecasting techniques for solar power production with no exogenous inputs. *Sol. Energy* **2012**, *86*, 2017–2028. [CrossRef]

8. Voyant, C.; Muselli, M.; Paoli, C.; Nivet, M.-L. Numerical weather prediction (NWP) and hybrid ARMA/ANN model to predict global radiation. *Energy* **2012**, *39*, 341–355. [CrossRef]

9. Monteiro, C.; Santos, T.; Fernandez-Jimenez, L.; Ramirez-Rosado, I.; Terreros-Olarte, M. Short-term power forecasting model for photovoltaic plants based on historical similarity. *Energies* **2013**, *6*, 2624–2643. [CrossRef]

10. Mellit, A.; Benghanem, M.; Kalogirou, S.A. An adaptive wavelet-network model for forecasting daily total solar-radiation. *Appl. Energy* **2006**, *83*, 705–722. [CrossRef]

11. Fryzlewicz, P.; van Bellegem, S.; von Sachs, R. Forecasting non-stationary time series by wavelet process modelling. *Ann. Inst. Stat. Math.* **2003**, *55*, 737–764. [CrossRef]

12. Catalão, J.P.S.; Pousinho, H.M.I.; Mendes, V.M.F. Short-term wind power forecasting in Portugal by neural networks and wavelet transform. *Renew. Energy* **2011**, *36*, 1245–1251. [CrossRef]

13. Domingues, M.O.; Mendes, O.; da Costa, A.M. On wavelet techniques in atmospheric sciences. *Adv. Space Res.* **2005**, *35*, 831–842. [CrossRef]

14. Mellit, A.; Pavan, A.M. A 24-h forecast of solar irradiance using artificial neural network: Application for performance prediction of a grid-connected PV plant at Trieste, Italy. *Sol. Energy* **2010**, *84*, 807–821. [CrossRef]

15. Azadeh, A.; Maghsoudi, A.; Sohrabkhani, S. An integrated artificial neural networks approach for predicting global radiation. *Energy Convers. Manag.* **2009**, *50*, 1497–1505. [CrossRef]
16. Chen, S.H.; Jakeman, A.J.; Norton, J.P. Artificial intelligence techniques: An introduction to their use for modelling environmental systems. *Math. Comput. Simul.* **2008**, *78*, 379–400. [CrossRef]
17. Kumar, L.; Skidmore, A.K.; Knowles, E. Modelling topographic variation in solar radiation in a GIS environment. *Int. J. Geogr. Inf. Sci.* **1997**, *11*, 475–497. [CrossRef]
18. Gueymard, C.A. Clear-sky irradiance predictions for solar resource mapping and large-scale applications: Improved validation methodology and detailed performance analysis of 18 broadband radiative models. *Sol. Energy* **2012**, *86*, 2145–2169. [CrossRef]

10

Error Assessment of Solar Irradiance Forecasts and AC Power from Energy Conversion Model in Grid-Connected Photovoltaic Systems

Gianfranco Chicco [1,*], Valeria Cocina [1], Paolo Di Leo [1], Filippo Spertino [1] and Alessandro Massi Pavan [2]

Academic Editor: Guido Carpinelli

[1] Energy Department, Politecnico di Torino, corso Duca degli Abruzzi 24, Torino 10129, Italy; valeria.cocina@polito.it (V.C.); paolo.dileo@polito.it (P.D.L.); filippo.spertino@polito.it (F.S.)

[2] Department of Engineering and Architecture, University of Trieste, Via Valerio 10, Trieste 34127, Italy; apavan@units.it

* Correspondence: gianfranco.chicco@polito.it

Abstract: Availability of effective estimation of the power profiles of photovoltaic systems is essential for studying how to increase the share of intermittent renewable sources in the electricity mix of many countries. For this purpose, weather forecasts, together with historical data of the meteorological quantities, provide fundamental information. The weak point of the forecasts depends on variable sky conditions, when the clouds successively cover and uncover the solar disc. This causes remarkable positive and negative variations in the irradiance pattern measured at the photovoltaic (PV) site location. This paper starts from 1 to 3 days-ahead solar irradiance forecasts available during one year, with a few points for each day. These forecasts are interpolated to obtain more irradiance estimations per day. The estimated irradiance data are used to classify the sky conditions into clear, variable or cloudy. The results are compared with the outcomes of the same classification carried out with the irradiance measured in meteorological stations at two real PV sites. The occurrence of irradiance spikes in "broken cloud" conditions is identified and discussed. From the measured irradiance, the Alternating Current (AC) power injected into the grid at two PV sites is estimated by using a PV energy conversion model. The AC power errors resulting from the PV model with respect to on-site AC power measurements are shown and discussed.

Keywords: photovoltaic systems; weather forecasts; photovoltaic (PV) conversion model; power profiles; error assessment; distributed generation; renewable energy; irradiance spike

---

1. Introduction

Photovoltaic (PV) generation strongly depends on weather conditions, in particular on solar irradiance and temperature. As such, availability of accurate weather forecast data is very important for PV system planning and operation. For grid-connected PV systems, the power injected into the grid is concentrated during sunlight hours, in which typically the maximum peak load occurs.

In the power system, the task of the Transmission System Operator (TSO) is to ensure a constant balance between supply and consumption within the grid. Actually, the presence of strong fluctuations of the irradiance increases the uncertainty on the PV generation and requires additional regulatory actions for the procurement of reserve services. This may cause an increase in the costs for ancillary services. The irradiance forecast is then useful for grid management, to obtain more accurate information on the expected weather conditions. This information may assist the operators in undertaking decisions concerning the energy market and to reduce the costs of energy imbalance.

The first research on solar irradiance forecasting was conducted more than twenty years ago [1], using the Model Output Statistics (MOS) technique [2]. This technique allows the prediction of a daily average value one or two days ahead. Concerning the forecasting on a short-time scale (a few hours), the effectiveness of a statistical approach based on the prediction of the motion of the clouds through images provided by satellites of the Meteosat constellation has been demonstrated in [3]. However, this method requires a huge computational effort. A multi-resolution decomposition technique applied to satellite images has been studied in [4], in order to obtain information on the local mean value and on the gradient of solar irradiance at different spatial scales. Other studies on short and very short time scales are available in the literature, taking into account the information provided by satellites. Currently, the weather forecasting tools are based on numerical techniques, which provide good results when applied to extended spatial scales. However, these tools are not able to address local variability of the weather conditions.

The forecasting field is rapidly evolving according to the growth of the PV market. A model to predict the power P produced by a PV plant can be written in the general form $P = f(X_1, X_2, ..., X_n)$ where X_i ($i = 1, 2, \ldots, n$) are n different physical quantities of influence. These quantities, including the solar irradiance, the PV module's temperature, the air temperature, the wind speed, the relative humidity, *etc.*, have to be provided by forecasting tools. The input of many models is the solar irradiance [5], while different methods are based on the irradiance and on the temperature of the air or of the PV module [6]. In general, as shown in [7] the models based on irradiance and temperature perform better than the ones where only the irradiance is considered. Conversely, the adoption of the models based on other working conditions does not necessarily improve the prediction accuracy.

Different techniques showing accurate results have been used to predict the power produced by a PV plant. Some examples are Artificial Neural Networks (ANN)-based techniques [7–9], regression model-based techniques [10], support vector machine [11], hybrid models [12] and PV system models [13]. The prediction methods may change depending on the availability of local data coming from a weather station specifically designed and installed to measure the operating conditions of the PV system under study. Most residential PV plants (as well as many of the commercial/industrial ones) are not equipped with any sensor of climate conditions. In this case, the usage of commercial weather data allows the implementation of simpler models for the estimation of the power produced by the PV plants [14].

Recently, a growing interest in using spatio-temporal forecasting methods has emerged, due to the availability of time series data over a large number of meteorological stations. Interesting results incorporating spatial-temporal forecasting method based on the vector autoregression framework have been presented in [15]. This framework combines observations collected by smart meters and distribution transformer controllers, to obtain 6 h-ahead forecasts at the residential PV and medium-voltage/low-voltage substation levels. Spatio-temporal information from satellite images has been used in [16] through an autoregressive approach to forecast the global horizontal irradiance at ground level.

This paper presents a procedure to assess the errors occurring in the 1 day-ahead solar irradiance estimation and in the model-based estimation of the Alternating Current (AC) power delivered to the grid by the PV system. This procedure is based on:

(1) The use of solar irradiance weather forecasts updated every a few hours from a provider. These data are interpolated with polynomial splines to obtain a higher number of estimated values during the day. The results are compared with the measurements gathered at 1-min intervals during a period of one year in order to calculate the estimation error.

(2) The application of a PV conversion model from solar irradiance to AC power, determining 1-min AC power estimates. From these values, 15-min averaged data are calculated and compared with the energy meter readings at 15-min intervals in order to calculate the error on AC power estimates.

The strengths of this paper are that:

(a)   The procedure presented can be used to select the best forecasting model or the best provider of weather forecasts in the location of interest.

(b)   The irradiance error estimation is particularly accurate, because the meteorological stations are equipped with pyranometers (secondary standards used as reference instruments) installed in the same sites of two operating PV plants. Moreover, the AC power error estimation is relevant because (i) the PV plants analyzed are located in the Italian region (Puglia) with the highest PV power density; and (ii) the measurements referring to the PV plants are taken from calibrated energy meters.

The results of the proposed procedure are presented in the form of duration curves of the positive and negative errors between estimated and measured values, determined during a period of one year. The evaluation of the estimation errors is helpful for the grid operator to estimate to what extent the estimation of the PV contribution as grid-connected local generation, calculated starting from 1 day-ahead forecast data, can be trustable. The occurrence of irradiance spikes due to the phenomenon of "broken clouds" is also presented and discussed as a source of possible irradiance peaks, in some cases exceeding the rated value of the AC power injected into the grid.

The next sections of this paper are organized as follows: the second section presents an overview of solar irradiance models and the PV conversion model. The third section introduces the proposed model for hourly classification in clear, variable and cloudy sky conditions, according to the clearness index values. Section 4 discusses the error calculation of the estimated irradiance with respect to the measured irradiance. The fifth section addresses the errors between the AC power estimates and the AC power values measured on two PV systems in operation. The last section contains the conclusions.

2. Solar Irradiance Models and PV System Description

2.1. Models for Extraterrestrial and Ground Level Irradiance

Solar irradiance, which is received by the Earth's surface, is the result of complex interactions between the irradiance from the Sun, the atmosphere and the Earth's surface. In general terms, it is subject to changes determined by the geometry of the Earth, its daily rotation and its annual revolution around the Sun. On a local scale, the main factor is the topography, especially the altitude, the slope, the exposure, and the shading caused by mountains or other natural obstacles. It is possible to determine, by means of appropriate geometric relationships, the amount of irradiance that reaches locally a plane parallel to the ground [17]. The first step is to determine the position of the Sun relative to an observer on the ground, by using a suitable reference system represented by the celestial coordinates. In this case, it is convenient to make use of local coordinates:

-    latitude ξ, in radians or degrees, with respect to the equator (>0 toward North);
-    longitude ζ, in radians or degrees, referred to the Greenwich line (>0 toward East);
-    solar declination δ, the angle between the Sun-Earth line and the equator plane (>0 North);
-    hour angle ψ, between the meridian plane passing through the observer and the meridian plane passing through the Sun (>0 West);
-    azimuth angle ϕ between the projection of the Sun-Earth line and the plane at the horizon with South direction (>0 West);
-    zenith angle z between the Sun-Earth line and the zenith direction;
-    solar height α, that is, the angle between the Sun-Earth line and the horizon plane.

The latitude ξ and the longitude ζ are the only parameters that do not require calculations, because they are known as the geographical coordinates of the place of observation.

The *hour angle* ψ of the Sun is a more complex parameter, because it depends on the position of the observer (longitude effect) and on the measurement of the local time. During the year, the time indicated from a meridian deviates periodically a few minutes with respect to the time indicated by a clock, which in Italy is normally referred to the Central European Time (CET). This difference is called "equation of time" τ, defined as the East or West component of the *analemma*, a curve representing the angular offset of the Sun from its mean position on the celestial sphere as viewed from the Earth. At any point in time, the solar irradiance incident on a horizontal plane outside the atmosphere is the normal solar irradiance G_0, given by ([18], pp. 37–41):

$$G_0 = G_{sc}\left(1 + 0.033\cos\frac{360d}{365}\right)\cos z \tag{1}$$

where G_{sc} is the solar constant, d is the day of the year, and $\cos z = \cos\xi\,\cos\delta\,\cos\psi + \sin\xi\,\sin\delta$.

By integrating this equation for an interval between the hour angles ψ_1 and ψ_2, which define an hour (where ψ_2 is the larger), the irradiation H_0 (in MJ/m^2) is obtained as:

$$H_0 = \frac{12 \times 3600}{\pi}G_{sc}\left(1 + 0.033\cos\frac{360d}{365}\right)[\cos\xi\,\cos\delta\,(\sin\psi_2 - \sin\psi_1) + (\psi_2 - \psi_1)\,\sin\xi\,\sin\delta] \tag{2}$$

The solar irradiance, in the path through the atmosphere towards the Earth surface, is subject to scatter, absorption, reflection, diffusion, meteorological conditions and air mass [19]. It is useful to define a standard "clear" sky and calculate the hourly and daily irradiance that would be received on a horizontal surface under these standard conditions ([18], pp. 85–95). In order to calculate the clear sky irradiance, several methods have been developed [20]. Among these methods, the Moon-Spencer model [21] provides the theoretical instantaneous values of the irradiance at clear sky on a surface orientated in any direction. The irradiance values obtained from the Moon-Spencer model refer to free view with respect to the horizon. The results may differ in case of obstacles hiding the visual landscape, for some parts of the day, in the directions in which the Sun should impact on the surface (e.g., presence of mountains or adjacent buildings). In these cases, the results of the Moon-Spencer model have to be adjusted to take into account the actual skyline seen from the surface.

The Moon-Spencer model was developed for the atmospheric conditions in the United States, but sometimes it over-estimates or under-estimates the global irradiance in case of geographical locations different from the United States. Therefore, this clear sky model cannot be considered as an ideal profile with minimum turbidity of the sky, but it represents an indicative daily evolution for the comparison with the weather forecasts. In order to take into account the information provided by the clear-sky model, a dedicated variable space has been created in [22], in which the time axis is normalized in such a way to map the time interval between the sunrise and the sunset in the (0, 1) interval; the irradiance values are normalized so that the unity value corresponds to the peak conditions at clear-sky from the Moon-Spencer model.

Another clear-sky model has been implemented in the online software PVGIS [23]. In [24] a comparison between the irradiance measured by the pyranometer and the simulated PVGIS clear-sky irradiance has been discussed for a particular day of July. Examining other days during the whole year, owing to higher air turbidity the measured values are, many times, lower than the ones obtained from the clear-sky model.

In this paper, the data are elaborated in order to create a partitioning into three types of sky conditions: *variable, clear,* and *cloudy*. Classifications with more types of sky conditions have been used in various references for different purposes [25,26].

2.2. Description of the Two Meteorological Stations and PV Systems

For our investigation, measurements have been collected by two meteorological stations, named for the sake of simplicity "Gi" and "Ma", near two grid-connected PV systems at latitude 40° North

(Puglia region, Figure 1). The distance between the two meteorological stations is 61 km. As discussed in [27], each meteorological station is provided with:

- A pyranometer (Secondary Standard according to ISO 9060 [28]) for measuring the horizontal global irradiance G_{pyr};
- Two reference solar cells in polycrystalline silicon (p-Si) with South orientation for measuring the 30° tilted global irradiance G_{tcell};
- One thermo-hygrometer for measuring the ambient temperature T_{amb}, relative humidity and wind speed w_s.

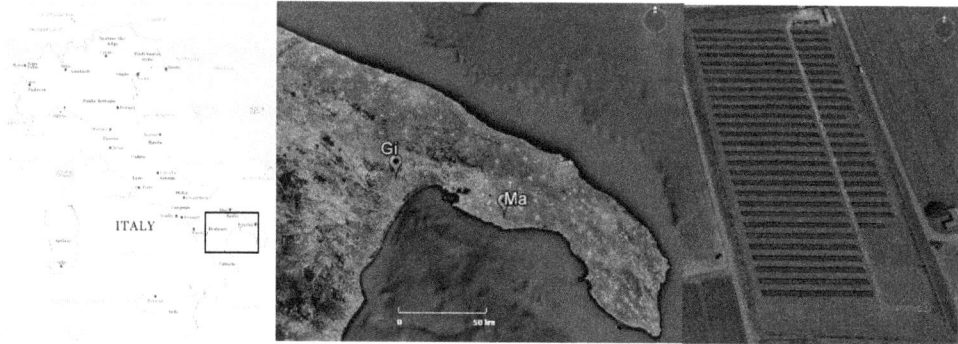

Figure 1. Location of the two grid-connected PV systems and satellite photo of the "Gi" site.

To obtain a correctly integrated value for the solar irradiance over the day, small time steps are recommended for data sampling. However, due to the response time of the pyranometer, the time step cannot be lower than 5 s. The sampling time steps may be chosen depending on the data collection system and on the calculation and update speed of the algorithms used. In our case, the pyranometer is connected to the meteorological station. Therefore, a time step of 10 s and an averaging time for the integrated values of 1 min are used. This averaging time is suitable for obtaining a sufficient number of data to be further averaged within longer time steps, such as the ones used in the electricity markets (e.g., 10, 15 min, 30 min, or 1 h). The global irradiance data from solar cells on the tilted plane are validated through the comparison with the pyranometer uncertainty [24], showing the possible effect of the measurement accuracy. The expanded uncertainty of a pyranometer having the same characteristics of the pyranometers used in the two sites is about 30 W/m^2 with confidence level 95% (coverage factor $k = 2$) [29]. This value is valid when the irradiance reaches its peaks up to 1200 W/m^2. A confidence level 99.7% (coverage factor $k = 3$) corresponds to an expanded uncertainty of about 45 W/m^2. At mid-level of irradiance (e.g., around 500 W/m^2) the expanded uncertainty for $k = 2$ is about 15 W/m^2.

The real grid-connected PV system has a power rating $P_{peak} = 993.6$ kW$_p$ for the site "Gi" and $P_{peak} = 997.3$ kW$_p$ for the site "Ma" at Standard Test Conditions (STC) with global irradiance $G_{STC} = 1$ kW/m^2, cell temperature $T_{STC} = 25$ °C and standard spectrum AM 1.5. The PV system in the site "Gi" is equipped with polycrystalline silicon modules of 230 W$_p$ each, tilted at 30° with South orientation. On the contrary, the PV system in the site "Ma" is equipped with mono-crystalline silicon modules of 230 W$_p$, 235 W$_p$ and 240 W$_p$ and with polycrystalline silicon modules of 230 W$_p$, 235 W$_p$ and 240 W$_p$, tilted at 30° with South orientation. The PV arrays of each site, placed on a metallic structure that permits the natural air circulation, feed two centralized inverters with high efficiency (transformerless option). These power conditioning units are slightly undersized, given that the 500-kVA inverter is supplied by a 552.0 kW$_p$ array for the site "Gi" and by a 542.0 kW$_p$ array for the site "Ma"; the 400-kVA inverter is supplied by a 441.6 kW$_p$ array for the site "Gi" and by a 455.3 kW$_p$ array for the site "Ma".

2.3. Definition of the PV Conversion Model

In the definition of the PV conversion model, it is important to take into account the efficiencies referring to the main loss factors affecting the PV system behavior. As mentioned in [30], the main loss factors are summarized in the efficiencies defined as follows:

- Efficiency η_{dirt}, due to losses for soiling and dirt (environmental pollution). To estimate the impact of dirt/soiling accumulation, a 10-day summer period without rain is considered. At the end of this period (10th day), the horizontal solar irradiation is calculated from the pyranometer and the solar cell. At the 11th day, the rain appears and naturally cleans the sensors. Finally, at the 12th day (clear-sky day), the solar irradiation is calculated in such a way as to practically have the same astronomical conditions of the 10th day. Therefore, the corresponding value of η_{dirt} for the PV plant, located in a relatively clean environment (*i.e.*, away from mines, landfills, *etc.*), is determined according to the following formula:

$$\eta_{dirt} = 100 \frac{(H_{a\_rain} - H_{b\_rain})}{H_{a\_rain}} \tag{3}$$

where H_{a_rain} and H_{b_rain} are the values of the daily irradiation in two clear-sky days, one after rain (12th day) and the other before rain (10th day), respectively. The corresponding value of η_{dirt} is generally in the range 0.97–0.98.

- Efficiency η_{refl}, due to reflection of the PV module glass; the value used is 0.971, taken from the PVGIS website [23].
- Efficiency η_{th}, due to the thermal losses l_{th} with respect to the STC, calculated as:

$$\eta_{th} = 1 - l_{th} = 1 - \gamma_{th}(T_C - T_{STC}) \tag{4}$$

where γ_{th} is the thermal coefficient of maximum power of the PV modules, depending on the PV technology (for crystalline silicon $\gamma_{th} = 0.5\%/°C$); T_C is the cell temperature (mean temperature in outdoor operation at $G_{NOCT} = 800$ W/m² and $T_{amb,NOCT} = 20$ °C), which can be calculated as a function of the ambient temperature T_{amb}, the cell irradiance on the tilted plane G_{tcell} and the Normal Operating Cell Temperature (*NOCT*) of 42–50 °C [31,32]:

$$T_C = T_{amb} + (NOCT - T_{amb,NOCT}) \frac{G_{tcell}}{G_{NOCT}} \tag{5}$$

- Efficiency η_{mism}, taking into account the current-voltage (*I-V*) mismatch losses, assuming that the bottleneck effect globally leads to 97% of the power rating declared by the manufacturer for all the modules in the PV array. This loss is a consequence of the weakest modules in the series connection inside the strings, and of the weakest strings in the parallel connection inside the PV array [33].
- Efficiency η_{cable}, including the DC cable losses, with the value 0.99 considered according to good design criteria [34].

Considering these efficiencies, the available power at the maximum power point is expressed as:

$$P_{mpp} = P_{rated}(G_{tcell} - G_{lim}) \eta_{dirt}\eta_{refl}\eta_{th}\eta_{mism}\eta_{cable} \tag{6}$$

where $G_{lim} = 17.7$ W/m² is the irradiance limit below which the output is vanishing, calculated by linear interpolation of the irradiance and power values declared by the manufacturer of the silicon modules installed in the PV array.

Finally, considering the efficiency η_{MPPT} of the maximum power point tracker, and thanks to the model of the power conditioning unit for grid connection, the AC power injected into the grid is calculated by solving the second-order equation [35]:

$$c_Q P_{AC}^2 + (1 + c_L) P_{AC} + P_0 - P_{DC} = 0 \qquad (7)$$

with:

$$P_{DC} = \eta_{MPPT} P_{mpp} \qquad (8)$$

where P_0 is the no-load power losses along the operation, while c_L and c_Q are the linear and quadratic loss coefficients, respectively.

Therefore, if the reference-cell data G_{tcell}, averaged on 15-min basis, are used as inputs of the above-described model, the power P_{AC} delivered to the grid can be compared with the power P_{meas} indicated by the energy meter of the PV plant.

For the error calculation of the AC power profiles compared to the experimental results of each PV plant in the two sites, the estimation error ΔP is defined as the difference between the estimated power to be delivered to the grid P_{AC} and the AC power P_{meas} measured by the energy meter of each PV plant:

$$\Delta P = P_{AC} - P_{meas} \qquad (9)$$

3. Clear, Variable and Cloudy Sky Classification

3.1. Determination of the Diffuse Contribution in the Global Irradiance

The amount of solar irradiance that reaches the ground, besides the daily and yearly apparent motion of the Sun, depends on the geographical location (latitude and longitude) and on the climatic conditions (e.g., cloud coverage). Many studies have proved that cloudiness is the main factor affecting the difference between the values of irradiance measured outside the atmosphere and on the Earth surface [36].

Let us define the global irradiance on a horizontal surface G_{th} (kW/m^2), composed of the diffuse component G_{dh} and beam component G_{bh}. Let us further consider the *hourly clearness index* $k_t = G_{th}/G_0$, calculated as the ratio of G_{th} (measured by the pyranometer) to the extra-atmospheric total irradiance G_0 (kW/m^2) defined in Equation (1). The ratio $k_d = G_{dh}/G_{th}$ can be expressed as a function of the clearness index, and permits to distinguish the sky conditions into clear, variable and cloudy. For this purpose, several correlations have been proposed to establish a relationship between the diffuse and the global horizontal irradiances. As explained in [37], some of the existing models [38,39] have been developed for Northern latitudes with high albedos and air masses. This is the reason for differences in diffuse irradiance values and an error source in modeling diffuse irradiance [40]. Only in recent years models to calculate the diffuse irradiance with respect to the global irradiance have been developed for the European Mediterranean area [41]. The hourly correlations considered in this context are represented by these expressions:

$$k_d = \begin{cases} 0.995 - 0.081 k_t & \text{for } k_t \leqslant 0.21 \\ 0.724 + 2.738 k_t - 8.32 k_t^2 + 4.967 k_t^3 & \text{for } 0.21 < k_t \leqslant 0.76 \\ 0.180 & \text{for } k_t > 0.76 \end{cases} \qquad (10)$$

The above-mentioned correlations permit the hourly classification of the sky conditions. In particular:

- for $k_t \leqslant 0.21$ a total *cloudy* sky condition occurs, and a linear expression of k_d is assumed;
- in the range $0.21 < k_t \leqslant 0.76$, a *variable* (*i.e.*, partially cloudy) sky condition occurs, in which the Sun is partially obscured by clouds, and the correlation is represented by a cubic polynomial expression;
- for $k_t > 0.76$ a *clear*-sky condition occurs, in which that the sunlight is not reduced by clouds, and the fraction of diffuse irradiance is assumed to be 18% of the global one.

The weather forecasts considered in this paper are provided by the meteorological service of Catalonia [42]. The available data are based on the Weather Research and Forecasting (WRF) [43] model: a next-generation mesoscale numerical weather forecast system designed to serve both atmospheric research and operational forecasting needs. It features two dynamical cores, a data assimilation system, and a software architecture allowing for parallel computation and system extensibility. The model serves a wide range of meteorological applications across scales ranging from meters to thousands of kilometers. As shown in Table 1, the coordinates of the points of the WRF model (WRF latitude and WRF longitude) are close to the real points (real latitude and real longitude) for the two sites under analysis.

Table 1. Real and WRF coordinates for the two sites (degrees).

Site	Real Latitude	Real Longitude	WRF Latitude	WRF Longitude
Ma	40.35	17.52	40.44	17.65
Gi	40.55	16.84	40.62	16.93

The data set taken from the meteorological provider consists of solar irradiance forecasts for the two meteorological stations during the whole year 2012 [44]. The forecast values are given in W/m^2 at four points in time (hours 7 a.m., 10 a.m., 1 p.m. and 4 p.m.). There is no indication on the uncertainty of these data. The maximum timespan is 72 h. The forecast data have been first interpolated by using polynomial splines [45], obtaining an estimated irradiance pattern that can be represented with a number of points per day higher than the four points available from the forecast data.

On the basis of the hourly correlations from Equation (10), each hour of a day has been classified belonging to clear, cloudy or variable sky conditions for both the estimated and measured data [44]. In particular, the value of the hourly clearness index of pyranometer k_{tp} has been calculated as the ratio of the solar irradiance from the pyranometer to the extra-atmospheric total irradiance G_0. Then, the hourly clearness index of forecasts k_{tf} has been calculated as the ratio between the 1 day-ahead forecast data (considered to be more than the 2 and 3 days-ahead forecasts, as confirmed by the results indicated in Section 4) and the abovementioned G_0.

3.2. Representation of Measured, Forecast and Estimated Data

In Figure 2, the daily evolution of the solar irradiance is represented for seven consecutive days of July 2012 (17–23 July) for the two sites "Gi" (a) and "Ma" (b). The 1, 2 and 3 days-ahead forecasts available every three hours from [42] are represented with circles, while the measurements from the pyranometer G_{pyr} are indicated with a black line. Furthermore, the lines obtained by polynomial spline interpolations are shown in different colors, corresponding to estimates obtained from the 1 day-ahead forecast (in red), the 2 days-ahead forecast (in blue) and the 3 days-ahead forecast (in green). These lines serve as references to identify the type of day on the basis of the actual measurements. Finally, the dashed line represents the outcomes of the Moon-Spencer model. By examining the daily evolution of measurements, for both sites there are four clear sky days, and three days with variable sky or cloudy sky conditions. In general, the weather forecasts are able to predict the clear-sky condition, but they are not able to predict the variable-sky condition: on 23 July for the "Gi" site only the 1 day-ahead forecast is closer to the measurements. In particular, the hours from 5 a.m. to 8 a.m. are classified by the algorithm related to the clearness index of measurement as cloudy, in which k_{tp} is between 0.04 and 0.16 and the value of the pyranometer irradiance does not reach 200 W/m^2. The calculation for the 1 day-ahead forecast of the same day classifies all the hours as variable, with k_{tf} in the range 0.36–0.67.

Actually, in variable-sky days, the phenomenon of broken clouds [46–48] (also called cloud enhancement [48,49], cloud edge [50] or overirradiance [51]) may affect the outcomes of the analysis. The presence of broken clouds appears when the sky is mainly clear, but the passage of clouds affects the irradiance evolution. The irradiance may reach values higher than the peak at clear-sky for short duration due to the clouds, surrounding the solar disc, which reflect a portion of irradiance in a small area of the ground. Obviously, on a longer period the clouds determine an irradiation lower than at clear-sky, because of the shield of the direct irradiance by the clouds. This phenomenon may cause fast positive and negative variations with the mentioned abnormal peak, which are revealed by the pyranometer on 1 min scale. In order to identify this effect in a synthetic way, let us introduce the acronym ISBC (Irradiance Spikes caused by Broken Clouds; no specific acronym has been found in the literature to identify this phenomenon. Thereby, the new acronym ISBC is introduced here to represent synthetically the occurrence of fast positive and negative irradiance variations in "broken clouds" conditions).

Averaging the irradiance data over a time period longer than 1 min (to obtain a smoother irradiance pattern) can reduce the relevance of the ISBC effect on the measured values. However, working with smoothed data would make the distinction among different sky conditions more challenging. In this respect, the ISBC effect is useful to provide highly variable real data characterizing *variable* or *cloudy* sky conditions, impacting on the increase of the forecasting errors. Hence, the polynomial splines obtained from the forecasted values are represented at 1-min time step as well in order to be compared with the measured values.

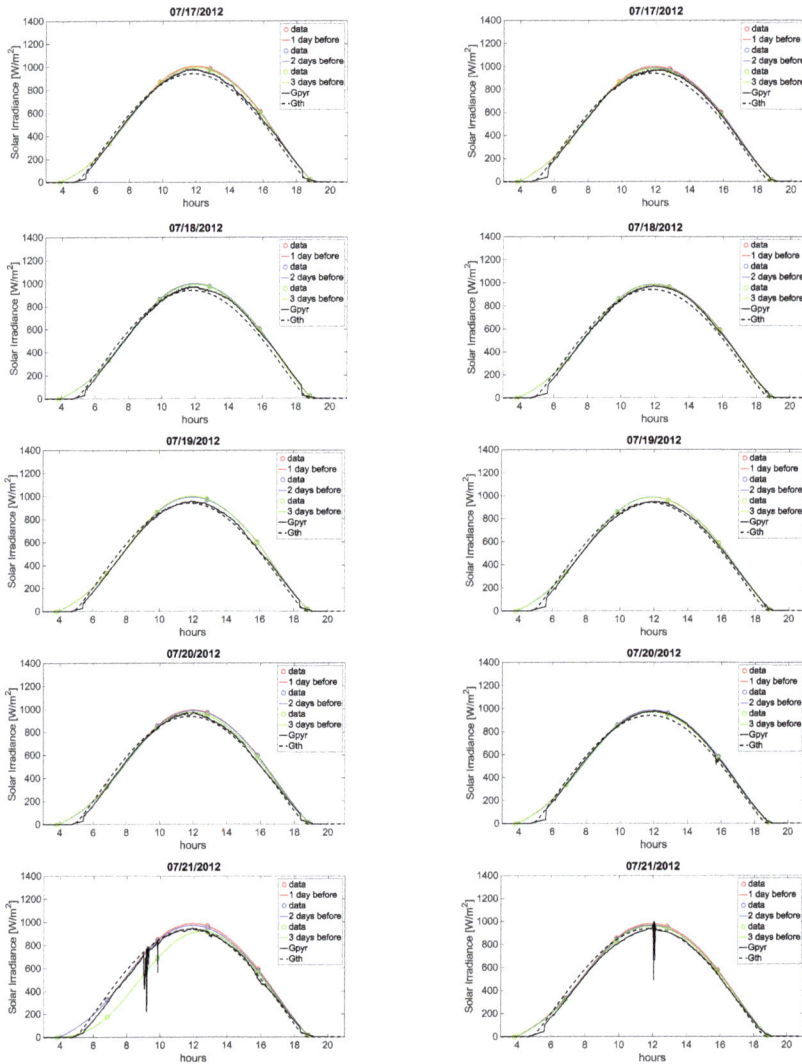

Figure 2. *Cont.*

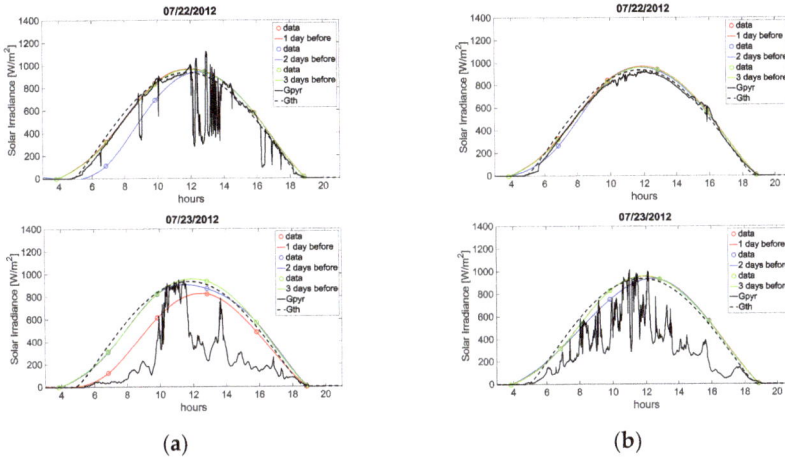

Figure 2. Solar irradiance values from: the Moon-Spencer model (G_{th}), pyranometer measurements (G_{pyr}), the 3-hour forecasts (dots) and the spline-approximated estimations for seven days of July 2012. (**a**) Site "Gi"; (**b**) Site "Ma".

A specific aspect can be observed in Figure 2b, where the measured data indicate a systematic irradiance reduction at the "Ma" site in the early morning. This is due to the location of the irradiance sensor in the meteorological station building. For this sensor, the direct irradiance is shadowed in the early morning in the summer period, causing a bias in the error calculations. This systematic error varies with the day. Its impact on the free-view curve can be determined by knowing the clear sky model of the days and the exact location of the shadowing obstacles. When the sensor shadowing occurs, the total irradiance is reduced to the diffuse irradiance. In the model used in this paper, for the periods in which the systematic error occurs, the maximum spline values can be reduced to the values of the diffuse irradiance calculated from the clear sky model.

3.3. Comparison between Estimated Values and Measurements in the Two PV Sites

On the basis of Equation (10), the clearness index determines the amount of diffuse hourly irradiation with respect to the global hourly irradiation, and thus the sky conditions. The assessment of *clear*, *variable* and *cloudy* sky conditions is carried out for the pyranometer measurements and for the polynomial splines representing the estimated values. In both cases, the values are averaged at each hour. When the assessment for measurement and estimate indicates the same type of sky conditions, the comparison result is marked as a "pass", otherwise it becomes a "fail".

Tables 2 and 3 report the comparison performance by showing the number of hours with correct classification (passes) and with incorrect classification (fails). The comparison is carried out with estimates referring to 1 day-ahead forecast, summarized for all the months of the year 2012 in the two sites. Only the hours having at least 30 min of sunlight are considered. The check on the sunlight hours is carried out for each day by using the clear sky model. The last row of each table reports the percentage of passes (or fails) per month related to the total number of sunlight hours per month.

For the site "Gi", in summer (in particular in July) the highest number of hourly passes occurs for the *variable* condition. The cases classified as *variable* from the measurements but *clear* from the estimation occur mainly in the middle hours of the day (from 10 a.m. to 3 p.m.) in the days with greater occurrence of the ISBC effect. On the other hand, in spring a great number of hourly fails occur for the *clear–variable* condition.

The low number of clear days, especially in spring and summer, can be explained by the actual air turbidity. In general, the air pollution can play a fundamental role. The number of clear-sky days at the site "Ma" is lower than at the site "Gi". In fact, the site "Ma" is close to high pollution areas, with the presence of fine dust in the air, produced by industrial steel mills.

Table 2. Number of passes for all the months of 2012.

(a) Site "Gi"

Measured	Estimated	January	February	March	April	May	June	July	August	September	October	November	December
Variable	Variable	192	88	136	80	112	139	148	130	140	120	178	185
Clear	Clear	8	30	51	98	108	129	54	109	86	77	5	5
Cloudy	Cloudy	7	31	18	33	12	29	10	16	31	45	31	24
Total Passes		207	149	205	211	232	297	212	255	257	242	214	214
Passes %		72%	48%	57%	53%	53%	67%	47%	60%	71%	72%	71%	72%

(b) Site "Ma"

Measured	Estimated	January	February	March	April	May	June	July	August	September	October	November	December
Variable	Variable	165	96	145	70	124	113	135	141	149	124	139	155
Clear	Clear	24	31	44	101	118	103	50	73	87	77	16	12
Cloudy	Cloudy	0	24	18	34	2	29	0	16	32	45	17	26
Total Passes		189	151	207	205	244	245	185	230	268	246	172	193
Passes %		65%	49%	57%	51%	56%	55%	41%	54%	74%	73%	57%	65%

Error Assessment of Solar Irradiance Forecasts and AC Power from Energy...

167

Table 3. Number of fails for all the months of 2012.

(a) Site "Gi"

Measured	Estimated	January	February	March	April	May	June	July	August	September	October	November	December
Variable	Clear	3	7	6	7	5	25	90	49	25	14	8	17
Variable	Cloudy	18	77	30	77	65	16	57	48	34	30	51	32
Clear	Variable	50	59	115	83	123	87	75	71	46	46	22	23
Clear	Cloudy	8	16	6	21	9	20	16	1	2	2	4	6
Cloudy	Variable	2	0	0	0	0	0	0	0	0	0	1	4
Cloudy	Clear	0	0	0	0	0	0	0	0	0	0	0	0
Total Fails		81	159	157	188	202	148	238	169	107	92	86	82
Fails %		28%	52%	43%	47%	47%	33%	53%	40%	29%	28%	29%	28%

(b) Site "Ma"

Measured	Estimated	January	February	March	April	May	June	July	August	September	October	November	December
Variable	Clear	12	6	11	12	12	70	120	67	27	32	38	31
Variable	Cloudy	30	75	18	77	75	30	66	48	27	25	85	49
Clear	Variable	51	66	116	83	96	74	62	76	41	31	2	15
Clear	Cloudy	7	10	10	18	7	24	16	2	1	0	0	8
Cloudy	Variable	0	0	0	4	0	0	0	0	0	1	2	2
Cloudy	Clear	0	0	0	0	0	0	0	0	0	0	1	0
Total Fails		100	157	155	194	190	198	264	193	96	89	128	105
Fails %		35%	51%	43%	49%	44%	45%	59%	46%	26%	27%	43%	35%

Comparing the number of passes with the number of fails, it is interesting to point out that in both sites only in two months (February and July) the number of passes is lower than the number of fails. In particular, in July the classification of the cloudy conditions in these sites is particularly challenging, especially when a cloudy day appears after a number of successive clear days.

3.4. Identification of the Irradiance Spikes caused by Broken Clouds (ISBC) Conditions

In order to estimate the number of occurrences of the ISBC effect in a day, an additional control has been performed. In general, the ISBC effect may happen at each time (minute, hour ...) of the day, when the irradiance spikes exceed the irradiance indicated by the reference model at clear sky. The maximum value of clear-sky irradiance changes in each day; the monthly values are shown in Table 4. The reference model considers a fixed plane (*i.e.*, with the same 30° tilt angle of the modules in the PV plant).

Table 4. Maximum clear-sky solar irradiance values with tilted angle of 30° for each month of 2012.

Month	Max Clear-Sky Irradiance (W/m^2) on a 30° Plane	
	Site "Gi"	Site "Ma"
January	852	856
February	971	977
March	1080	1100
April	1100	1110
May	1060	1070
June	1060	1060
July	1020	1030
August	1080	1080
September	1040	1040
October	1000	1020
November	906	905
December	856	853

In this paper, a spike is defined by the occurrence of an increase in the global irradiance G_{tcell} on the tilted solar cells, from one minute to the successive minute, higher than a threshold value. The threshold is set to 45 W/m^2, that is, the maximum expanded uncertainty of the pyranometer with 99.7% confidence indicated in Section 2.2. The occurrence of the ISBC effect has been determined by showing two different outcomes:

(i)   the number of irradiance spikes for which the measured irradiance exceeds the irradiance of the reference model at the same minute;

(ii) the number of irradiance spikes for which the measured irradiance is so high to exceed the maximum irradiance G_{max} indicated by the reference model of the corresponding day. The rationale of this choice is that for irradiance values higher than the maximum value established at clear-sky conditions the PV system may inject in the electrical network a power that could be even higher than the rated power of the PV plant.

For the two sites, the global irradiance G_{tcell} is gathered with a resolution of 1 min. For the "Gi" site, in the year 2012 the ISBC effect occurs for a number of minutes corresponding to about 2 days and 17 h. The details by month are shown in Table 5a. The occurrence of the ISBC effect in 2012 at the "Ma" site is similar to what happens at the "Gi" site, with a number of minutes in which the ISBC effect occurs corresponding to 2 days and 16 h. The details by month are shown in Table 5b. The months with a higher occurrence of the ISBC effect are in spring (April–May) and in autumn (September).

Table 5. Number of ISBC events for each month, considering G_{tcell} with time step of 1 min, exceeding the minute-by-minute points and the daily peak of the clear sky model.

| Month | Number of ISBC Events (Year 2012) | | | |
| | (a) Site "Gi" | | (b) Site "Ma" | |
	Exceeding the Minute-by-Minute Points of the Clear Sky Model	Exceeding the Daily Peak of the Clear Sky Model	Exceeding the Minute-by-Minute Points of the Clear Sky Model	Exceeding the Daily Peak of the Clear Sky Model
January	378	190	357	172
February	228	117	231	117
March	232	89	430	217
April	552	283	469	221
May	646	343	397	190
June	252	108	202	73
July	227	104	167	57
August	154	60	34	10
September	493	207	663	311
October	345	161	416	166
November	245	121	214	79
December	174	86	289	140
Total year	3926	1869	3869	1753

4. Accuracy of the Estimated Values

4.1. Error Indices to Compare the Irradiance Estimates with the Measurements

In order to compare the estimated quantities with the measured ones, the estimation error ε_G is the difference between the estimated irradiance G_{est} and the measured irradiance G_{meas} [52]:

$$\varepsilon_G = G_{est} - G_{meas} \tag{11}$$

Different statistical parameters [53] have been calculated on a daily basis, by considering a generic error ε:

- the root mean square error (RMSE):

$$RMSE = \sqrt{\frac{1}{N}\sum_{i=1}^{N} \varepsilon_i^2} \tag{12}$$

- the mean bias error (MBE), representing the systematic part (bias) of the error [54]:

$$MBE = \bar{\varepsilon} = \frac{1}{N}\sum_{i=1}^{N} \varepsilon_i \tag{13}$$

- the mean absolute error (MAE):

$$MAE = \frac{1}{N}\sum_{i=1}^{N} |\varepsilon_i| \tag{14}$$

The estimation error is calculated by using Equation (11) with G_{est} given by the estimated data interpolated from the 1, 2 and 3 days-ahead irradiance forecasts, and G_{meas} equal to the irradiance G_{pyr} measured from the pyranometer. Figures 3–5 show the RMSE, the MBE and the MAE calculated errors, respectively, expressed in kW/m^2 for the days of July 2012 in the sites "Gi" and "Ma". For example, it can be observed that on 23 July there is a significant increase of the three errors with respect to the preceding days, at both sites. This situation can be explained by looking at the last graphs of Figure 2, in which the sky on 23 July becomes variable after a sequence of clearer days. A similar situation occurs on 24 July at both sites and on 25 July at the site "Gi".

On the average, the 1-day ahead estimate is the most accurate, with lower errors compared to the 2-days and 3-days ahead estimates. In particular, considering the RMSE errors in Figure 3, the average value of the 1-day ahead estimates give figures around 119 W/m^2 for the site "Gi" and around 107 W/m^2 for the site "Ma"; whereas the maximum value is around 266 W/m^2 for the site "Gi" and around 280 W/m^2 for the site "Ma". Furthermore, the minimum value is around 81 W/m^2 for the site "Gi" and around 82 W/m^2 for the site "Ma". From Figure 4, it can be pointed out that the 1-day ahead estimates give MBE figures around 50 W/m^2 for the site "Gi" and around 40 W/m^2 for the site "Ma". In the best results the MBE index decreases down to 20 W/m^2 for the site "Gi" and down to 10 W/m^2 for the site "Ma", and in the worst results the MBE raises up to 140 W/m^2 for the site "Gi" and up to 160 W/m^2 for the site "Ma". The presence of a positive bias (with estimates higher than the measured values) may be associated with air pollution. From Figure 5, the MAE for the 1-day ahead estimates is 90 W/m^2 for the site "Gi" and 80 W/m^2 for the site "Ma", the minimum value is about 60 W/m^2 for both sites and the maximum value is 180 W/m^2 for the site "Gi" and 190 W/m^2 for the site "Ma".

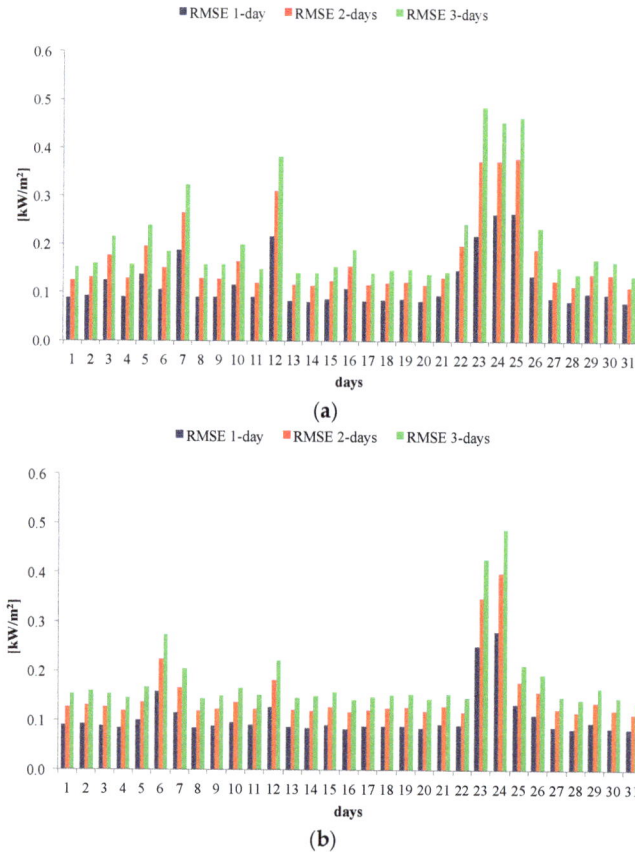

Figure 3. The RMSE in kW/m^2 in July. (a) Site "Gi"; (b) Site "Ma".

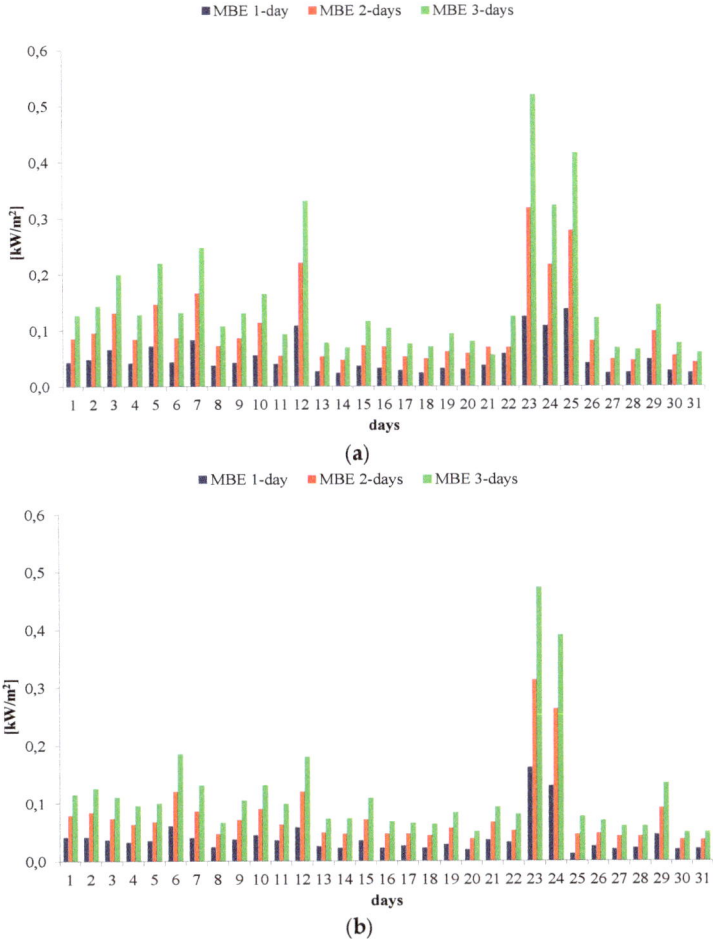

Figure 4. The MBE in kW/m^2 in July. (**a**) Site "Gi"; (**b**) Site "Ma".

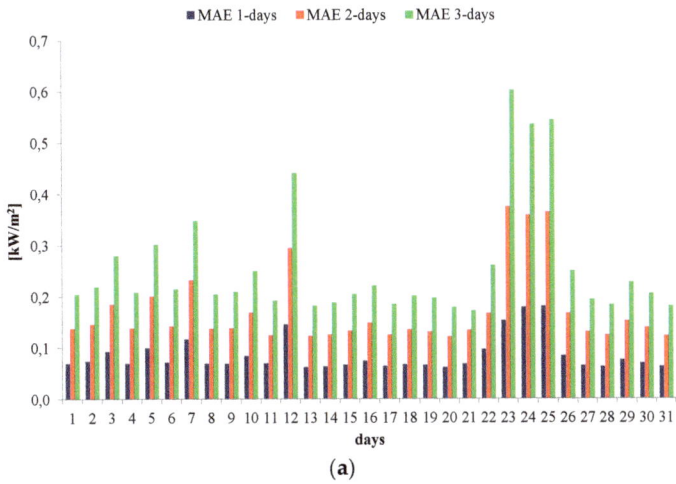

Figure 5. *Cont.*

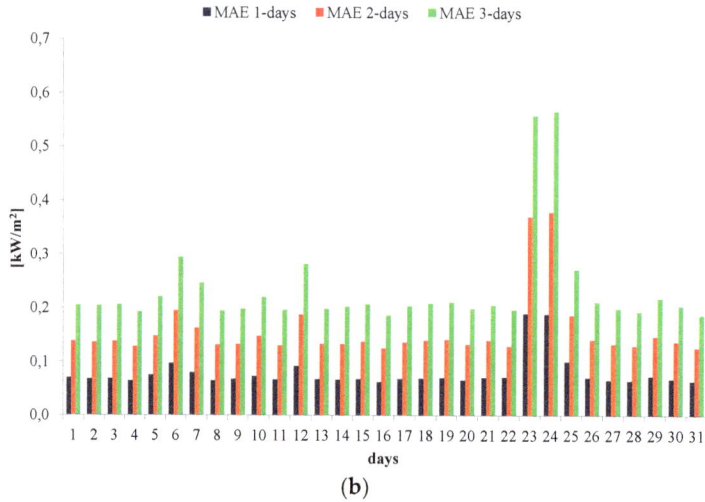

(b)

Figure 5. The MAE in kW/m^2 in July. (**a**) Site "Gi"; (**b**) Site "Ma".

In order to provide an overall view of the estimation errors during the year 2012, Figure 6 shows the monthly average values of the daily RMSE, MAE and MBE obtained from the 1 day-ahead estimates. At both sites, relatively high average errors occur in April and May, while generally the months with lower errors are from October to January. These results are consistent with the occurrence of the most relevant ISBC effects in April and May, as shown in Table 5.

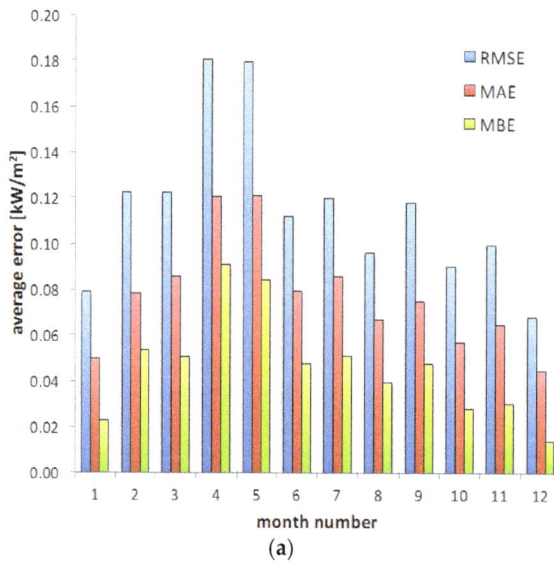

(a)

Figure 6. *Cont.*

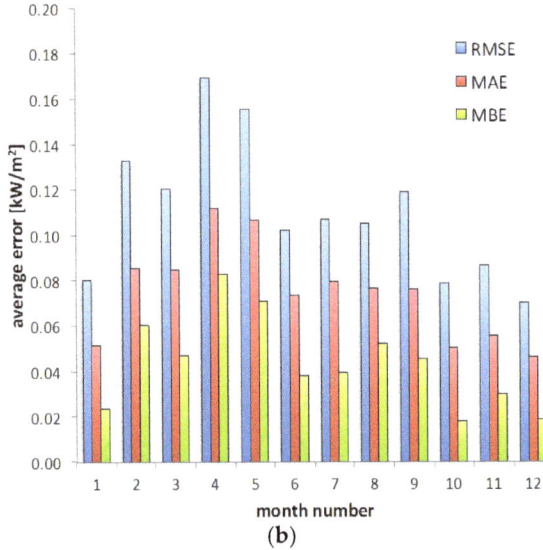

Figure 6. Monthly average values of the daily RMSE, MAE and MBE obtained from the 1-day ahead estimates at the two sites. (**a**) Site "Gi"; (**b**) Site "Ma".

4.2. Duration Curves of Bias and Absolute Error

The figures reported in this section show the error duration curves for the sites "Gi" and "Ma", representing the bias and the absolute errors of the estimates referring to the 1 day-ahead forecast with respect to the irradiance measurements. The data are averaged at 15-min time steps for one year. As renewable sources are rapidly gaining a larger share of the market, the importance of forecasting their productivity is essential for grid operators who have always to balance electricity supply and demand. The evaluation of the forecast errors is helpful for the grid operator, for example, to estimate to what extent the contribution of PV plants to the grid-connected local generation (calculated by using the 1 day-ahead forecast data) can be trustable.

Figure 7 shows the error duration curves of positive and negative bias, where positive biases correspond to estimates higher than the measurements, and *vice versa*. The errors are expressed in per units by considering a reference value of 1 kW/m^2. For both sites, the number of positive bias entries is higher than the number of negative bias entries, indicating a major trend to overestimate the irradiance values in these cases. This overestimation issue deserves some attention. In fact, using the estimated irradiance values as input data to estimate the PV power generation, an overestimation of the PV production creates an undue imbalance between the actual generation and the load. This imbalance impacts on the need to procure further reserves for supporting the grid operation.

Figure 8 indicates that the positive bias exceeded by 5% of the quarters of hour during the year is about 0.49 for the site "Gi", and about 0.44 for the site "Ma". Figure 9 shows that the negative bias exceeded by 5% of the quarters of hour during the year is about -0.24 for the site "Gi", and about -0.27 for the site "Ma". Furthermore, Figure 10 shows the duration curves for the absolute error. In the zoom of Figure 11 it is shown that the absolute error exceeded by 5% of the quarters of hour during the year is about 0.43 for the site "Gi" and about 0.38 for the site "Ma".

The bias and error values obtained are compared with the pyranometer uncertainty indicated in Section 2.2. The absolute error exceeded in 5% of the cases is indicatively one order of magnitude higher than the expanded uncertainty (45 W/m^2, that is, 0.045 per units) of the pyranometer at confidence level 99.7%. Thereby, the measurement uncertainty plays a minor role in the interpretation of the results.

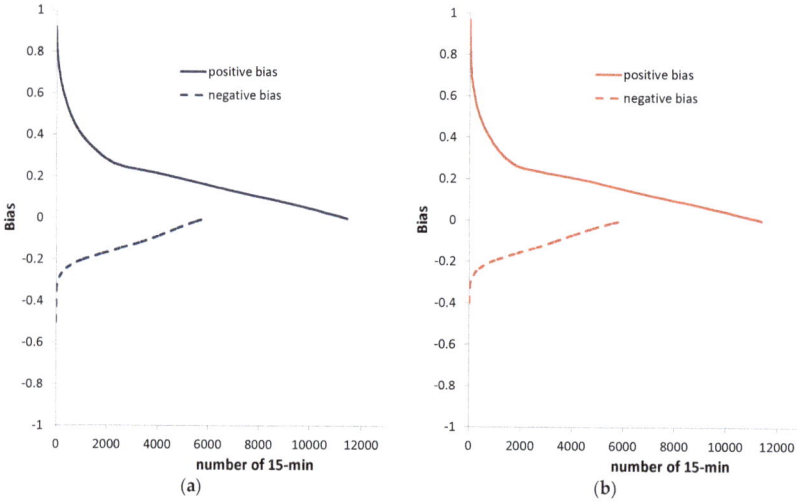

Figure 7. Duration curve of positive and negative bias for the solar irradiance estimated data with respect to measurements for the year 2012. (**a**) Site "Gi"; (**b**) Site "Ma".

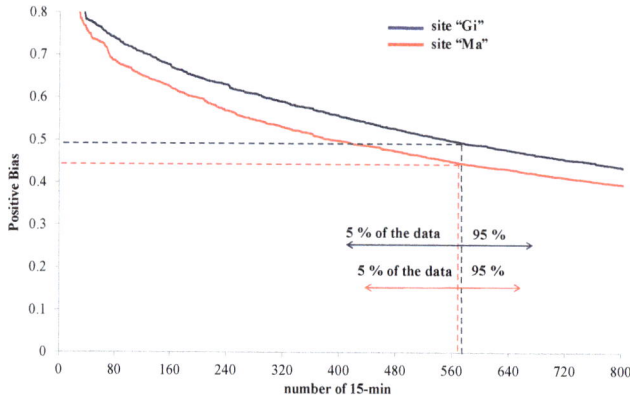

Figure 8. Zoom of the duration curves of positive bias for the solar irradiance estimated data with respect to measurements for the year 2012 at the sites "Gi" and "Ma".

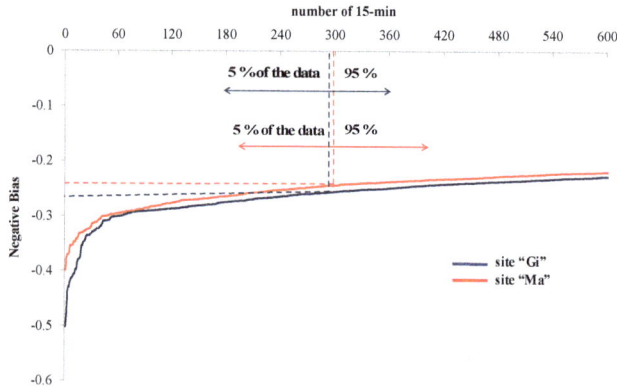

Figure 9. Zoom of the duration curves of negative bias for the solar irradiance estimated data with respect to measurements for the year 2012 at the sites "Gi" and "Ma".

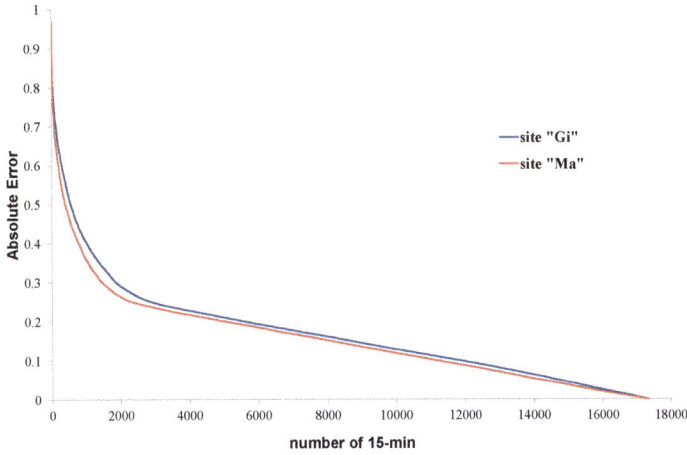

Figure 10. Duration curve of the absolute errors for the solar irradiance estimated data with respect to measurements for the year 2012.

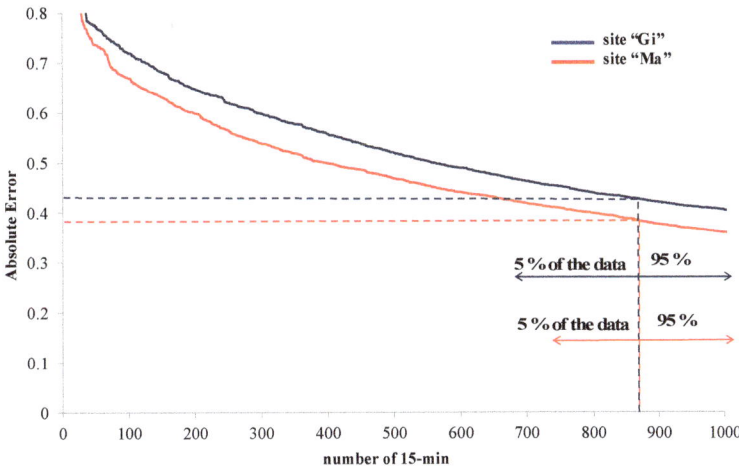

Figure 11. Zoom of the duration curves of the absolute error for the solar irradiance estimated data with respect to measurements for the year 2012 at the sites "Gi" and "Ma".

5. AC Power Estimations Compared with Experimental Results

Figure 12 shows the application of the PV conversion model to the reference-cell irradiance data for seven consecutive days of July 2012 in the sites "Gi" and "Ma". For example, low deviations occur on 23 July (an extremely variable day), meaning that the model is able to follow also huge irradiance variations. Moreover, thanks to the days with clear sky, since the deviations are proportional to the solar irradiance, it is possible to detect the failure of a portion of the PV arrays, becoming evident by the occurrence of large and regular deviations between the measured values and the outputs of the model, as reported in [20]. Thereby, a salient characteristic of the model is that it can be useful for fault diagnosis.

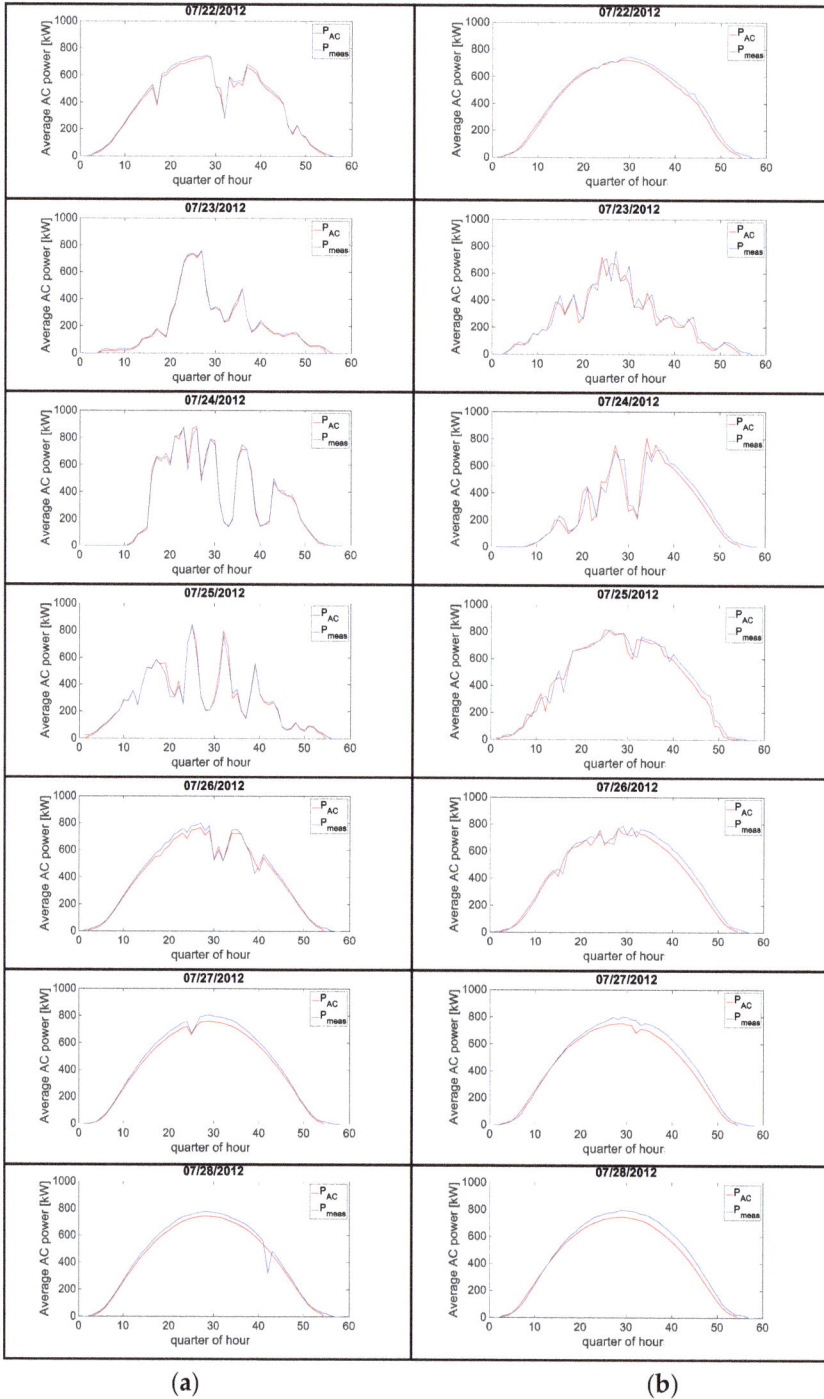

Figure 12. Comparison between PV power measurements and simulations for seven consecutive days of July 2012. (**a**) Site "Gi"; (**b**) Site "Ma".

The differences appearing in Figure 12 may be also attributed to the fact that average values have been used in the representation of the sources of losses and efficiencies, with the scope of estimating both clear sky and variable conditions without varying the efficiencies in function of the sky conditions.

For the calculation of the power P_{AC} to be used in Equation (9), the reference-cell irradiance data gathered on the tilted plane G_{tcell}, averaged on 15-min basis, are used as inputs of the model previously described. The results of error calculation for the estimated power profiles are reported in the following figures. The error duration curves calculated for ΔP are represented with averaging time step of 15 min and on an annual basis. Figure 13 shows the duration curves of positive and negative ΔP errors. At both sites, the number of negative errors ΔP^{-} is higher than the number of positive errors ΔP^{+}, but the positive errors may be quantitatively higher. This aspect is confirmed by comparing the results shown in the zoom of Figure 14 (for positive estimation errors) and Figure 15 (for negative estimation errors). The corresponding errors exceeded by 5% of the number of quarters of hour in one year are of about 120–130 kW for the positive errors at the two sites, and of about -60 kW and -100 kW for the negative errors at the sites "Gi" and "Ma", respectively.

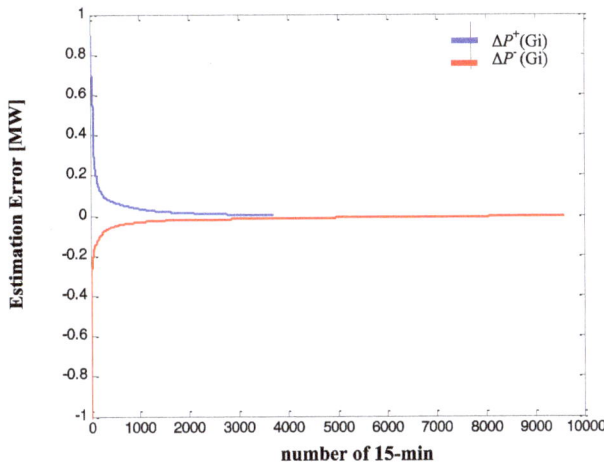

(a)

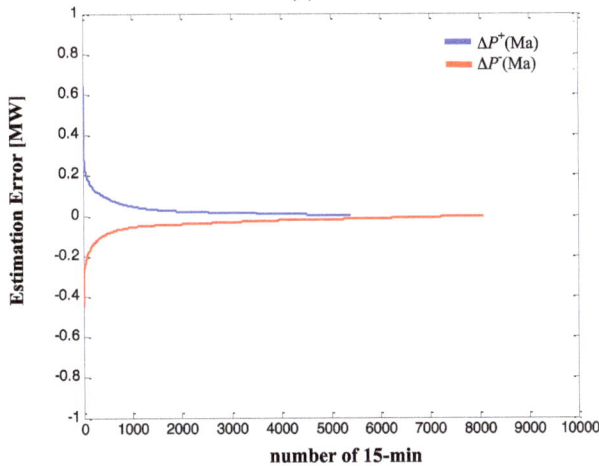

(b)

Figure 13. Duration curve of positive and negative estimation errors of the AC power profiles with respect to experimental results for the year 2012. (a) Site "Gi"; (b) Site "Ma".

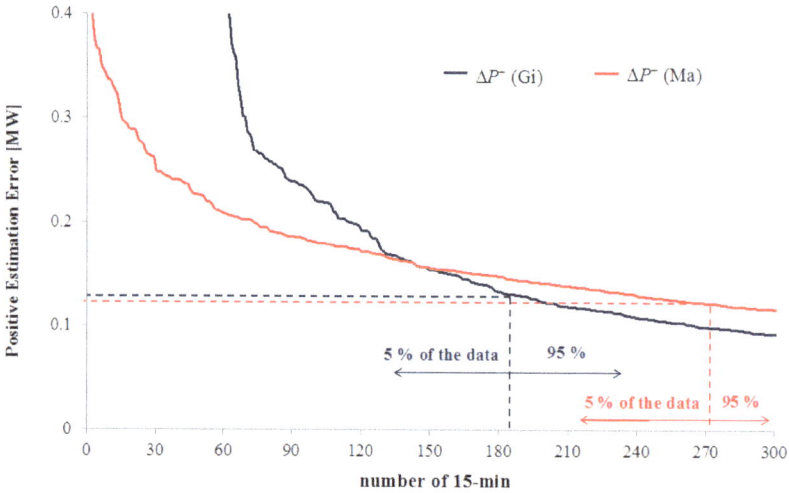

Figure 14. Zoom of the duration curves of positive estimation errors of the AC power profiles with respect to experimental results for the year 2012 at the sites "Gi" and "Ma".

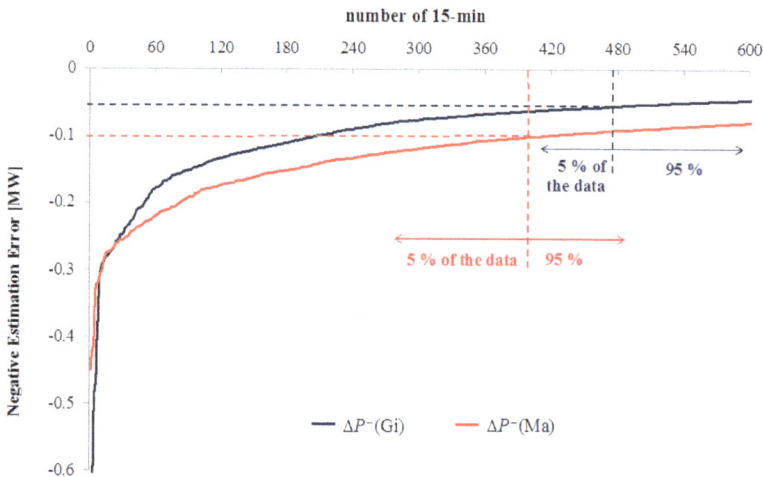

Figure 15. Zoom of the duration curves of negative estimation errors of the AC power profiles with respect to experimental results for the year 2012 at the sites "Gi" and "Ma".

6. Conclusions

This paper has presented the comparison of irradiance and AC power estimates with respect to the experimental results gathered from meteorological stations and energy meters in grid-connected PV systems. The data used refer to two PV sites located in Southern Italy. The solar irradiance forecasts up to 3 days ahead on the horizontal plane, available from a weather forecast provider at geographic coordinates close to the PV plants, have been used to obtain estimated data patterns. It has been established to what extent the estimated data referring to the 1 day-ahead forecast are better than the estimated data determined from the 2 and 3 days-ahead forecasts, by calculating some classical average errors. A method to classify each hour of a day by using three categories (*variable*, *cloudy*, or *clear*) has been implemented. Examining the results month by month, it has been possible to determine the number of successful and unsuccessful classifications provided by the 1 day-ahead estimated data with respect to the pyranometer measurements. The low number of clear-sky days, especially in spring and summer, can be explained by the air turbidity, e.g., due to pollution deriving from human activities. Deep cloudy weather cases can be reproduced when the WRF forecasts indicate low irradiance values and the polynomial spline connecting these points remains well below the clear sky conditions.

The results obtained for the ISBC effect have shown that this particular effect contributes to determine variable sky conditions. Assessing the ISBC effect is useful to explain the presence of power production peaks even higher than the rated power specified at STC of irradiance and cell temperature. The ISBC effect, even if noticeable on 1-min scale, is smoothed on the 15-min scale and considering the aggregation of more locations.

For the comparison between estimated and measured data, the statistical indicators RMSE, MAE and MBE have been calculated. Considering positive and negative MBE, the error duration curves have been obtained for 15-min averaged irradiance values on an annual basis. Finally, the combination of irradiance estimation and PV conversion model provides interesting results to boost the PV penetration into the grid. Considering the error of the AC power calculated from the PV model with respect to the AC power measured by the meters on the real grid-connected PV system, the error duration curve allows us determine which positive or negative errors occur for an established percentage of the data analyzed.

The categorization of the types of sky for each period of the day, associated with the short-term estimation of the weather conditions, is a specific information that can be used to quantify the additional reserve necessary to balance the fluctuations of the PV generation in periods in which high fluctuations are expected, without requiring such reserve to be continuously available [55,56]. The connection of PV power to the grid, like in the case of wind power, requires additional reserve with respect to the normal reserve required for the balance control of the grid [57]. For this purpose, the information on the PV forecasting uncertainty can be handled to assist the assessment of the amount of reserves needed to integrate the uncertain PV generation into the electrical system. For this purpose, persistence models, Markov chains and neural networks can be applied [56–58]. These results are also useful to estimate the contribution of PV in the definition of capacity value and capacity credit of renewable energy sources [59–61].

Acknowledgments: Part of the research leading to these results has received funding from the European Union Seventh Framework Programme FP7 under grant agreement No. 309048, project SiNGULAR (Smart and Sustainable Insular Electricity Grids Under Large-Scale Renewable Integration).

Author Contributions: The authors contributed equally to the work.

Conflicts of Interest: The authors declare no conflict of interest.

References

1. Jensenius, J.S. Solar Resources. In *Insolation Forecasting*; Massachusetts Institute of Technology (MIT) Press: Cambridge, MA, USA, 1989; pp. 335–349.

2. Glahn, H.R.; Lowry, D.A. The use of model output statistics (MOS) in objective weather forecasting. *Appl. Meteorol.* **1972**, *11*, 1203–1211. [CrossRef]

3. Kaifel, A.K.; Jesemann, P. An adaptive filtering algorithm for very short-range forecast of cloudiness applied to meteosat data. In Proceedings of the 9th Meteosat Scientific Users Meeting, Locarno, Switzerland, 15–18 September 1992.

4. Beyer, H.G.; Costanzo, C.; Heinemann, D.; Reise, C. Short range forecast of PV energy production using satellite image analysis. In Proceedings of the 12th European Photovoltaic Solar Energy Conference, Amsterdam, The Netherlands, 11–15 April 1994; pp. 1718–1721.

5. Kang, B.O.; Tam, K. New and improved methods to estimate day-ahead quantity and quality of solar irradiance. *Appl. Energy* **2015**, *137*, 240–249. [CrossRef]

6. Huld, T.; Amillo, A. Estimating PV module performance over large geographical regions: The role of irradiance, air temperature, wind speed and solar spectrum. *Energies* **2015**, *8*, 5159–5181. [CrossRef]

7.   Mellit, A.; Massi Pavan, A. Performance prediction of 20 kW$_P$ grid-connected photovoltaic plant at Trieste (Italy) using artificial neural network. *Energy Convers. Manag.* **2010**, *51*, 2431–2441. [CrossRef]

8.   Mellit, A.; Massi Pavan, A. A 24-h forecast of solar irradiance using artificial neural network: Application for performance prediction of a grid-connected PV plant at Trieste, Italy. *Sol. Energy* **2010**, *84*, 807–821. [CrossRef]

9.   Izgi, E.; Oztopal, A.; Yerli, B.; Kaymak, M.K.; Sahin, A.D. Short-mid-term solar power prediction by using artificial neural networks. *Sol. Energy* **2012**, *86*, 725–733. [CrossRef]

10.   Da Silva Fonseca, J.G., Jr.; Oozeki, T.; Takashima, T.; Koshimizu, G.; Uchida, Y.; Ogimoto, K. Use of support vector regression and numerically predicted cloudiness to forecast power output of a photovoltaic power plant in Kitakyushu, Japan. *Progress Photovolt. Res. Appl.* **2011**, *20*, 874–882. [CrossRef]

11.   Shi, J.; Lee, W.; Liu, Y.; Yang, Y.; Wang, P. Forecasting power output of photovoltaic systems based on weather classification and support vector machines. *IEEE Trans. Ind. Appl.* **2012**, *48*, 1064–1069. [CrossRef]

12.   Bouzerdoum, M.; Mellit, A.; Massi Pavan, A. A hybrid model (SARIMA–SVM) for short-term power forecasting of a small-scale grid-connected photovoltaic plant. *Sol. Energy* **2013**, *98*, 226–235. [CrossRef]

13.   Pelland, S.; Galanis, G.; Kallos, G. Solar and photovoltaic forecasting through post-processing of the global environmental multiscale numerical weather prediction model. *Progress Photovolt. Res. Appl.* **2011**, *21*, 284–296. [CrossRef]

14.   Cai, T.; Duan, S.; Chen, C. Forecasting power output for grid-connected photovoltaic power system without using solar radiation measurement. In Proceedings of the 2nd IEEE International Symposium on Power Electronics for Distributed Generation Systems (PEDG), Hefei, China, 16–18 June 2010.

15.   Bessa, R.J.; Trindade, A.; Miranda, V. Spatial-temporal solar power forecasting for smart grids. *IEEE Trans. Ind. Inform.* **2015**, *11*, 232–241. [CrossRef]

16.   Dambreville, R.; Blanc, P.; Chanussot, J.; Boldo, D. Very short term forecasting of the global horizontal irradiance using a spatio-temporal autoregressive model. *Renew. Energy* **2014**, *72*, 291–300. [CrossRef]

17.   Iqbal, M. *An introduction to solar radiation*; Academic Press: Toronto, ON, Canada, 1983.

18.   Duffie, J.A.; Beckman, W.A. *Solar Engineering of Thermal Processes*, 2nd ed.; Wiley Interscience: New York, NY, USA, 1991.

19.   Şen, Z. *Solar Energy Fundamentals and Modeling Techniques*; Springer: Berlin, Germany, 2008; pp. 70–71.

20.   The American Society of Heating, Refrigerating and Air-Conditioning Engineers (ASHRAE). *Handbook of Fundamentals, American Society of Heating, Refrigeration and Air-Conditioning Engineers*; ASHRAE: Atlanta, GA, USA, 1993.

21.   Moon, P.; Spencer, D.E. Illumination from a non uniform sky. *Illum. Eng.* **1942**, *37*, 707–726.

22.   Chicco, G.; Cocina, V.; Spertino, F. Characterization of solar irradiance profiles for photovoltaic system studies through data rescaling in time and amplitude. In Proceedings of the 49th International Universities' Power Engineering Conference (UPEC 2014), Cluj-Napoca, Romania, 2–5 September 2014.

23.   Joint Research Centre of the European Commission, Photovoltaic Geographical Information System (PVGIS). Available online: http://re.jrc.ec.europa.eu/pvgis/apps4/pvest.php (accessed on 16 December 2015).

24.   Spertino, F.; di Leo, P.; Cocina, V. Accurate measurements of solar irradiance for evaluation of photovoltaic power profiles. In Proceedings of the IEEE Conference Powertech, Grenoble, France, 16–20 June 2013; pp. 1–5.

25.   Czekalski, D.; Chochowski, A.; Obstawski, P. Parameterization of daily solar irradiance variability. *Renew. Sustain. Energy Rev.* **2012**, *16*, 2461–2467. [CrossRef]

26.   Badosa, J.; Haeffelin, M.; Chepfer, H. Scales of spatial and temporal variation of solar irradiance on Reunion tropical island. *Sol. Energy* **2013**, *88*, 42–56. [CrossRef]

27.   Chicco, G.; Cocina, V.; di Leo, P.; Spertino, F. Weather forecast-based power predictions and experimental results from photovoltaic systems. In Proceedings of the IEEE Conference Speedam, Ischia, Italy, 18–20 June 2014.

28.   International Organization for Standardization (ISO). *Solar Energy—Specification and Classification of Instruments for Measuring Hemispherical Solar and Direct Solar Radiation*; ISO 9060:1990; ISO: Geneva, Switzerland, 1990.

29.   Carullo, A.; Ferraris, F.; Vallan, A.; Spertino, F.; Attivissimo, F. Uncertainty analysis of degradation parameters estimated in long-term monitoring of photovoltaic plants. *Measurement* **2014**, *55*, 641–649. [CrossRef]

30. Reinders, A.H.M.E.; van Dijk, V.A.P.; Wiemken, E.; Turkenburg, W.C. Technical and economic analysis of grid-connected PV systems by means of simulation. *Progress Photovolt. Res. Appl.* **1999**, *7*, 71–82. [CrossRef]

31. Markvart, T. *Solar Electricity*, 2nd ed.; John Wiley & Sons: Hoboken, NJ, USA, 2000.

32. International Electrotechnical Commission (IEC). *Crystalline Silicon Photovoltaic (PV) Array. On-Site Measurement of I-V Characteristics*; IEC 61829; British Standards Institution: Geneva, Switzerland, 1998.

33. Spertino, F.; Sumaili Akilimali, J. Are manufacturing I-V mismatch and reverse currents key factors in large photovoltaic arrays? *IEEE Trans. Ind. Electron.* **2009**, *56*, 4520–4531. [CrossRef]

34. Spertino, F.; Corona, F. Monitoring and checking of performance in photovoltaic plants: A tool for design, installation and maintenance of grid-connected systems. *Renew. Energy* **2013**, *60*, 722–732. [CrossRef]

35. Spertino, F.; Corona, F.; di Leo, P. Limits of advisability for master-slave configuration of DC-AC converters in photovoltaic systems. *IEEE J. Photovolt.* **2012**, *2*, 547–554. [CrossRef]

36. Tapakis, R.; Charalambides, A.G. Equipment and methodologies for cloud detection and classification: A review. *Sol. Energy* **2013**, *95*, 392–430. [CrossRef]

37. De Miguel, A.; Bilbao, J.; Aguiar, R.; Kambezidis, H.; Negro, E. Diffuse solar irradiation model evaluation in the North Mediterranean Belt area. *Sol. Energy* **2001**, *70*, 143–153. [CrossRef]

38. Erbs, D.G.; Kein, S.A.; Duffie, J.A. Estimation of the diffuse radiation fraction for hourly, daily and monthly-average global radiation. *Sol. Energy* **1982**, *28*, 293–302. [CrossRef]

39. Orgill, J.F.; Hollands, K.G.T. Correlation equation for hourly diffuse radiation on a horizontal surface. *Sol. Energy* **1977**, *19*, 357–359. [CrossRef]

40. Iqbal, M. Prediction of hourly diffuse solar radiation from measured hourly global radiation on a horizontal surface. *Sol. Energy* **1980**, *24*, 491–503. [CrossRef]

41. Lorenz, E.; Heinemann, D. Prediction of solar irradiance and photovoltaic power. *Compr. Renew. Energy* **2012**, *1*, 239–292.

42. Meteorological Service of Catalonia, Meteo.Cat. Available online: http://www.meteo.cat (accessed on 16 December 2015).

43. The Weather Research & Forecasting Model. Available online: http://www.wrf-model.org (accessed on 16 December 2015).

44. Cocina, V. Economy of Grid-Connected Photovoltaic Systems and Comparison of Irradiance/Electric Power Predictions vs. Experimental Results. Ph.D. Thesis, Politecnico di Torino, Torino, Italy, 2014.

45. Sluiter, R. Interpolation methods for climate data. Literature review. KNMI (Koninklijk Nederlands Meteorologisch Instituut) intern rapport IR 2009-04. De Bilt, The Netherlands, 19 November 2008.

46. Estupiñán, J.G.; Raman, S.; Crescenti, G.H.; Streicher, J.J.; Barnard, W.F. Effects of clouds and haze on UV-B radiation. *J. Geophys. Res.* **1996**, *101*, 16807–16816. [CrossRef]

47. Craig, F.; Bohren, C.F.; Clothiaux, E.E. *Fundamentals of Atmospheric Radiation*; Wiley-VCH: Weinheim, Germany, 2006.

48. Morf, H. A stochastic solar irradiance model adjusted on the Ångström-Prescott regression. *Sol. Energy* **2013**, *87*, 1–21. [CrossRef]

49. Luoma, J.; Kleissl, J.; Murray, K. Optimal inverter sizing considering cloud enhancement. *Sol. Energy* **2012**, *86*, 421–429. [CrossRef]

50. Wirth, G.; Lorenz, E.; Spring, A.; Becker, G.; Pardatscher, R.; Witzmann, R. Modeling the maximum power output of a distributed PV fleet. *Progress Photovolt. Res. Appl.* **2015**, *23*, 1164–1181. [CrossRef]

51. Yordanov, G.H.; Midtgård, O.-M.; Saetre, T.O.; Nielsen, H.K.; Norum, L.E. Overirradiance (Cloud Enhancement) events at high latitudes. *IEEE J. Photovolt.* **2013**, *3*, 271–277. [CrossRef]

52. Lorenz, E.; Hurka, J.; Heinemann, D.; Beyer, H.G. Irradiance forecasting for the power prediction of grid-connected photovoltaic systems. *IEEE J. Sel. Top. Appl. Earth Obs. Remote Sens.* **2009**, *2*, 2–10. [CrossRef]

53. IEA (International Energy Agency). Photovoltaic and solar forecasting: State of the art. In *Report IEA Photovoltaic Power Systems Programme (PVPS) T14*; IEA: St. Ursen, Switzerland, 2013; Volume 1, pp. 1–36.

54. Diagne, M.; David, M.; Lauret, P.; Boland, J.; Schmutz, N. Review of solar irradiance forecasting methods and a proposition for small-scale insular grids. *Renew. Sustain. Energy Rev.* **2013**, *27*, 65–76. [CrossRef]

55. Nijhuis, M.; Rawn, B.; Gibescu, M. Prediction of power fluctuation classes for photovoltaic installations and potential benefits of dynamic reserve allocation. *IET Renew. Power Gener.* **2014**, *8*, 314–323. [CrossRef]

56. Yan, X.; Francois, B.; Abbes, D. Operating power reserve quantification through PV generation uncertainty analysis of a microgrid. In Proceedings of the IEEE PowerTech, Eindhoven, The Netherlands, 29 June–2 July 2015.
57. Brouwer, A.S.; van den Broek, M.; Seebregts, A.; Faaij, A. Impacts of large-scale intermittent renewable energy sources on electricity systems, and how these can be modeled. *Renew. Sustain. Energy Rev.* **2014**, *33*, 443–466. [CrossRef]
58. Tabone, M.D.; Callaway, D.S. Modeling variability and uncertainty of photovoltaic generation: A hidden state spatial statistical approach. *IEEE Trans. Power Syst.* **2015**, *30*, 2965–2973. [CrossRef]
59. Perez, R.; Taylor, M.; Hoff, T.; Ross, J.P. Reaching consensus in the definition of photovoltaics capacity credit in the USA: A practical application of satellite-derived solar resource data. *IEEE J. Sel. Top. Appl. Earth Obs. Remote Sens.* **2008**, *1*, 28–33. [CrossRef]
60. Simoglou, C.K.; Biskas, P.N.; Bakirtzis, E.A.; Matenli, A.N.; Petridis, A.I.; Bakirtzis, A.G. Evaluation of the capacity credit of RES: The Greek case. In Proceedings of the IEEE PowerTech, Grenoble, France, 16–20 June 2013.
61. Munoz, F.D.; Mills, A.D. Endogenous assessment of the capacity value of solar PV in generation investment planning studies. *IEEE Trans. Sustain. Energy* **2015**, *6*, 1574–1585. [CrossRef]

11

Molecular Level Factors Affecting the Efficiency of Organic Chromophores for *p*-Type Dye Sensitized Solar Cells

Svitlana Karamshuk [1], Stefano Caramori [2,*], Norberto Manfredi [1], Matteo Salamone [1], Riccardo Ruffo [1], Stefano Carli [2], Carlo A. Bignozzi [2] and Alessandro Abbotto [1,*]

Academic Editor: Claudia Barolo

[1] Department of Materials Science and Milano-Bicocca Solar Energy Research Center—MIB-Solar, University of Milano-Bicocca, INSTM Unit, Via Cozzi 55, 20125 Milano, Italy; svitlpetrova@gmail.com (S.K.); norberto.manfredi@unimib.it (N.M.); matteo.salamone@unimib.it (M.S.); riccardo.ruffo@unimib.it (R.R.)

[2] Department of Chemistry, University of Ferrara, Via L. Borsari 46, 44121 Ferrara, Italy; stefano.carli@unife.it (S.C.); carloalberto.bignozzi@unife.it (C.A.B.)

* Correspondence: stefano.caramori@unife.it (S.C.); alessandro.abbotto@unimib.it (A.A.)

Abstract: A series of mono- and di-branched donor-π-acceptor charge-separated dyes incorporating triphenylamine as a donor and either Dalton's or benzothiadiazole group as strong acceptors was synthesized and its fundamental properties relevant to the sensitization of nanocrystalline NiO investigated. The dyes exhibited an intense visible absorption band with a strong charge transfer character favorable to NiO sensitization, shifting the electron density from the donor to the acceptor branches. Nevertheless, the computed exciton binding energy is *circa* twice that of a common literature standard (P1), suggesting a more difficult charge separation. When tested in *p*-type dye-sensitized solar cells the dyes successfully sensitized NiO electrodes, with photocurrent densities about half than that of the reference compound. Being recombination kinetics comparable, the larger photocurrent generated by P1 agrees with the superior charge separation capability originating by its smaller exciton binding energy.

Keywords: dyes; heteroaromatic; visible absorption; dipolar; donor–acceptor; triphenylamine; branched; Suzuki coupling; DFT-TDDFT

1. Introduction

With the increasing demand of clean, secure, cost-effective, and renewable energy sources, the exploitation of solar light as a major source has clearly emerged as a key strategic priority. Following the first publication of Grätzel and O'Regan in 1991 [1], dye-sensitized solar cells (DSSCs) were recognized as a relatively cheap and easy-to-scale approach to direct solar-to-electrical power conversion. Furthermore, their transparency, versatile design and wide color palette offer unique structural and architectural possibilities in the emerging field of building integration [2,3], for example through the realization of photovoltaic windows and façades. The key feature of DSSCs is the sensitization of a porous wide-band-gap semiconductor thin film with a photoactive dye, which, following excitation, is able to transfer either electrons (*n*-type sensitization) or holes (*p*-type sensitization) to a semiconductor substrate. *n*-type DSSCs have been intensively investigated over the last two decades, recently reaching optimized power conversion efficiencies up to 13% with a single photoactive junction (dye/*n*-TiO$_2$) [4]. Despite this recent progress, power conversion efficiencies in *n*-type DSSCs seem to have reached a plateau, urging the search for new viable approaches to further increase DSSCs efficiencies. A particularly promising strategy to produce DSSCs with significantly enhanced power conversion efficiencies is the connection of an n-type photoelectrode (n-dye/TiO$_2$) with a p-type photoelectrode (p-dye/NiO), affording a tandem cell composed by two serially connected photoactive electrodes, each contributing to the total photovoltage delivered by the cell. Applying such a concept, organic-based photovoltaic devices with up to 40% conversion efficiency could be theoretically achievable [5]. Unfortunately, thus far, sensitized p-type systems have been much less investigated and much lower photocurrents, compared to their n-type counterparts, have been reported [6]. One of the main limitations in p-type systems, commonly based on NiO as a hole semiconductor [7], arises from the fast charge recombination [8] between the photoinjected hole in NiO and the reduced dye. Therefore, it is essential to develop new p-type chromophores, which could produce a long-lived charge separated state and minimize back recombination.

Several families of p-type dyes have been so far synthesized for this purpose. Those based on coumarin, porphyrin and peryleneimide scaffolds showed incident monochromatic photon-to-current conversion efficiencies (IPCE) only up to 4% and overall efficiencies lower than 0.2% [9–12]. One of the best examples of improved charge separation through dye design is represented by push–pull systems based on a di-branched D-(π-A)$_2$ (where D = electron-donor group, π = π-spacer, and A = electron-acceptor group) structures, like **P1**, firstly reported in 2008 by Sun and coworkers [13]. This prototypical p-type dye had a carboxylic anchoring group on the triphenylamine donor moiety bridged to a dicyanovinyl acceptor by a thienylene ring acting as a π-linker. The di-branched architecture, constituted by one donor and two π-spacer and acceptor units, follows the same design strategy used for recent n-type DSSC sensitizers [14–16]. Using the same general and successful principles, more complex dyes based on much more elaborate triphenylamine/oligothiophene dyes have soon followed [17–22]. Nevertheless, even in the best cases, the power conversion efficiencies obtained by p-type sensitizers are still about one order of magnitude lower than then the average efficiency delivered by n-type devices.

We were thus triggered to explore new structures for potentially efficient chromophores for p-type devices, by considering that the intramolecular charge transfer, at the basis of efficient charge separation in donor–acceptor dyes, is strongly dependent on the electron-withdrawing ability of the acceptor. We have therefore designed and investigated new p-type dyes where **P1**-like structures have been optimized by inserting stronger and more efficient electron-acceptor groups compared to the conventional dicyanovinyl moiety. First, we have designed the **SK2** dye, where the dicyanovinyl group of **P1** has been replaced by the much stronger 3-cyano-2-(dicyanomethylene)-4,5,5-trimethyl-2,5-dihydrofuran (Dalton's acceptor) group, widely used in other materials science fields [23]. Second, mono- and di-branched dyes (**SK3** and **SK4**), characterized by a D-π-A-π architecture, were realized. In the latter dyes, an additional benzothiadiazole-based acceptor-spacer unit was introduced both in linear (**SK3**) and branched (**SK4**) geometrical motifs for the manifold purpose of increasing the transition dipole moment, improving the spatial separation between hole and electron, and favoring the electron transfer to the electron mediator (I^-/I_3^-), resulting in improved interfacial charge separation.

Herein, we report the synthesis of the new p-type chromophores as well as their computational, electrochemical and photoelectrochemical properties in comparison with the literature standard **P1**. We have investigated the new dyes (Figure 1) as p-type sensitizers in p-type DSSC and the results were compared with the reference dye **P1**.

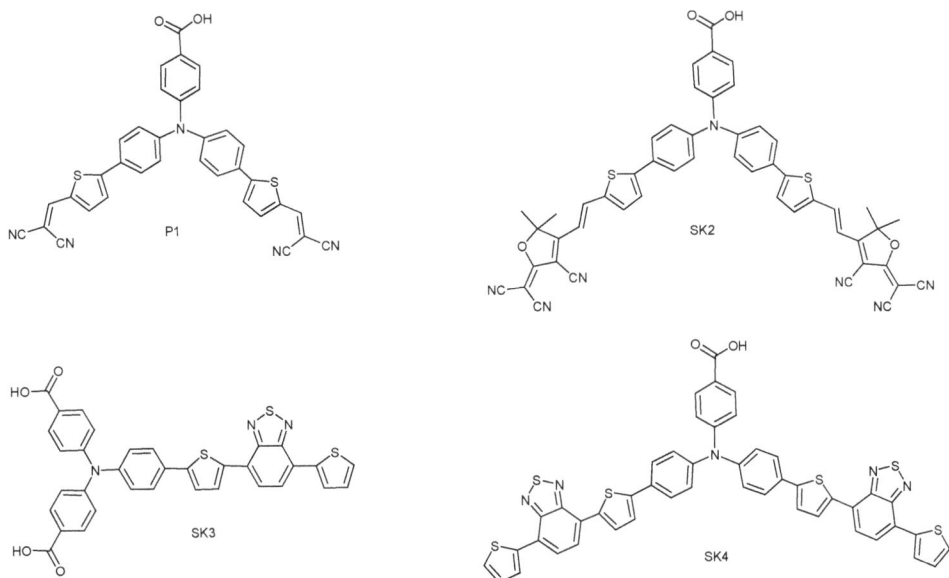

Figure 1. Structure of the investigated and reference (P1) dyes.

2. Results and Discussion

2.1. Synthesis

SK2, SK3 and SK4 were obtained following a synthetic pathway shown in Scheme 1. 4-[*Bis*-(4-bromophenyl)amino]benzaldehyde (1) was oxidized to the corresponding benzoic acid 2 by standard silver mirror reaction and then submitted to the Suzuki coupling with 5-formyl-2-thienylboronic acid for introducing the two thienyl linkers [13]. Knoevenagel reaction between the resulting bis-aldehyde 3 and 2-(3-cyano-4,5,5-trimethylfuran-2(5H)-ylidene)malononitrile afforded the desired chromophore SK2. It should be noted that the Knoevenagel condensation of 3 could be successfully accomplished only in acidic conditions at relatively high temperatures (>75 °C), while the reaction under conventional basic conditions (piperidine/EtOH) did not afford the condensation product, likely for the presence of the terminal COOH substituent on the donor core.

Preparation of SK3 and SK4 started from bromination of 4,7-di(thiophen-2-yl)benzo[c][1,2,5] thiadiazole (4) to the mono-bromide 5 using *N*-bromosuccinimide (NBS) in presence of a 1:1 solution of CH$_2$Cl$_2$ and acetic acid [24]. Since such reaction gave mono- and di-substituted products possessing close polarities, the separation of these two compounds by column chromatography had to be performed with care in order to afford the mono-derivative in moderate yields. Borination of 5 with *bis*(pinacolato)diboron resulted in the key intermediate boronic ester 6. Different reaction conditions (solvent, base, and temperature) were investigated for Suzuki condensation between 6 and triphenylamine derivatives. The best choice in both cases was dimethoxyethane (DME) as a solvent and aqueous solution of K$_2$CO$_3$ as a base. Under these conditions, the coupling reaction started immediately after mixing up the reagents and within a considerably short time afforded the desired targets 7 and 8 with fairly good yields. SK3 and SK4 were then obtained by following the previously described procedure for SK2.

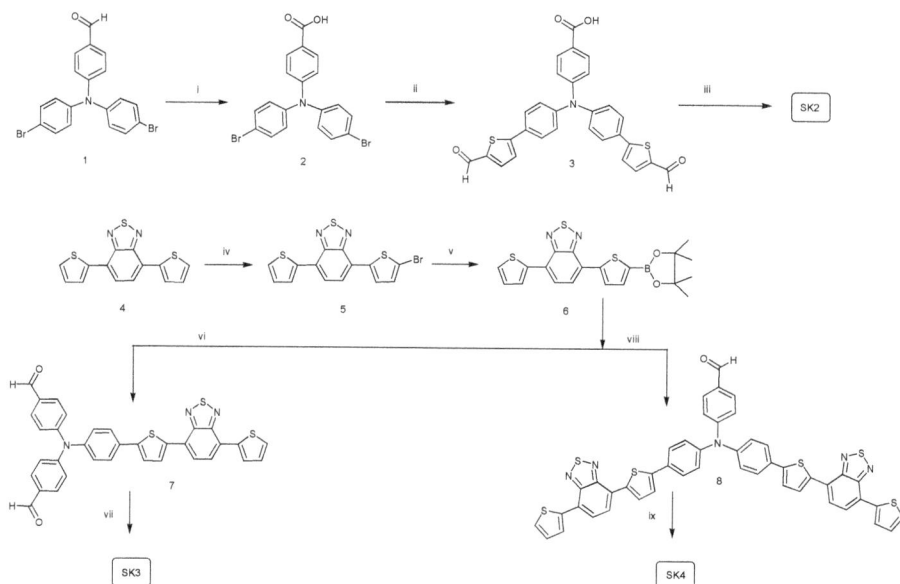

Scheme 1. Synthesis of **SK2**, **SK3** and **SK4**. Reagents and conditions: (**i**) Ag₂O, EtOH, NaOH, r.t.; (**ii**) Pd(dppf)Cl₂, 5-formyl-2-thienylboronic acid, K₂CO₃, toluene/MeOH, microwave; (**iii**) 2-(3-cyano-4,5,5-trimethylfuran-2(5H)-ylidene)malononitrile, NH₄OAc/AcOH, EtOH, reflux; (**iv**) NBS, CH₂Cl₂/AcOH, r.t.; (**v**) bis(pinacolato)diboron, Pd(dppf)Cl₂, KOAc, dioxane, reflux; (**vi**) 4-bromo-N,N-bis(4-formylphenyl)aniline, Pd(dppf)Cl₂, aq. K₂CO₃, DME, reflux; (**vii**) Ag₂O, EtOH, NaOH, r.t.; (**viii**) 4-(bis(4-bromophenyl)amino)benzaldehyde, Pd(dppf)Cl₂, aq. K₂CO₃, DME, reflux; and (**ix**) Ag₂O, EtOH, NaOH, r.t.

2.2. Spectroscopic and Electrochemical Properties

The absorption spectra of **SK2**, **SK3** and **SK4** in dimethyl sulfoxide (DMSO) are depicted in Figure 2 together with that of the reference dye **P1**. The normalized absorption and emission spectra of the SK dye series are reported in Figure S1 (Supplementary Materials). All dyes under investigations exhibited similar general spectral features, summarized by two intense ($\varepsilon > 3 \times 10^4$ M^{-1}·cm^{-1}) relatively broad absorption bands extending the light harvesting up to 700 nm in the case of **SK2**, where the presence of an efficient intramolecular charge transfer to the strong electron-withdrawing groups resulted in a significant bathochromic shift (*ca.* 70 nm) of the visible absorption maximum. **SK3**, **SK4**, and **P1** are *ca.* 100 nm blue-shifted with respect to **SK2**, showing a visible absorption maximum at 480–490 nm and an absorption onset at *ca.* 600 nm. In all cases, the second band is located in the UV region, with an absorption peak in the interval 350–370 nm, where **SK3** and his di-branched analogue **SK4**, incorporating the same donor–acceptor side arm, showed the sharpest and most intense band. The main optical and electrochemical parameters together with HOMO/LUMO energies estimated by cyclic voltammetry in 0.1 M TBAClO₄ in DMSO are collected in Table 1.

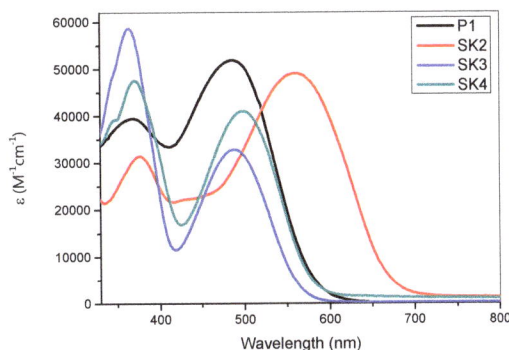

Figure 2. Absorption spectra of **P1**, **SK2**, **SK3** and **SK4** in dimethyl sulfoxide (DMSO).

Table 1. Optical and electrochemical parameters of the dyes. [a]

Dye	λ_{abs}(nm) (ε) $(10^4 \text{ M}^{-1}\cdot\text{cm}^{-1})$	λ_{em} (nm)	$E^{\circ\circ}$ (eV) [b]	$E_{(HOMO)}$ *vs.* NHE(V) (*vs.* Vacuum) (eV) [c]	$E_{(LUMO)}$ *vs.* NHE (V) (*vs.* Vacuum) (eV) [c]
P1	372 (3.97); 489 (5.17)	611	2.14	1.34(−6.0)	−0.62(−4.0)
SK2	373 (3.12); 559 (4.90)	674	1.91	1.26(−5.9)	−0.22(−4.4)
SK3	357 (5.87); 488 (3.28)	609	2.21	1.17(−5.8)	−0.85(−3.8)
SK4	365 (4.76); 498 (4.11)	621	2.17	1.16(−5.8)	−0.86(−3.7)

[a] in DMSO; [b] Calculated from the onset of the normalized absorption spectra of the dyes; [c] Evaluated from the DPV oxidation and reduction peak potentials in the presence of ferrocene as an internal reference ($E_{1/2}$ Fc$^+$/Fc = 0.68 V *vs.* NHE) [25] and using a potential value of −4.6 eV for NHE *vs.* vacuum [26].

As often observed with organic D-π-A dyes [27,28], the electrochemical behavior (Figure S2a, Supplementary Materials) of the series under investigation is dominated by irreversible processes that complicate a rigorous thermodynamic evaluation of the redox levels relevant to NiO sensitization. For these reasons, we have used the Differential Pulsed Voltammetry (DPV) to evaluate orbital energies. DPV shows higher sensitivity to Faradaic currents compared to other techniques (Figure S2b, Supplementary Materials). At anodic potentials, all chromophores presented a similar oxidation behavior resulting in an oxidation wave having a peak potential at *ca.* 0.5 V *vs.* Fc$^{+/\circ}$, which is related to the oxidation of the electron-rich TPA group [29]. The reductive behavior is generally more complex, showing for all dyes a first, weak and irreversible wave likely originated by the reductive chemisorption of the acidic protons onto the electrode surface [30]. The following current wave corresponds to the injection of the electron into the LUMO orbital and was used to estimate the corresponding energy level. In agreement with the structural similarity of their acceptor group, **SK3** and **SK4** showed close LUMO energies at −3.8 eV. This estimate is in reasonable agreement with the quasi-reversible wave observed at −1.55 V *vs.* Fc$^{+/\circ}$. The reduction of **SK2** dye, bearing a stronger acceptor group, is comparatively more anodically shifted with a lower energy LUMO at −4.5 eV. Thus, no thermodynamic limitation to hole transfer to NiO are expected: in all cases, the HOMO energies are similar to the **P1** dye and sufficiently lower than the upper valence band edge of NiO, located at 0.5–0.6 V *vs.* NHE [17,31]. Electron transfer to I_3^-, relevant to dye regeneration, is in all cases, hexoergonic, with ΔG° values larger than −0.2 eV, allowing to predict favorable kinetics.

2.3. Computational Investigation

To gain insights into the electronic and optical properties of the investigated sensitizers, we performed Density Functional Theory (DFT) calculations. All the calculations have been performed using the Gaussian 09 program package [32]. The dye structures at the ground state were optimized at the B3LYP level employing the 6-31G* basis set, starting from a semi-empirical PM3 pre-optimized geometry. Similar structural features were observed throughout the whole series. The nitrogen atom of the triphenylamine core gives rise to a distorted tetrahedral motif, resulting in a dihedral angle between the branches of the order of 40°. The π-A unit, constituted by thienylene bridges and dicyanovinly groups in **P1** and **SK2** and benzothiadiazole fragments in **SK3** and **SK4** (Figure 3), is essentially planar in agreement with the presence of an extended π-electron delocalization. Each branch is linked to the electron donating group via a benzene ring, with a dihedral angle of about 20°–27°. Such twisting angle is expected to be beneficial for decoupling holes and electrons, once the initial charge separation is achieved.

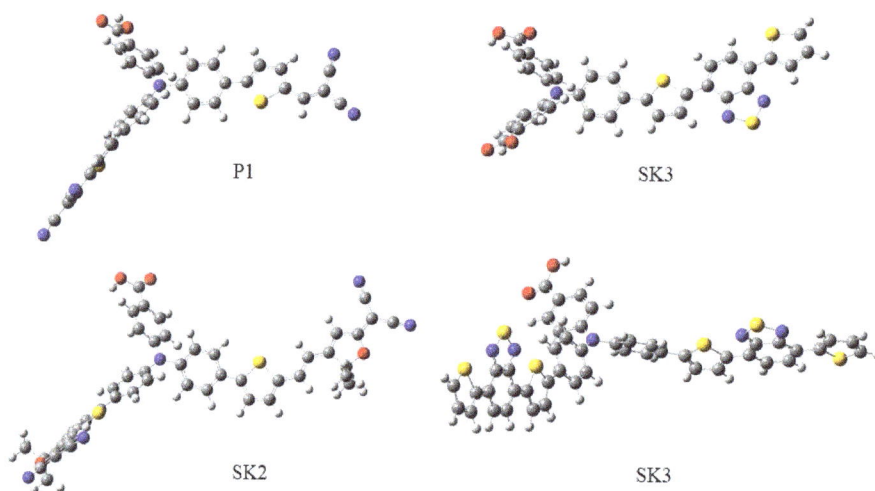

Figure 3. Equilibrium geometries of the *p*-type dyes under investigation calculated at the DFTB3LYP6-31G* level.

It is notoriously difficult for Time-Dependent DFT (TDDFT) methods to reliably describe charge transfer states, particularly those involving spatially separated orbitals and long range excitations as in the case of our *p*-type systems [33,34]. Therefore, in the attempt to drive reasonable insights on the electronic structure and excited state energetics of the dye series under investigation, the excitation energies (Table 2), resulting from calculations with B3LYP [35] and BH & H (Half and Half) [36,37] functionals were compared. Frontier orbitals isodensity maps (isovalue of 0.02) of the selected dyes are depicted in Figure S3 (Supplementary Materials).

Table 2. Comparison of experimental and calculated lowest absorption maximum of the investigated dyes in DMSO.

Dye	E_{exp}(eV) [λ_{max}(nm)]	$E^{\circ P}_{BH\ \&\ H}$(eV) [λ_{max}(nm)]	$E^{\circ P}_{B3LYP}$ (eV) [λ_{max}(nm)]	$|\Delta E_{BH\ \&\ H}|$(eV)	$|\Delta E_{B3LYP}|$(eV)
P1	2.53 [489]	2.79 [443]	2.30 [540]	0.26	0.23
SK2	2.22 [559]	2.86 [433]	2.08 [595]	0.64	0.14
SK3	2.54 [488]	2.62 [473]	2.14 [580]	0.08	0.40
SK4	2.49 [498]	2.48 [501]	1.92 [614]	0.01	0.57

The BH & H is considerably successful in predicting the optical transition energies of dyes **SK3** and **SK4**, where deviations from the experimentally measured spectrum are as low as 0.08 and 0.01 eV, but less successful with dyes **P1** and **SK2**, where the calculated $E^{\circ P}$ is higher than the experimental value by

0.26 and 0.64 eV, respectively. The lowest transition has, in all cases, a major HOMO-LUMO component (55%–87%) in the Kohn–Sham basis, with minor components arising from higher energy excitations, typically HOMO-1→LUMO, HOMO-1→LUMO + 1 and HOMO→LUMO + 2. The description of the lowest singlet state (S1) by the B3LYP calculation is roughly similar to that of the BH & H, although the HOMO-LUMO component to the excitation is largely predominating (>93%). The lowest transition in **P1** and **SK2** is identified by B3LYP with a lower absolute error (0.21 and 0.11 eV) than the BH & H functional, whereas larger errors are found with **SK3** and, particularly with **SK4**, which shows long range charge transfers with a high degree of spatial charge separation, as indicated by Electron Density Difference Maps (EDDM) (Scheme 2). Both the B3LYP and BH & H functional agree in the description of the EDDMs, showing a shift in electron density from the triphenylamine core to the electron acceptor branches, where the maximum density is localized in the proximity of the cyano groups (**P1** and **SK2**) or on the benzothiadiazole acceptor (**SK3** and **SK4**). It can also be appreciated that, in branched dyes (**P1**, **SK2**, and **SK4**), the electronic excitation involves the two acceptor branches simultaneously. Thus, the electronic distribution resulting from the lowest, and usually most intense, electronic transition is favorable to a successful charge separation. The hole density is localized in the immediate proximity of the surface interacting with NiO via COOH groups. This causes photoexcitation of the electron to occur predominantly on the farthest from the NiO part of the molecule where scavenging of the electron by I_3^- is more favorable.

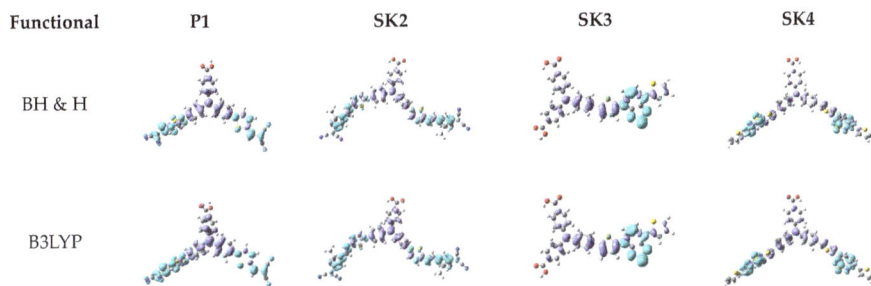

Scheme 2. Electron Density Difference Maps (EDDMs) (isovalue = 0.001) of the main transition in the visible region calculated with a 6311 G,d,+ basis set obtained with the BH & H (top row) and B3LYP (bottom row) functionals. DMSO solvent was described as a polarizable continuum model (PCM). Light blue and violet indicate an increased and a depleted electron density respectively.

The exciton binding energy (EBE) [38] was calculated [39] from the B3LYP data set, since BH & H was found to give unrealistically high HOMO-LUMO gaps (Table S1, Supplementary Materials) as well as binding energies (>1 eV). For the **SK** dyes, similar binding energies of the order of 0.33–0.38 eV are found, being comparable to that of some efficient n-type push pull charge separators [40] recently reported in the literature. **P1** displays the smallest EBE within the series with a value as small as 0.17 eV, comparable to some of the best n-type charge separators reported [41]. Smaller EBE favor separation of the electron–hole pair, and, consequently, charge injection into the semiconductor. The largest binding energy value of 0.38 eV is found for the dye **SK3**, which means that a higher energy, usually provided by the local electric potential at the semiconductor/dye interface, or by collision with phonons, is needed in order to promote charge injection. The larger EBE of the **SK** series may constitute a disadvantage compared to the reference dye **P1** for obtaining hole injection with high quantum yields.

2.4. Photoelectrochemical Investigation in DSSCs

The **SK** dyes together with reference compound **P1** were tested as photosensitizers in p-type DSSCs. NiO photocathodes were fabricated by blading a NiO colloidal paste, obtained by dispersing commercial 25-nm NiO nanoparticles in terpineol, with ethylcellulose as an organic binder/densificating agent, followed by high temperature sintering on FTO (Fluorine-doped Tin Oxide) coated glass. The resulting photocurrent (J)/photovoltage (V) curves under AM 1.5 illumination (1 sun) are shown in Figure 4A and their main efficiency parameters are listed in Table 3.

Table 3. Photovoltaic parameters of investigated chromophores in p-type dye-sensitized solar cells (DSSCs).

Dye	J_{sc} (mA/cm^2)	V_{oc} (mV)	FF	PCE (%)
P1	1.14	95	0.32	0.035
SK2	0.51	81	0.33	0.014
SK3	0.54	82	0.33	0.015
SK4	0.43	134	0.32	0.018

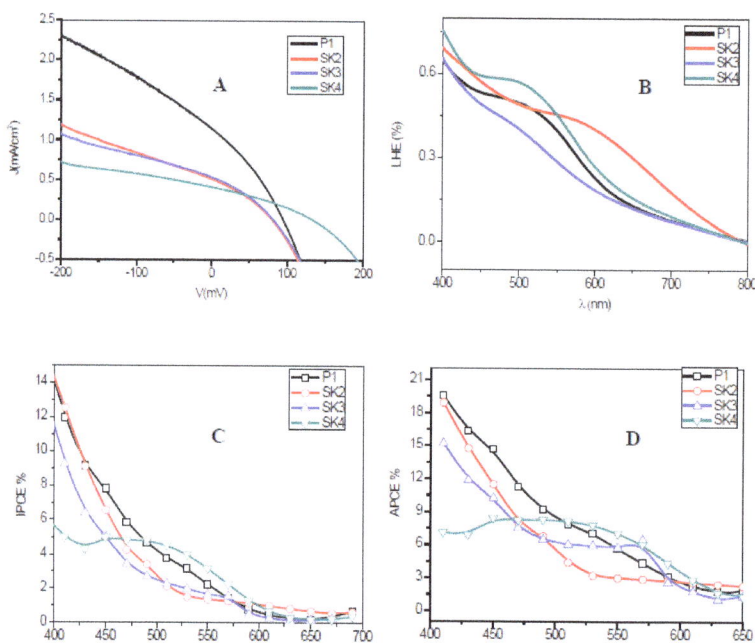

Figure 4. Photoelectrochemical and spectral properties of p-type DSSCs in the presence of 1 M LiI and 0.1 M I$_2$ in acetonitrile: (**A**) current–voltage characteristics; (**B**) Light Harvesting Efficiency (LHE) (LHE(λ) = $1 - 10^{-A(\lambda)}$), where A(λ) is the background-subtracted absorbance of the dyed NiO films; (**C**) photoaction spectra (IPCE $vs.$ λ); and (**D**) absorbed photon conversion efficiency spectra (APCE $vs.$ λ).

The photoelectrochemical investigation was carried out with the iodide/triiodide redox couple (solution of 1.0 M LiI and 0.1 M I$_2$ in CH$_3$CN). Dye **SK4** produces the highest open circuit voltage (133 mV), that is $ca.$ 40 mV higher than that of device sensitized by the other reference and novel dyes. The **SK2** and **SK3** molecules produced comparable J_{sc} ($ca.$ 0.5 mA/cm^2), which was, however, approximately half that of the reference **P1** ($ca.$ 1.1 mA/cm^2). As a comparison, blank NiO cells without the sensitizer generated $ca.$ 0.25 mA/cm^2 at 0 V, that is less than half of the cells sensitized by the SK series and $ca.$ ¼ of the reference P1 (Figure S4, Supplementary Materials). In all cases the sloped J/V characteristic at reverse (negative) bias is consistent with a small shunt resistance of the NiO films, causing charge leaks and losses by recombination. For instance, in the case of **P1** the photoinjected charge extracted from the photocathode under short circuit conditions (0 V) is roughly one half that available at −0.2 V. Such a photocurrent loss upon moving from negative to positive potential is less marked for the **SK** series, which, giving rise to a smaller extent of hole injection, result in fewer charge leaks to the electrolyte. The overall PCE of **P1**-sensitized cells is about twice as large as that of the SK-based devices, mainly due to their lower photocurrents. In general, although not optimized, the efficiencies of the NiO sensitized solar cells recorded in this work are in agreement with the average performances reported so far in the literature for similar types of sensitized p-type electrodes.

To understand the nature of photocathodic current, the IPCE spectra of the p-type cells, under short circuit conditions were measured (Figure 4 C). Consistent with their spectral properties, all dyes displayed a strongest response in the blue region of the spectrum for λ < 450 nm, reaching the best values of *ca.* 12%–14% for **P1** and **SK2**. The higher IPCE in this narrow wavelength Vis region is motivated by both a direct contribution from the I^-/I_3^- photochemistry and by a more efficient collection of the charge carriers generated in closer proximity of the electron collector by photons having a comparatively short penetration length into the film. The photoconversion then decreases by moving from the blue to the red region of the visible spectrum, showing a shoulder the 500–600 nm region, with values of 4%–5% in the case of **P1** and **SK4**. Despite the opacity of the NiO electrodes, causing significant light scattering, and the strong competitive absorption of the iodine based electrolyte, in general the photoconversion appeared to be in a relatively good agreement with the Light Harvesting Efficiency (LHE) (Figure 4B) of the sensitized NiO photoelectrodes, confirming that the excited state of the dye is responsible for the observed photocurrent in the visible region of spectrum.

The LHE plot, with values of 50%–60% in the 550–450 nm region, showed that **SK4** has here the best light harvesting performance, consistent with its relatively good IPCE response at these wavelengths. The absorbed monochromatic photon-to-current conversion efficiencies (APCE) spectra, obtained by dividing the corresponding IPCE spectra by the LHE of the respective photocathodes, are shown in Figure 4D. The low energy response of **P1** and **SK4** are similar, with APCE ~10% at 500 nm. Interestingly, **SK4** is able to yield a slightly more efficient conversion for λ ⩾ 550 nm, while **P1** prevails below 480 nm. As for the IPCE, the best APCE response is observed in the blue region of the spectrum, with values of ~18%–20% for **P1** and **SK2**.

Electrochemical impedance spectroscopy (EIS) was used to further investigate the comparative behavior of the chromophores in photovoltaic devices. All dyes were stable during potentiostatic measurements, which were performed at potentials between −10 mV and the V_{oc} of the cells. The J/V curves after EIS measurements showed essentially the same performances found before the measurements, which allows to rule out substantial dye or cell degradation affecting the results. The EIS spectra (Figure S5, Supplementary Materials) could be well fitted with the equivalent circuit, comprising the serial resistance, the counter-electrode interface, the diffusion element accounting for electrolyte transport impedance in a thin layer cell, and the transmission line [42] (DX1 in Zview) which describes the transmission network of the NiO, applied by Zhang *et al.* [17,18] to p-type DSSCs, comprising both the transport and the charge transfer resistance (R_{ct} (NiO)) (recombination resistance) at the NiO/electrolyte interface.

The impedance response of the NiO based DSSC exhibited similar features, appearing as a small high frequency semicircle, due essentially to the counter-electrode electrochemical interface, followed by a much larger loop at lower frequencies, which bears the main contribution of the NiO/electrolyte interface. Indeed, the J/V curves are essentially dominated by the charge transfer resistance of the NiO, as can be appreciated by the comparison of the derivative ($\frac{\partial i}{\partial V}$) of the J/V curves and the inverse of the interfacial charge transfer resistance $(R_{ct}(NiO))^{-1}$ (Figure 5B). This agreement is particularly good in the case of the **SK** series, for which the charge transfer resistance was the largely predominant contribution to the total cell resistance at each potential and varied, typically between 1000 and 200 Ω, depending on the voltage. Obviously, the lowest R_{ct} (NiO) are in all cases found at the open-circuit voltage, due to hole accumulation into the mesoscopic film under open-circuit conditions.

The NiO capacitance (Figure 5A), calculated according to $C = CPE(\omega)^{n-1}$, where n is the exponent of the CPE admittance from the transmission line, in all cases variable between 0.9 and 0.84, and ω is the frequency at the maximum of the large low frequency loop, was reasonably linear on a logarithmic scale, indicating a chemical capacitance [43] behavior determined by empty states in the valence bands or in surface states following a Boltzmann distribution near the valence band edge. The inspection of Figure 5 reveals that **P1** is the dye which exhibits the lowest charge transfer resistance and highest capacitance, indicating a superior capability of hole injection.

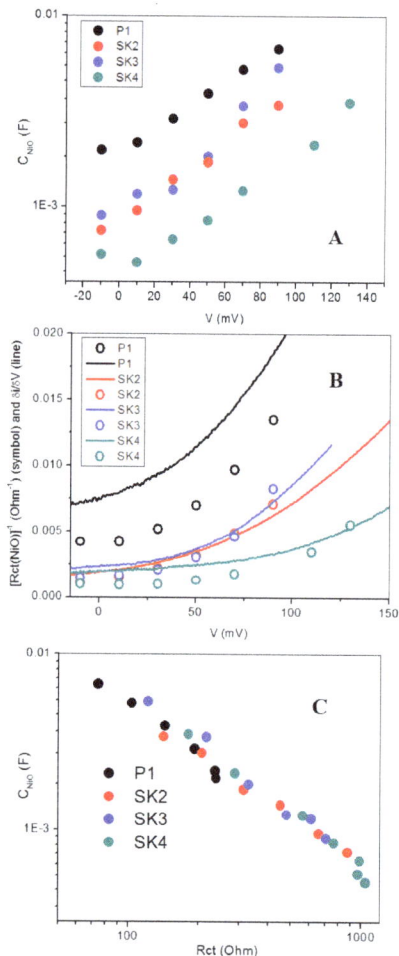

Figure 5. (**A**) Capacitance of sensitized NiO films at increasing potentials in the forward direction. (**B**) Reciprocal of the charge transfer resistance at the NiO/electrolyte interface (squares) compared to the derivative of the J/V curve (solid lines). (**C**) Chemical capacitance of NiO *vs.* interfacial charge transfer resistance in *p*-type DSSCs under study.

The representation of $R_{ct}(NiO)$ *vs.* $C(NiO)$ on a logarithmic scale indicates that all the dyes behave intrinsically similarly to each other, as far as recombination kinetics are concerned, when compared at the same chemical capacitance of the NiO. In other terms, the charge loss pathways involving direct dye^-/h^+ and hole recapture by I^- occurred at comparable rates for all the dyes under investigation. Accordingly, the higher photovoltage observed for the **SK4**-sensitized device is likely due to a positive shift of the valence band edge of the NiO by *ca.* 30–40 mV, induced by the adsorption of the dye sensitizer, as can be observed by comparing the different voltages at which the similar capacitances are found, rather than by slower interfacial recombination kinetics. The voltage shift observed for **SK4** in EIS under illumination is quantitatively consistent with the more positive onset of the anodic current observed for this moiety in the dark (Figure S6, Supplementary Materials).

3. Experimental Section

3.1. General Remarks

NMR spectra were recorded with an instrument operating at 500.13 (^{1}H) and 125.77 MHz (^{13}C). Flash chromatography was performed with silica gel 230–400 mesh (60 Å). Reactions conducted under a nitrogen atmosphere were performed in oven-dried glassware and monitored by thin-layer chromatography by using UV light (254 and 365 nm) as a visualizing agent. 4-[Bis-(4-bromo-phenyl)-amino]-benzaldehyde and 4,4'-(4-bromophenylazanediyl)dibenzaldehyde were synthesised following procedure known in literature [44,45]. All other reagents were obtained from commercial suppliers at the highest purity and used without further purification. Anhydrous solvents were purchased from commercial suppliers and used as received. Extracts were dried with anhydrous Na_2SO_4 and filtered before removal of the solvent by evaporation.

3.2. Synthesis of the p-Type Sensitizers

3.2.1. 4-[Bis-(4-bromo-phenyl)-amino]-benzoic Acid (2)

A procedure reported in the literature was adapted with some modifications [13]. The suspension of sodium hydroxide (62 mmol, 2.5 g) in 45 mL of ethanol was vigorously stirred for 20 min and then silver (I) oxide (9.5 mmol, 2.19 g). Resulting mixture was stirred for another 20 min before 4-[bis-(4-bromo-phenyl)-amino]-benzaldehyde (1) (1.2 mmol, 550 mg) in 4 mL of toluene was slowly added. The resulting suspension was stirred at room temperature for 24 h and then neutralized with HCl (15%, 25 mL) until formation of grey silver chloride precipitate was observed. The resulting solution was decanted and extracted with ethyl acetate (3 × 40 mL). The organic phase was washed with water and dried. Thereafter the solvent was removed by rotary evaporation giving the desired acid 2 (0.8 g, 78%) as white solid, which was used without further purification. ^{1}H NMR (500 MHz, DMSO-d_6): δ 6.90 (d, J = 8.5 Hz, 2H), 6.93 (d, J = 8.8 Hz, 4H), 7.44 (d, J = 8.8 Hz, 4H), 7.79 (d, J = 8.5 Hz, 2H).

3.2.2. 4-{Bis-[4-(5-formyl-thiophen-2-yl)-phenyl]-amino}-benzoic Acid (3)

The solution of benzoic acid 2 (0.4 mmol, 200.0 mg) and Pd(dppf)Cl$_2$ (0.1 mmol, 79 mg) in toluene (5 mL) was mixed with the solution of 5-formyl-2-thienylboronic acid (1.2 mmol, 187 mg) and K$_2$CO$_3$ (2.5 mmol, 355 mg) in methanol (5 mL). The resulting mixture was irradiated in the microwave oven at 80 °C for 20 min and then poured into a saturated solution of ammonium chloride (60 mL). After extraction with ethyl acetate (3 × 100 mL), the combined organic phases were washed with brine, dried and then filtered over Celite and concentrated under reduced pressure. The crude residue was purified by column chromatography over silica gel using dichloromethane (DCM)/methanol 1:9 mixture as the eluent to give the product (145 mg, 62%) in form of yellow solid. m.p. 137–139 °C, ^{1}H NMR (500 MHz, acetone-d_6): δ 7.21 (d, J = 8.8 Hz, 2H), 7.28 (d, J = 8.7 Hz, 4H), 7.66 (d, J = 4.0 Hz, 2H), 7.83 (d, J = 8.7 Hz, 4H), 7.98 (d, J = 4.0 Hz, 2H), 8.00 (d, J = 8.8 Hz, 2H), 9.94 (s, 2H), 11.14 (s, 1H).

3.2.3. 4-(Bis(4-(5-((E)-2-(4-cyano-5-(dicyanomethylene)-2,2-dimethyl-2,5-dihydrofuran-3-yl)vinyl) thiophen-2-yl)phenyl)amino)benzoic Acid (SK2)

To a solution of the intermediate 4 (100 mg, 0.2 mmol) and 2-(3-cyano-4,5,5-trimethylfuran-2 (5H)-ylidene)malononitrile (77.2 mg, 0.4 mmol) in ethanol (20 mL) was added catalytic amount of ammonium acetate (8 mg, 0.1 mmol) in 2 mL of acetic acid. Reaction mixture was refluxed under nitrogen atmosphere for 100 h. Afterwards, organic solvent was evaporated under reduced pressure and resulted crude product was washed with ethyl acetate to give the desired dye SK2 (75 mg, 74%) in form of dark violet solid. m.p. 252–253 °C. ^{1}H NMR (500 MHz, DMSO-d_6): δ 12.75 (s, 1H), 8.17 (d, J = 16 Hz, 2H), 7.92 (d, J = 8 Hz, 2H), 7.87 (d, J = 4 Hz, 2H), 7.84 (d, J = 8.0 Hz, 4H), 7.73 (d, J = 4 Hz, 2H), 7.2 (d, J = 9 Hz, 4H), 7.16 (d, J = 9 Hz, 2H), 6.83 (d, J = 16 Hz, 2H), 1.81 (s, 12 H). ^{13}C NMR (acetone-$d6$): δ 26.7, 54.9, 98.3, 99.2, 100.2, 112.4, 113.4, 114.2, 114.3, 123.9, 126.5, 127.2, 129, 129.6, 132.5,

139.5, 140.3, 141.7, 148.2, 151.3, 152.2, 168.1, 175.9, 178.2. $\nu_{max}(cm^{-1})$ 3583, 3476, 3007, 2310, 1738, 1590, 1373, 1217. Found: C, 70.11; H, 4.19; N, 11.60. $C_{51}H_{33}N_7O_4S_2$ requires: C, 70.25; H, 3.81; N, 11.24.

3.2.4. 4-(5-Bromothiophen-2-yl)-7-(thiophen-2-yl)benzo[1,2,5]thiadiazole (5)

To a solution of 4,7-di-2-thienyl-2,1,3-benzothiadiazole (4) [46] (350 mg, 1.17 mmol) in a 100 mL of acetic acid-CH_2Cl_2 1:1 mixture, was added N-bromosuccinimide (NBS) (229 mg, 1.28 mmol) in DCM by small portions over 30 min. The resulting reaction mixture was stirred overnight, then poured into water (100 mL) and the separated organic layer was dried, filtered and concentrated in vacuum. Obtained crude product was purified by flash column chromatography on silica gel (DCM-hexane = 1:4) giving a target benzothiadiazole derivative (5) in form of bright orange solid with 70% yield (310 mg, 0.8mmol). ^{1}H NMR (500 MHz, CDCl$_3$): δ 8.12 (d, 1H,), 7.88 (d, 1H), 7.8 (m, 2H), 7.46 (d, 1H), 7.22 (d, 1H), 7.15 (d, 1H).

3.2.5. 4-(5-(4,4,5,5-Tetramethyl-1,3,2-dioxaborolan-2-yl)thiophen-2-yl)-7-(thiophen-2-yl)benzo[c][1,2,5]thiadiazole (6)

Schlenk flask was charged with 4,7-di(thiophen-2-yl)benzo[c][1,2,5]thiadiazole (5) (550 mg, 1.45 mmol), potassium acetate (440 mg, 4.48 mmol) and bis(pinacolato)diboron (490 mg, 1.93 mmol) together with the catalytic amount of Pd(dppf)Cl$_2$(0.05 mmol, 39 mg) and 15 mL of dry dioxane. Resulting mixture was heated at reflux under argon atmosphere for 24 h then poured in water, extracted with ethyl acetate (3 × 50 mL) and dried. After organic solvent was evaporated under reduced pressure, crude product was purified on silica gel using CH_2Cl_2-cyclohexane 1:4 as eluent. Product was obtained in form of brownish orange solid (318 mg, 55% yield). m.p. 168 °C–170 °C. ^{1}H NMR (500 MHz, CDCl$_3$): δ = 8.18 (d, $J = 4$ Hz, 1H) 8.13 (m, 1H), 7.92 (d, $J = 8$ Hz, 1H), 7.87 (d, $J = 8$ Hz, 1H), 7.72 (d, $J = 3$ Hz, 1H), 7.45 (m, 1H), 7.21 (m, 1H), 1.38 (s, 12H). ^{13}C NMR (CDCl$_3$): δ = 152.66 (CH), 152.63 (CH), 145.89 (CH), 139.32 (CH), 137.89 (C), 128.65 (C), 128.03 (C), 127.67 (C), 126.97 (C), 126.48 (CH), 126.37 (C), 125.83 (CH), 125.75 (C), 84.26 (CH3), 24.79 (C) ppm. Found C, 56.39; H, 4.60; N, 6.29. $C_{20}H_{19}BN_2O_2S_3$ requires: C, 56.34; H, 4.49; N, 6.57.

3.2.6. 4,4'-(4-(5-(7-(Thiophen-2-yl)benzo[c][1,2,5]thiadiazol-4-yl)thiophen-2-yl)phenylazanediyl) dibenzaldehyde (7)

4,4'-(4-Bromophenylazanediyl)dibenzaldehyde (114 mg, 0.3 mmol), benzothiadaizole derivative 6 (130 mg, 0.31 mmol), a catalytic amount of Pd(dppf)Cl$_2$ (0.019 mmol, 18 mg) and a solution of K$_2$CO$_3$ (400 mg, 2.9 mmol) in water (3 mL) were charged into Schlenk flask and then dissolved in 15 mL of dry dimethoxyethane (DME). Reaction mixture was refluxed under argon atmosphere for 15 h and after cooling down to the room temperature it was washed with water, dried and concentrated under vacuum. Crude product was purified by column chromatography on silica gel using ethyl acetate-cyclohexane = 2:7 as the eluent. The pure intermediate 7 (55 mg, 30% yield) was obtained in form of bright red solid. m.p. 173 °C–175 °C. ^{1}H NMR (500 MHz, CDCl$_3$): δ = 9.91 (s, 2H), 8.12 (dd, $J = 8.0$; 15 Hz, 2H), 7.86 (s, 1H), 7.81 (d, $J = 17.0$ Hz, 4H), 7.68 (d, $J = 17.0$ Hz, 2H), 7.45 (d, $J = 8.0$ Hz, 1H), 7.39 (d, $J = 8.0$ Hz, 1H), 7.24 (d, $J = 17.0$ Hz, 4H), 7.21 (d, $J = 8.0$ Hz, 1H), 7.17 (d, $J = 17.0$ Hz, 2H).^{13}C NMR (CDCl$_3$): δ= 191.4, 153.5, 153.4, 152.6, 145.8, 140.18, 139.9, 132.7, 132.4, 132.2, 129.5, 128.9, 128.5, 128.2, 127.89, 127.83, 126.9, 126.6, 126.5, 126.3, 125.2, 123.9 ppm. Found: C, 67.84; H, 3.63; N, 7.33. $C_{34}H_{21}N_3O_2S_3$ requires: C, 68.09; H, 3.53; N, 7.01.

3.2.7. 4-(Bis(4-(5-(7-(thiophen-2-yl)benzo[c][1,2,5]thiadiazol-4-yl)thiophen-2-yl)phenyl)amino) benzaldehyde (8)

4-[Bis-(4-bromo-phenyl)-amino]-benzaldehyde (1) (100 mg, 0.232 mmol) and benzothiadaizole derivative 6 (213 mg, 0.5 mmol) and K$_2$CO$_3$ (691 mg, 5 mmol)were dissolved in 5 mL of water and charged into Schlenk flask together with the catalytic amount of Pd(dppf)Cl$_2$ (0.02 mmol, 19 mg) and then dissolved in 15 mL of dry DME. Reaction mixture was refluxed under argon atmosphere for

15 h. After cooling to the room temperature it was washed with water, dried and concentrated under vacuum. Crude product was washed with ethyl acetate and filtered. The desired intermediate **8** was obtained as dark red solid (70 mg, yield 45%). m.p. 189 °C–190 °C (decomp.). ^{1}H NMR (500 MHz, CDCl$_3$): δ = 9.88 (s, 1H), 8.14 (d, J = 8.0, 4H), 7.9 (s, 1H), 7.77 (d, J = 17.0 Hz, 2H), 7.69 (d, J = 17.0 Hz, 4H), 7.47 (d, J = 8.0 Hz, 2H), 7.41 (d, J = 8.0 Hz, 2H), 7.23 (d, J = 17.0 Hz, 4H), 7.22 (d, J = 8.0 Hz, 2H), 7.19 (d, J = 17.0 Hz, 2H). ^{13}C NMR (CDCl$_3$): δ = 191.4, 153.6, 153.5, 153.4, 146.4, 145.3, 140.2, 139.5, 132.3, 131.7, 130.9, 129.6, 128.9, 128.4, 127.9, 127.8, 127.7, 127.0, 126.8, 126.6, 126.2, 124.9, 121.6 ppm. Found C 64.50; H 2.88; N 7.91. C$_{47}$H$_{27}$N$_5$OS$_6$ requires: C, 64.87; H, 3.13; N, 8.05.

3.2.8. 4,4′-(4-(5-(7-(Thiophen-2-yl)benzo[c][1,2,5]thiadiazol-4-yl)thiophen-2-yl)phenylazanediyl) dibenzoic Acid (**SK3**)

A suspension of sodium hydroxide (600 mg, 15 mmol) in 10 mL of ethanol was vigorously stirred for 15 min before Ag$_2$O (208 mg, 0.8 mmol) was added. After another 15 min, dibenzaldehyde **7** (50 mg, 0.08 mmol) in toluene was added dropwise to the solution. Reaction mixture was stirred overnight at room temperature and quenched by the slow addition of 1 M HCl solution, then it was extracted with CH$_2$Cl$_2$ (3 × 75 mL). Combined organic layers were washed with water and dried. After evaporation of solvent under reduced pressure resulting crude product was washed with diethyl ether to give final compound **SK3** in form of dark red solid (40 mg. 87% yield). m.p. 215 °C–217 °C. ^{1}H NMR (500 MHz, DMSO-$d6$): δ = 8.20 (t, J = 9 Hz, 2H), 8.17 (d, J = 8 Hz, 1H), 8.12 (d, J = 8.0 Hz, 1H), 7.92 (d, J = 8.7 Hz, 4H), 7.80 (d, J = 8.5 Hz, 2H), 7.78 (d, J = 5.1 Hz, 1H), 7.67 (d, J = 4.0 Hz, 1H), 7.29 (d, J = 8.7 Hz, 1H), 7.22 (d, J = 8.5 Hz, 2H) 7.15 (d, J = 8.7 Hz, 4H). ^{13}C NMR (DMSO-$d6$): δ = 168.1, 153.0, 152.9, 151.4, 146.7, 145.7, 139.8, 138.9, 132.5, 131.4, 130.1, 129.7, 129.5, 128.8, 128.3, 127.7, 127.2, 126.9, 126.4, 126.3, 126.0, 125.9, 124.0 ppm. ν$_{max}$ (cm^{-1}) 3456, 2970, 2281, 1738, 1494, 1366, 1217. Found: C, 64.36; H, 3.74; N, 6.94. C$_{34}$H$_{21}$N$_3$O$_4$S$_3$ requires: C, 64.64; H,3.35; N, 6.65.

3.2.9. 4-(*Bis*(4-(5-(7-(thiophen-2-yl)benzo[c][1,2,5]thiadiazol-4-yl)thiophen-2-yl)phenyl)amino) benzoic Acid (**SK4**)

To the vigorously stirred suspension of NaOH (300 mg, 7.5 mmol) in 10 mL of ethanol was added silver (I) oxide (205 mg, 0.8 mmol). Mixture was stirred for 15 min before adding benzaldehyde **8** (50 mg, 0.057 mmol) in CH$_2$Cl$_2$. Reaction mixture was stirred overnight at the room temperature and quenched by the slow addition of 1 M HCl solution, extracted with ethyl acetate (3 × 75 mL) and then combined organic layers were washed with water and dried. After evaporation of solvents, crude product was washed with dichloromethane. **SK4** dye was obtained in form of black-red solid (40 mg, 90% yield). m.p. > 285 °C (decomp.). ^{1}H NMR (500 MHz, CDCl$_3$): δ = 8.22 (d, J = 3.9 Hz, 2H), 8.20 (d, J = 3.9 Hz, 2H), 8.18 (s, 2H), 8.14 (d, J = 8.3 Hz, 2H), 7.88 (d, J = 8.7 Hz, 2H), 7.81 (d, J = 8.5 Hz, 4H), 7.79 (d, J = 5.5 Hz, 2H), 7.68 (d, J = 4.1 Hz, 2H), 7.31 (t, J = 8.8 Hz, 2H), 7.23 (d, J = 8.5 Hz, 4H), 7.11 (d, J = 8.7 Hz, 2H). ^{13}C NMR (DMSO-d_6): δ = 169.07, 153.5, 153.3, 151.8, 151.7, 146.97, 145.8, 142.2, 132.0, 131.5, 130.8, 129.6, 128.9, 128.2, 127.8, 127.7, 126.9, 126.7, 126.6, 126.3, 126.1, 124.6, 122.3 ppm. ν$_{max}$ (cm^{-1}) 3472, 2976, 2110, 1738, 1546, 1372, 1254. Found C 63.88, H 3.44, N 7.68. C$_{47}$H$_{27}$N$_5$O$_2$S$_6$ requires: C, 63.70; H, 3.07; N, 7.90.

3.3. Solar Cell Assembly

NiO films were prepared by grounding 7.5 g of NiO nanoparticles (Inframat, nominal size 25 nm) in a mortar in the presence of 200 mL of acetilacetone. 100 mL of terpineol were then added, followed by 10 g of ethlycellulose dissolved in 110 mL of absolute ethanol. The resulting mixture was homogenized by stirring and sonication in an ultrasonic bath. Ethanol was evaporated under reduced pressure leaving a dense paste constituted by terpineol, ethylcellulose and NiO nanoparticles. The resulting paste could be spread by blading onto well cleaned FTO glass to form, after drying and high temperature sintering, the NiO thin films. The temperature program adopted for drying and sintering in air was the following: 10 min at 120 °C, followed by a ramp (15 °C/min) to 450 °C which were maintained for 30 min. A subsequent ramp (10 °C/min) brought the temperature from 450 °C to 550 °C which were kept constant for further 20 min. Cooling occurred naturally, by interrupting the heating. Sensitization of the resulting NiO electrodes, having an active surface of 0.25 cm², was carried out overnight in the dark, by immersion in the DMSO solutions at the concentration of 2×10^{-4} M of the selected sensitizers. Solar cells were fabricated by clamping the sensitized NiO photocathode with a platinum coated FTO counter electrode (Chimet, Badia al Pino, Italy). The electrolyte was 1 M LiI/0.1 M I_2 in acetonitrile.

3.4. Spectroscopic, Electrochemical, and Photoelectrochemical Measurements

UV-Vis spectra of sensitized NiO thin films, collected in transmission mode, were recorded on a JASCO V570 UV-Vis-NIR spectrophotometer (Jasco Analytical instruments, Easton, MD, USA). UV-Vis spectra in solution were obtained with a V-570 JASCO spectrophotometer (Jasco Analytical instruments, Easton, MD, USA). Emission spectra were recorded FP6200JASCO spectrofluorometer. TBAlClO₄ 0.1 DMSO.

Cyclic voltammetry of the dye dissolved in 0.1 M solution of tetrabutylammonium perchlorate (Fluka, electrochemical grade, 99.0%) in anhydrous DMSO (Sigma-Aldrich) as the supporting electrolyte was carried out at a glassy carbon working electrode with a PARSTA2273 potentiostat in a two-chamber three-electrodeelectrochemical cell with a scan rate of 50 mVs⁻¹. Potentials are referred to the ferrocenium/ferrocene (Fc⁺/Fc) couple as internal reference. An Ecochemie PGSTAT 302/N electrochemical workstation equipped with the FRA 2 Frequency Response Analyzer and running under either GPES or Nova software was used for collecting both the JV characteristics and the EIS response of the p-DSSCs. Solar cells were illuminated under simulated solar conditions (Am 1.5 G 100 mW/cm²) generated by an ABET sun simulator. J/V curves were recorded by linear scan voltammetry, using a scan rate of 5 mV/s. Every cell was left to equilibrate under illumination until a superimposable J/V response was obtained upon subsequent scans. Curves reported in this work, representative of the average performance of three cells, are those measured after steady state was achieved. Potentiostatic EIS data were collected on a single cell at various voltages under illumination by applying a 10 mV sinusoidal frequency variable from 10⁵ to 10⁻² Hz. Impedance data were fitted with Zview.

IPCE data were collected under monochromatic illumination, having a band pass of 10 nm, generated by an Applied Photophysics monochromator coupled to a 175 W Luxtel Xe arc lamp. Short circuit photocurrents were recorded on an Agilent 34401A multimeter. Incident irradiance was measured with a calibrated Centronic OSD 7Q silicon photodiode having an active area of 1 cm².

3.5. Computational Studies

Optimized ground state geometry of the p-type dyes were obtained at the DFT-B3LYP level by using a 631G* basis set. The DFT calculation was carried out on pre-optimized structures at the PM6 level. Time dependent (TDDFT) calculations in the presence of DMSO solvent, treated as continuum polarizable medium, were carried out on the optimized structures by considering the two different B3LYP and BH & H functionals with the 6311 G* basis set incorporating also diffuse functions (+). All calculations were carried out with Gaussian 09 A02 [32] on multi-core computers. Structures and isodensity surfaces were graphically visualized with Gaussview 5. Electron Density Difference Maps (EDDMs) were generated with GaussSum 2.2 [47].

4. Conclusions

New p-type photosensitizers based on organic linear and di-branched donor–acceptor systems bearing tryphenylamine as a donor and strong electron-acceptor Dalton's (**SK2**) or benzothiadiazole (**SK3** and **SK4**) groups were synthesized and characterized by steady state spectroscopic, electrochemical, and computational studies. All the dyes under investigation exhibited strong charge transfer bands in the visible regions with ground and excited state energetic which are favorable to the sensitization of NiO electrodes. The computational investigation revealed a clear directionality of the lowest excited state exhibiting a marked charge transfer character, shifting the electron density from the donor core to the acceptor branches, an electronic situation that is favorable to the hole injection in p-type semiconductors such as NiO. However, the exciton binding energy, *i.e.*, the energy of the bound electron-hole pair, is about twice as large as that of a known literature standard (**P1**), suggesting that a more difficult charge separation in the new dyes might occur.

When tested as photosensitizers in p-type DSSCs, the **SK** series was able to successfully sensitize NiO electrodes, resulting in photocurrents that are about half that of **P1**-based cells. The charge recombination kinetics, probed by considering the charge transfer resistance at the NiO/electrolyte interface at a comparable chemical capacitance, showed that the dyes behaved similarly under this respect and that the higher photovoltage observed for the device based on **SK4** dye is seemingly due to a positive shift of the valence band edge, consistent with the shift in the anodic threshold measured in the dark. The fact that similar recombination resistances were found at a comparable photohole density in the NiO corroborates the indication, gained by DFT studies, that the superior performance, particularly in photocurrent, of **P1** may be ascribed to a superior charge separation capability originating by its smaller exciton binding energy. Future designs of dyes for p-type sensitization will take this parameter into consideration for achieving a more effective charge separation.

Acknowledgments: The authors thank the Ministero dell'Università e della Ricerca (MIUR-PRIN) (grants No. 2008CSNZFR and 20104XET32) for financial support.

Author Contributions: Norberto Manfredi and Alessandro Abbotto conceived and designed the molecules and the synthesis; Svitlana Karamshuk performed the synthesis; Matteo Salamone and Riccardo Ruffo designed and performed electrochemical analysis; Stefano Carli and Stefano Caramori performed the computational calculations; Stefano Caramori and Carlo-Alberto Bignozzi designed and prepared the photovoltaic devices.

Conflicts of Interest: The authors declare no conflict of interest.

References

1. O'Regan, B.; Gratzel, M. A low-cost, high-efficiency solar cell based on dye-sensitized colloidal TiO_2 films. *Nature* **1991**, *353*, 737–740. [CrossRef]

2. International Energy Agency (IEA). *Potential for Building Integrated Photovoltaics*; Report IEA PVPS Task 7; NET Ltd.: St. Ursen, Switzerland, 2002.

3. Giordano, F.; Guidobaldi, A.; Petrolati, E.; Vesce, L.; Riccitelli, R.; Reale, A.; Brown, T.M.; di Carlo, A. Realization of high performance large area Z-series-interconnected opaque dye solar cell modules. *Prog. Photovolt. Res. Appl.* **2013**, *21*, 1653–1658. [CrossRef]

4. Mathew, S.; Yella, A.; Gao, P.; Humphry-Baker, R.; CurchodBasile, F.E.; Ashari-Astani, N.; Tavernelli, I.; Rothlisberger, U.; NazeeruddinMd, K.; Grätzel, M. Dye-sensitized solar cells with 13% efficiency achieved through the molecular engineering of porphyrin sensitizers. *Nat. Chem.* **2014**, *6*, 242–247. [CrossRef] [PubMed]

5. Green, M.A. *Third Generation Photovoltaics: Advanced Solar Energy Conversion*; Springer-Verlag: Heidelberg, Germany, 2003.

6. Hagfeldt, A.; Boschloo, G.; Sun, L.; Kloo, L.; Pettersson, H. Dye-Sensitized Solar Cells. *Chem. Rev.* **2010**, *110*, 6595–6663. [CrossRef] [PubMed]

7. Lide, D.R. *Handbook of Chemistry and Physics*, 90th ed.; CRC Press: Boca Raton, FL, USA, 2010.

8. Odobel, F.; Pellegrin, Y.; Gibson, E.A.; Hagfeldt, A.; Smeigh, A.L.; Hammarström, L. Recent advances and future directions to optimize the performances of p-type dye-sensitized solar cells. *Coord. Chem. Rev.* **2012**, *256*, 2414–2423. [CrossRef]

9. Morandeira, A.; Boschloo, G.; Hagfeldt, A.; Hammarström, L. Coumarin 343-NiO Films as Nanostructured Photocathodes in Dye-Sensitized Solar Cells: Ultrafast Electron Transfer, Effect of the I^{3-}/I^- Redox Couple and Mechanism of Photocurrent Generation. *J. Phys. Chem. C* **2008**, *112*, 9530–9537. [CrossRef]

10. Zhu, H.; Hagfeldt, A.; Boschloo, G. Photoelectrochemistry of Mesoporous NiO Electrodes in Iodide/Triiodide Electrolytes. *J. Phys. Chem. C* **2007**, *111*, 17455–17458. [CrossRef]

11. Borgström, M.; Blart, E.; Boschloo, G.; Mukhtar, E.; Hagfeldt, A.; Hammarström, L.; Odobel, F. Sensitized Hole Injection of Phosphorus Porphyrin into NiO: Toward New Photovoltaic Devices. *J. Phys. Chem. B* **2005**, *109*, 22928–22934. [CrossRef] [PubMed]

12. Gibson, E.A.; Smeigh, A.L.; le Pleux, L.; Fortage, J.; Boschloo, G.; Blart, E.; Pellegrin, Y.; Odobel, F.; Hagfeldt, A.; Hammarström, L. A *p*-Type NiO-Based Dye-Sensitized Solar Cell with an Open-Circuit Voltage of 0.35 V. *Angew. Chem. Int. Ed.* **2009**, *48*, 4402–4405. [CrossRef] [PubMed]

13. Qin, P.; Zhu, H.; Edvinsson, T.; Boschloo, G.; Hagfeldt, A.; Sun, L. Design of an Organic Chromophore for *P*-Type Dye-Sensitized Solar Cells. *J. Am. Chem. Soc.* **2008**, *130*, 8570–8571. [CrossRef] [PubMed]

14. Abbotto, A.; Manfredi, N.; Marinzi, C.; de Angelis, F.; Mosconi, E.; Yum, J.H.; Zhang, X.X.; Nazeeruddin, M.K.; Gratzel, M. Di-branched di-anchoring organic dyes for dye-sensitized solar cells. *Energy Environ. Sci.* **2009**, *2*, 1094–1101. [CrossRef]

15. Abbotto, A.; Leandri, V.; Manfredi, N.; de Angelis, F.; Pastore, M.; Yum, J.H.; Nazeeruddin, M.K.; Gratzel, M. Bis-Donor-Bis-Acceptor Tribranched Organic Sensitizers for Dye-Sensitized Solar Cells. *Eur. J. Org. Chem.* **2011**, *2011*, 6195–6205. [CrossRef]

16. Manfredi, N.; Cecconi, B.; Abbotto, A. Multi-Branched Multi-Anchoring Metal-Free Dyes for Dye-Sensitized Solar Cells. *Eur. J. Org. Chem.* **2014**, *2014*, 7069–7086. [CrossRef]

17. Nattestad, A.; Mozer, A.J.; Fischer, M.K.R.; Cheng, Y.B.; Mishra, A.; Bauerle, P.; Bach, U. Highly efficient photocathodes for dye-sensitized tandem solar cells. *Nat. Mater.* **2010**, *9*, 31–35. [CrossRef] [PubMed]

18. Zhang, X.L.; Zhang, Z.; Chen, D.; Bauerle, P.; Bach, U.; Cheng, Y.-B. Sensitization of nickel oxide: Improved carrier lifetime and charge collection by tuning nanoscale crystallinity. *Chem. Commun.* **2012**, *48*, 9885–9887. [CrossRef] [PubMed]

19. Li, L.; Gibson, E.A.; Qin, P.; Boschloo, G.; Gorlov, M.; Hagfeldt, A.; Sun, L. Double-Layered NiO Photocathodes for *p*-Type DSSCs with Record IPCE. *Adv. Mater.* **2010**, *22*, 1759–1762. [CrossRef] [PubMed]

20. Qin, P.; Linder, M.; Brinck, T.; Boschloo, G.; Hagfeldt, A.; Sun, L. High Incident Photon-to-Current Conversion Efficiency of *p*-Type Dye-Sensitized Solar Cells Based on NiO and Organic Chromophores. *Adv. Mater.* **2009**, *21*, 2993–2996. [CrossRef]

21. Zhu, L.; Yang, H.B.; Zhong, C.; Li, C.M. Rational design of triphenylamine dyes for highly efficient *p*-type dye sensitized solar cells. *Dyes Pigment.* **2014**, *105*, 97–104. [CrossRef]

22. Yen, Y.-S.; Chen, W.-T.; Hsu, C.-Y.; Chou, H.-H.; Lin, J.T.; Yeh, M.-C.P. Arylamine-Based Dyes for *p*-Type Dye-Sensitized Solar Cells. *Org. Lett.* **2011**, *13*, 4930–4933. [CrossRef] [PubMed]

23. Melikian, G.; Rouessac, F.P.; Alexandre, C. Synthesis of Substituted Dicyanomethylendihydrofurans. *Synth. Commun.* **1995**, *25*, 3045–3051. [CrossRef]

24. Yassar, A.; Videlot, C.; Jaafari, A. Synthesis and photovoltaic properties of mono-substituted quaterthiophenes bearing strong electron-withdrawing group. *Sol. Energy Mater. Sol. Cells* **2006**, *90*, 916–922. [CrossRef]

25. Barrette, W.C.; Johnson, H.W.; Sawyer, D.T. Voltammetric evaluation of the effective acidities (pKa') for Broensted acids in aprotic solvents. *Anal. Chem.* **1984**, *56*, 1890–1898. [CrossRef] [PubMed]

26. Bockris, J.O.M.; Khan, S.U.M. *Surface Electrochemistry—A Molecular Level Approach*; Kluwer Academic/Plenum Publishers: New York, NY, USA, 1993.

27. Scrascia, A.; Pastore, M.; Yin, L.; Anna Picca, R.; Manca, M.; Guo, Y.-C.; de Angelis, F.; Della Sala, F.; Cingolani, R.; Gigli, G.; *et al.* Organic Dyes Containing A Triple Bond Spacer for Dye Sensitized Solar Cells: A Combined Experimental and Theoretical Investigation. *Curr. Org. Chem.* **2011**, *15*, 3535–3543. [CrossRef]

28. Mba, M.; D'Acunzo, M.; Salice, P.; Carofiglio, T.; Maggini, M.; Caramori, S.; Campana, A.; Aliprandi, A.; Argazzi, R.; Carli, S.; *et al.* Sensitization of Nanocrystalline TiO$_2$ with Multibranched Organic Dyes and Co(III)/(II) Mediators: Strategies to Improve Charge Collection Efficiency. *J. Phys. Chem. C* **2013**, *117*, 19885–19896. [CrossRef]

29. Ji, Z.Q.; Natu, G.; Huang, Z.J.; Wu, Y.Y. Linker effect in organic donor-acceptor dyes for *p*-type NiO dye sensitized solar cells. *Energy Environ. Sci.* **2011**, *4*, 2818–2821. [CrossRef]

30. Barolo, C.; Nazeeruddin, M.K.; Fantacci, S.; Di Censo, D.; Comte, P.; Liska, P.; Viscardi, G.; Quagliotto, P.; de Angelis, F.; Ito, S.; *et al.* Synthesis, Characterization, and DFT-TDDFT Computational Study of a Ruthenium Complex Containing a Functionalized Tetradentate Ligand. *Inorg. Chem.* **2006**, *45*, 4642–4653. [CrossRef] [PubMed]

31. Qin, P.; Wiberg, J.; Gibson, E.A.; Linder, M.; Li, L.; Brinck, T.; Hagfeldt, A.; Albinsson, B.; Sun, L.C. Synthesis and Mechanistic Studies of Organic Chromophores with Different Energy Levels for *p*-Type Dye-Sensitized Solar Cells. *J. Phys. Chem. C* **2010**, *114*, 4738–4748. [CrossRef]

32. Petersson, G.A.; Nakatsuji, H.; Caricato, M.; Li, X.; Hratchian, H.P.; Izmaylov, A.F.; Bloino, J.; Zheng, G.; Sonnenberg, J.L.; Hada, M.; *et al. Gaussian 09, revision A.02.*; Gaussian, Inc.: Wallingford, CT, USA, 2009.

33. Autschbach, J. Charge-Transfer Excitations and Time-Dependent Density Functional Theory: Problems and Some Proposed Solutions. *Chemphyschem* **2009**, *10*, 1757–1760. [CrossRef] [PubMed]

34. Ziegler, T.; Seth, M.; Krykunov, M.; Autschbach, J.; Wang, F. Is charge transfer transitions really too difficult for standard density functionals or are they just a problem for time-dependent density functional theory based on a linear response approach. *J. Mol. Struct. THEOCHEM* **2009**, *914*, 106–109. [CrossRef]

35. Becke, A.D. Density-functional thermochemistry. III. The role of exact exchange. *J. Chem. Phys.* **1993**, *98*, 5648–5652. [CrossRef]

36. Becke, A.D. A new mixing of Hartree-Fock and local density-functional theories. *J. Chem. Phys.* **1993**, *98*, 1372–1377. [CrossRef]

37. Koch, W.; Holthausen, M. *A Chemist's Guide to Density Functional Theory*, 2nd ed.; Wiley-VCH: Weinheim, Germany, 2001.

38. Gregg, B.A. Excitonic Solar Cells. *J. Phys. Chem. B* **2003**, *107*, 4688–4698. [CrossRef]

39. Zhen, C.-G.; Becker, U.; Kieffer, J. Tuning Electronic Properties of Functionalized Polyhedral Oligomeric Silsesquioxanes: A DFT and TDDFT Study. *J. Phys. Chem. A* **2009**, *113*, 9707–9714. [CrossRef] [PubMed]

40. Kim, B.-G.; Zhen, C.-G.; Jeong, E.J.; Kieffer, J.; Kim, J. Organic Dye Design Tools for Efficient Photocurrent Generation in Dye-Sensitized Solar Cells: Exciton Binding Energy and Electron Acceptors. *Adv. Funct. Mater.* **2012**, *22*, 1606–1612. [CrossRef]

41. Choi, H.; Baik, C.; Kang, S.O.; Ko, J.; Kang, M.-S.; Nazeeruddin, M.K.; Grätzel, M. Highly Efficient and Thermally Stable Organic Sensitizers for Solvent-Free Dye-Sensitized Solar Cells. *Angew. Chem. Int. Ed.* **2008**, *47*, 327–330. [CrossRef] [PubMed]

42. Bisquert, J.; Fabregat Santiago, F. Impedance spectroscopy: A general introduction and application to dye-sensitized solar cells. In *Dye Sensitized Solar Cells*; Kalyanasundaram, K., Ed.; CRC Press: Boca Raton, FL, USA, 2010.

43. Bisquert, J. Chemical capacitance of nanostructured semiconductors: Its origin and significance for nanocomposite solar cells. *Phys. Chem. Chem. Phys.* **2003**, *5*, 5360–5364. [CrossRef]

44. Lefebvre, J.F.; Sun, X.Z.; Calladine, J.A.; George, M.W.; Gibson, E.A. Promoting charge-separation in *p*-type dye-sensitized solar cells using bodipy. *Chem. Commun.* **2014**, *50*, 5258–5260. [CrossRef] [PubMed]

45. Yang, Z.; Zhao, N.; Sun, Y.; Miao, F.; Liu, Y.; Liu, X.; Zhang, Y.; Ai, W.; Song, G.; Shen, X.; *et al.* Highly selective red- and green-emitting two-photon fluorescent probes for cysteine detection and their bio-imaging in living cells. *Chem. Commun.* **2012**, *48*, 3442–3444. [CrossRef] [PubMed]

46. Lee, D.H.; Lee, M.J.; Song, H.M.; Song, B.J.; Seo, K.D.; Pastore, M.; Anselmi, C.; Fantacci, S.; de Angelis, F.; Nazeeruddin, M.K.; *et al.* Organic dyes incorporating low-band-gap chromophores based on π-extended benzothiadiazole for dye-sensitized solar cells. *Dyes Pigments* **2011**, *91*, 192–198. [CrossRef]

47. O'Boyle, N.M.; Tenderholt, A.L.; Langner, K.M. cclib: A library for package-independent computational chemistry algorithms. *J. Comput. Chem.* **2008**, *29*, 839–845. [CrossRef] [PubMed]

12

Reconfiguration of Urban Photovoltaic Arrays Using Commercial Devices

Sergio Ignacio Serna-Garcés [1,*], Juan David Bastidas-Rodríguez [2] and Carlos Andrés Ramos-Paja [3]

Academic Editor: Narottam Das

[1] Departamento de Electrónica y Telecomunicaciones, Instituto Tecnológico Metropolitano, Cra. 31 No. 54-10, Medellín 050013, Colombia
[2] Escuela de Ingenierías Eléctrica, Electrónica y de Telecomunicaciones, Universidad Industrial de Santander, Cra. 27 Calle 9, Bucaramanga 680002, Colombia; jdbastir@uis.edu.co
[3] Departamento de Energía Eléctrica y Automática, Universidad Nacional de Colombia, Cra. 80 No. 65-223, Medellín 050041, Colombia; caramosp@unal.edu.co
* Correspondence: sergioserna@itm.edu.co

Abstract: A recent approach to mitigate the adverse effects of photovoltaic (PV) arrays operating under mismatching conditions is the dynamic electrical reconfiguration of the PV panels. This paper introduces a procedure to determine the best configuration of a PV array connected in a series-parallel structure without using complex mathematical models. Such a procedure uses the experimental current *vs.* voltage curves of the PV panels, which are composed of multiple PV modules, to construct the power *vs.* voltage curves of all of the possible configurations to identify the optimal one. The main advantage of this method is the low computational effort required to reconstruct the power *vs.* voltage curves of the array. This characteristic enables one to implement the proposed solution using inexpensive embedded devices, which are widely adopted in industrial applications. The proposed method, and its embedded implementation, were tested using a hardware-in-the-loop simulation of the PV system. Finally, the real-time operation and benefits of the proposed solution are illustrated using a practical example based on commercial devices.

Keywords: optimal configuration; reconfiguration; series-parallel arrays; embedded systems; partial shading; mismatching conditions; hardware in the loop

1. Introduction

Urban photovoltaic (PV) systems are formed, in general, by a few PV panels driven by a grid-connected inverter. Moreover, the PV panels are connected in a series-parallel (SP) configuration [1–3], which enables one to reach the large DC voltages required by the inverter. Despite the existence of other array configurations, such as total cross-tied (TCT), bridge-linked (BL) or honey-comb (HC) [4], the SP configuration is the most widely adopted in commercial systems due to its simplicity and reduced number of connections. In such a way, the manufacturers of commercial inverters commonly describe the device ratings in terms of SP arrays, e.g., see Sunny Boy 1200 [5] or the Solar Edge SE6000 datasheets [6]. The PV inverter operates to achieve two objectives: deliver the PV power to the grid with an unitary power factor and, at the same time, track the optimal operating condition to achieve the maximum power production in the PV array [7,8].

In uniform conditions, *i.e.*, all of the PV panels exhibiting the same parameters, temperature and solar irradiance, the electrical characteristics of the PV array are proportional to the ones of any panel, just scaled in voltage by the number of panels in series and scaled in current by the number of strings in parallel. Hence, there exists a single optimal operating condition despite the connection between the PV panels [9–11]. However, if some (or all) PV panels exhibit different parameters, e.g., different irradiance due to shade, the PV array produces a different maximum power depending on the electrical connections between the panels. Moreover, in those mismatched conditions, the PV array exhibits multiple maximum power points (MPP) instead of a single one, as in uniform conditions [12–14].

Mismatched conditions commonly occur in urban PV installations due to shade generated by objects adjacent to the array, e.g., trees, buildings, posts, other PV arrays, *etc.* Figure 1 shows some commercial PV systems affected by urban shade: since the Sun moves along the day, a PV panel can be fully irradiated, partially shaded or completely shaded, depending on the hour of the day.

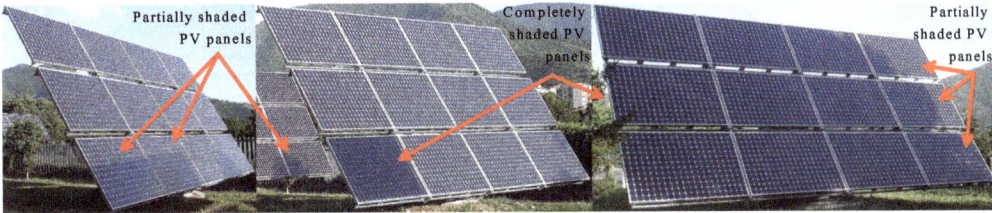

Figure 1. Shading profiles affecting commercial PV systems in urban environments.

The reconfiguration of PV systems consists of modifying the electrical connections between the panels to increase the power production of a PV array operating under mismatched conditions [15,16]. Such a technique has been analyzed in the literature concerning the reconfiguration of PV modules [17,18], which are the basic units forming the commercial PV panels. However, taking into account that commercial panels are formed by two or more PV modules connected in series, e.g., the BP585 PV panel, this paper focuses on a more realistic scenario: reconfiguring panels instead of modules. In addition, a large amount of solutions use analytical models to evaluate the best set of connections for a given shading profile [19–22]. Nevertheless, since costly processing devices, such as full-size computers, must be avoided in commercial applications to improve the profitability of the system, this paper is aimed at implementing a reconfiguration algorithm in commercial-friendly (and inexpensive) embedded systems, like FPGAs or DSPs. Therefore, the algorithm proposed in this paper is based on simple manipulations of experimental measurements to avoid the adoption of complex computational models, providing a fast and accurate solution.

The paper is organized as follows. Section 2 presents the basic concepts of PV systems' reconfiguration. Then, Section 3 presents the proposed solution to reconfigure the connections between PV panels based on experimental data. The simulation platform used to evaluate the proposed method is presented in Section 4, and the performance of this solution is illustrated in Section 5. Finally, the conclusions close the paper.

2. Reconfiguration of PV Systems

In an SP array, multiple panels are connected in series to form strings, and the strings are connected in parallel to form the array. Considering that each panel is formed by two or more modules connected in series, a string is composed of a number of series-connected modules. Moreover, each module has a diode connected in antiparallel, named the bypass diode, which is used to protect the cells forming the module. The bypass diode provides an alternative path for the string current when a PV module is operating under mismatching conditions; thus, avoiding not only the dissipation of power in the cells of the mismatched PV module [23], but also the reduction of the current in the string [15].

Figure 2 illustrates the operation of the bypass diodes under uniform and mismatching conditions for a simple string of two PV modules. Under uniform conditions, the bypass diodes are reverse biased, and no current flows through them (see Figure 2a). If the module M_2 is shaded, its maximum current ($I_{pv2,max}$) is lower than the maximum current of M_1 ($I_{pv2,max}$). Then, if the string current (I_{st}) is lower than $I_{pv2,max}$ (Figure 2b), no current flows through the bypass diode of M_2; however, if $I_{st} > I_{pv2,max}$, the difference between I_{st} and $I_{pv2,max}$ flows through the bypass diode of M_2 to avoid the string current limitation and the power dissipation in the cells of M_2.

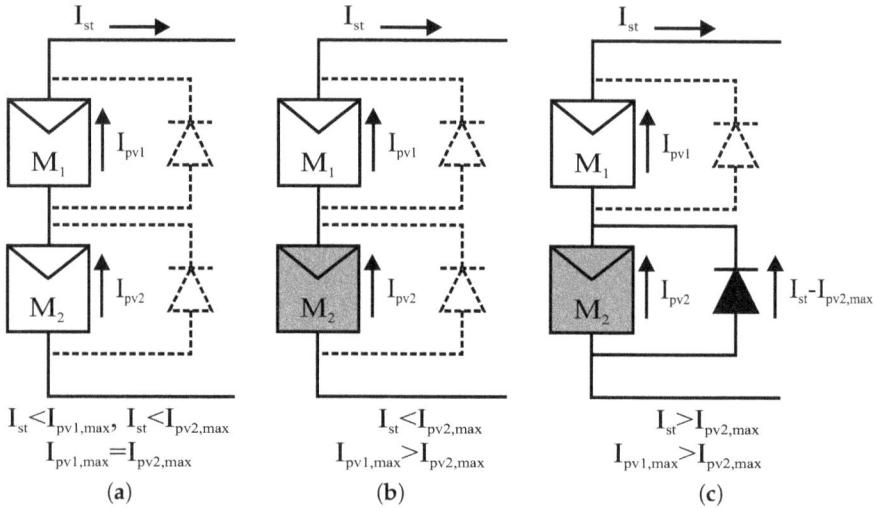

Figure 2. Series-connected PV modules with bypass diodes under uniform and mismatching conditions. (**a**) Uniform conditions: $I_{pv1,max} = I_{pv2,max}$; (**b**) Mismatching conditions: $I_{st} < I_{pv2,max}$; (**c**) mismatching conditions: $I_{st} > I_{pv2,max}$.

The activation and deactivation of the bypass diodes produce multiple maximum power points in the power *vs.* voltage (P-V) curves of the PV arrays [15,24,25]. Therefore, if the panels in an SP PV array are organized in different ways (*i.e.*, different configurations), it is possible to obtain different P-V curves from the same PV panels involved. Moreover, one of those configurations will have a global MPP higher than the other configurations.

The classical reconfiguration of PV systems consist of modifying the electrical connections between the modules of the PV array. In SP arrays, such a procedure consists of moving, electrically, a module from one string to another one. To illustrate this concept, Figure 3 shows a PV array formed by two strings, St_1 and St_2, where each string is formed by two modules. Therefore, four modules (M_1–M_4) form the PV array. Such modules can be connected in three different configurations CF_1, CF_2 and CF_3, as described in Figure 3. In this way, if the PV array originally operates in CF_1, the electrical connections between the modules can be modified to configure the array in CF_2 or CF_3. Such a procedure is performed using commercial controlled switches, e.g., relays or MOSFETs, which are organized in matrices to reduce the required space [26,27].

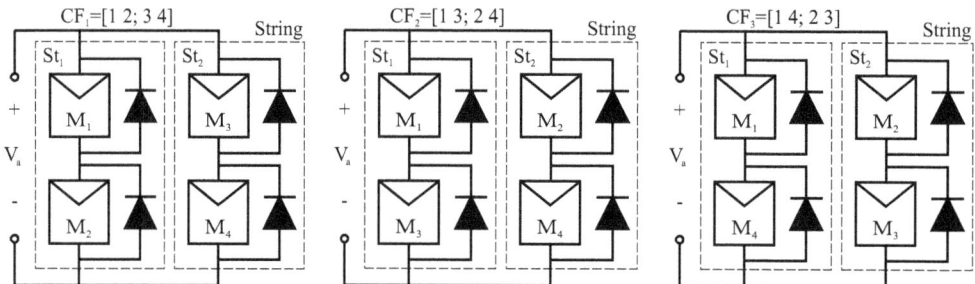

Figure 3. Possible configurations of a series-parallel (SP) array with two strings of two modules each.

Figure 4 shows the experimental current-voltage (I-V) curves of one module from a BP 2150S PV panel [28] operating under different shading conditions. Four BP 2150S PV panels under different mismatched conditions were used to simulate the three configurations CF_1, CF_2 and CF_3, obtaining the power curves presented at Figure 4: CF_1 and CF_3 produce almost the same maximum power (57.7 W at 21.84 V), while CF_2 produces 16.18% more power (67 W at 21.47 V) just by connecting the four modules in a different way. This example puts in evidence the advantages of PV reconfiguration systems: provide higher power in comparison with static arrays; it requires simple switching elements with negligible power losses; and its cost is low in comparison with distributed maximum power point techniques that require one power converter for each PV panel (e.g., the Power optimizer from Solar Edge [29]).

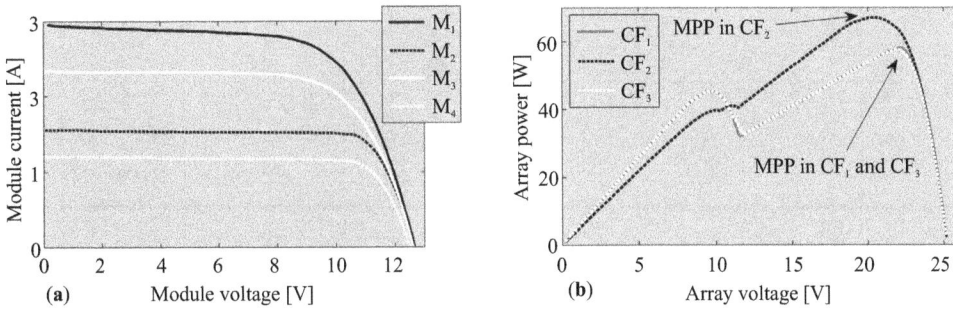

Figure 4. (a): I-V curves of M_1, M_2, M_3 and M_4. (b): P-V curves of CF_1, CF_2 and CF_3.

A large amount of reconfiguration solutions proposed in literature, e.g., [30], are focused on structures similar to the one reported in Figure 3, *i.e.*, reconfigure PV modules and not PV panels. However, commercial PV systems are constructed using PV panels formed by two or more PV modules; hence, the reconfiguration of modules using commercial devices is not possible. Therefore, the following section proposes a procedure to reconfigure PV panels based on experimental data.

3. Reconfiguration of Panels Based on Experimental Data

Several reconfiguration solutions are based on processing models to evaluate the array configuration that provides the highest power in the operating conditions to which the array is subjected, e.g., [15,31]. However, such an approach requires a large amount of calculations to estimate the maximum power of each configuration due to the complex non-linear systems required to model a PV array.

Instead, this paper proposes to reconstruct the I-V curves of the PV array from the I-V curves of the PV panels to ensure the identification of the best configuration. Therefore, it is required to acquire the I-V curve of each panel in the array by performing a current/voltage sweep to measure both the voltage and current. This process can be done by using DC/DC converters, as reported in [32]; however, that device could be complex and costly; hence, this paper proses a simpler solution.

In a similar way, some commercially-available PV optimizers, designed for small PV installations (4, 8, 16 or 24 panels), acquire the I-V curve of each panel in the array for reconfiguration and monitoring purposes. This is the case of the ENDANA PV optimizer [33]. Moreover, the ENDANA device also enables disconnecting PV panels from the array for either maintenance purposes or to exclude shading panels from a PV string. Therefore, it is possible to assume that the inclusion of a device for tracing the I-V curve of each panel and the inclusion of switches to dynamically change the array electrical connections do not significantly degrade the economic viability of small PV installations.

3.1. Structure of the Reconfiguration System

The structure of the proposed system is presented in Figure 5a. It is composed of the PV panels, a switches' matrix, a sweep device, an embedded controller and a commercial PV inverter. The switches' matrix, the sweep device and the controller are described in the following paragraphs.

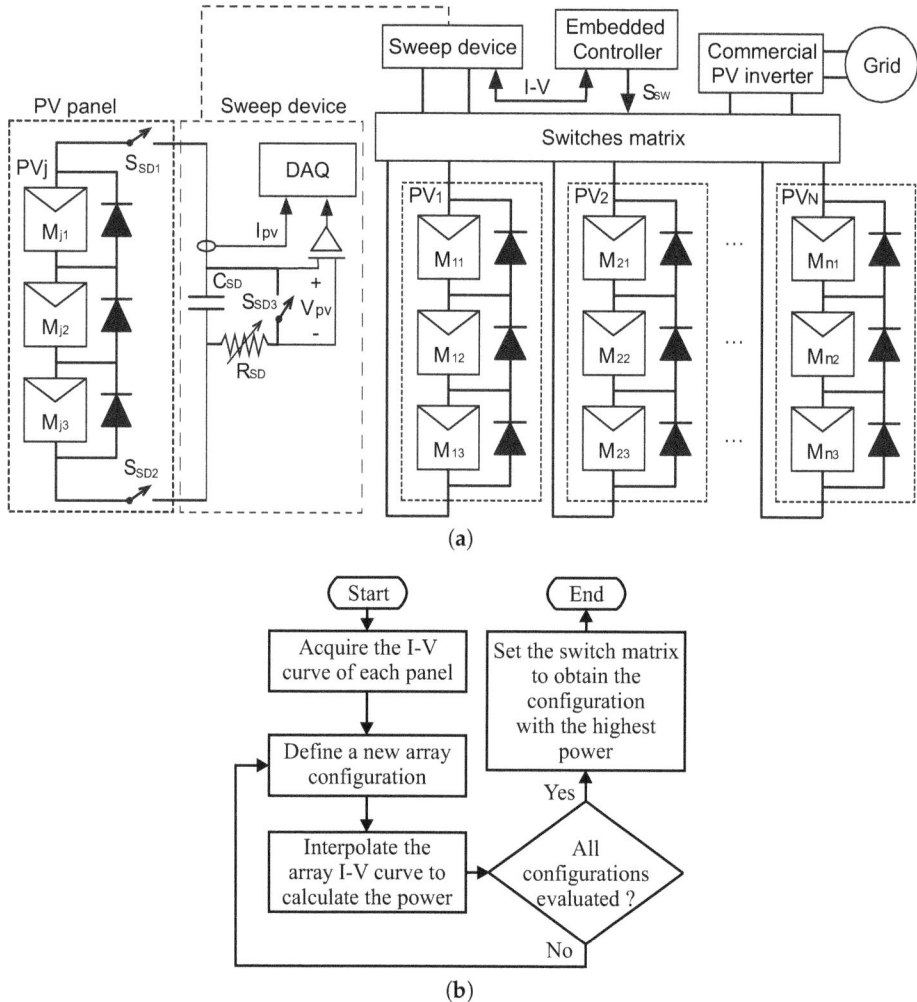

(a)

(b)

Figure 5. Proposed reconfiguration system. (**a**) General structure; (**b**) Controller algorithm.

The PV panels are connected to the switches' matrix, which is implemented by a set of interconnected relays to obtain low power losses, as reported in [15,21,33,34]. The switches' matrix has two main functions defined by the controller algorithm: connect each PV panel to some string of the array, and connect each PV panel to the sweep device.

The switches' matrix can be implemented in a centralized structure, as show in [15,21], or in a distributed structure, as proposed in [33]. In a centralized structure, each PV panel must be wired to the switches' matrix; therefore, it needs to be located close to the PV panels to avoid high ohmic losses and high installation costs caused by long cables. A distributed switches' matrix is composed of multiple units (remote units) located close to the PV panels. In this case, a set of PV panels (e.g., four PV panels in [33]) are connected to each remote unit, which is able to connect any PV panel to any string of the array or to an I-V tracing device located in a central unit ([33]). Hence, a distributed switches' matrix reduces considerably the cabling requirements and, as a consequence, the ohmic losses and the installation costs in comparison with a centralized matrix. The design and implementation of different switches matrices are reported in [15,21,34,35].

On the other hand, the circuit proposed to perform the voltage sweep (sweep device) in each panel only requires a capacitor, a set of switches, current/voltage sensors and an optional resistor to discharge the capacitor. Moreover, since the solution proposed in this paper is designed to be implemented in commercial-friendly (and inexpensive) embedded controllers, the voltage/current data from each panel are acquired using analog-to-digital converters (ADC) available in most embedded devices, such as DSP or FPGA.

The structure of the sweep device is depicted in Figure 5a: the capacitor C_{SD} is connected in parallel to the PV panel under test by means of the switches S_{SD1} and S_{SD2}. The capacitor C_{SD} starts discharging, since the switch S_{SD3} is closed at the beginning of the test. Then, switch S_{SD3} is set open, and the PV panel starts charging the capacitor from 0 V to the panel open-circuit voltage ($V_{OC,P}$), performing a sweep on the I-V curve. The control of the switches is performed by the embedded controller used to implement the reconfiguration algorithm. Moreover, the voltage and current data of the panel are acquired by means of ADCs available in that embedded controller. When the panel current is zero, switch S_{SD3} is closed to discharge the capacitor, so that a new test can be performed for another panel. The time required to perform a sweep of an I-V curve depends on the value of C_{SD} and on the panel current. Therefore, the longest sweep time is obtained at the lower irradiance condition. With the aim of providing an equation for the time required to perform the voltage sweep, the ideal single-diode model [36], given in Equation (1), is used. Taking into account that the PV current charges the capacitor, which in turn defines the PV voltage, the differential equation given in Equation (2) describes the evolution of the PV voltage.

$$i_{pv} = i_{SC} - A \times \exp\left(B \times v_{pv}\right) \tag{1}$$

$$\frac{d\,v_{pv}}{dt} = \frac{1}{C_{SD}} \times \left[i_{SC} - A \times \exp\left(B \times v_{pv}\right)\right] \tag{2}$$

Solving Equation (2) requires numerical methods and the identification of the parameters of the PV model; therefore, an estimation of the capacitor charge time is obtained by considering a PV current equal to $I_{SC,min}$, which corresponds to the minimum short-circuit current to be tested. In the light of such an assumption, the time required to perform a voltage sweep T_{SD} is approximated by Equation (3). It must be highlighted that Equation (3) must be used for selecting the value of C_{SD} and to set the acquisition time of the ADCs in the embedded controller. Therefore, expression Equation (3) is not used in the normal operation of the reconfiguration system.

$$T_{SD} = \frac{C_{SD} \times V_{OC,P}}{I_{SC,min}} \tag{3}$$

The discharge of the capacitor C_{SD} can be done by means of an additional resistor R_{SD} when switch S_{SD3} is closed. The design of R_{SD} depends on the desired discharge time T_{dis} and minimum C_{SD} voltage accepted to start a new test δv_{pv}, as given in Equation (4). A fast discharge can be imposed by reducing R_{SD} (even to a short-circuit) at the expense of increasing the stress in C_{SD}. However, C_{SD} can be implemented with a parallel array of multiple smaller capacitors to reduce the stress in each individual capacitor and, at the same time, to increase the reliability of the sweep device.

$$R_{SD} = -\frac{T_{dis}}{C_{SD} \times \ln\left(\delta v_{pv}/V_{OC,P}\right)} \tag{4}$$

Each panel of the PV array must be sequentially connected to the sweep device to obtain the I-V curve of all of the PV panels. Therefore, the time T_{sweep} required to acquire all of the I-V curves is given in Equation (5), where M stands for the number of strings in the array and N stands for the number of panels in each string; hence, $N \times M$ represents the number of panels in the array. Moreover, T_{SW} represents the time required to open and close switches S_{SD1}, S_{SD2} and S_{SD3} in the following sequence: S_{SD1} and S_{SD2} are closed, and S_{SD3} is open at the same time of performing the voltage sweep, while S_{SD1} and S_{SD2} are open and S_{SD3} is closed at the same time of discharging the capacitor. Therefore, T_{SW} is equal to the time required to open and close a switch. It must be clarified that S_{SD1} and S_{SD2} are part of the switches' matrix associated with each panel; hence, the connection to the sweep device is considered as an additional string in the switches' matrix.

$$T_{sweep} = (N \times M) \times (T_{SD} + T_{dis} + T_{SW}) \tag{5}$$

Finally, the embedded controller has two main functions: first, it processes the reconfiguration algorithm; and second, it coordinates the operation of the switches' matrix to set the new configuration and to trace the I-V curve of all of the panels using the sweep device. The controller algorithm is described in the flowchart of Figure 5b. It starts acquiring the I-V curve of all of the PV panels, then it defines an array configuration and use interpolation to generate the array I-V curve to identify the global maximum power point (GMPP). Such a process is repeated for all of the possible configurations to identify the one that provides the highest GMPP (best configuration). Subsequently, the embedded controller defines the state of each switch (open/closed) in the switches' matrix to set the best configuration in the PV array. This process is done using digital signals wired to the switches' control inputs. The vector S_{SW} represents those signals in Figure 5a.

The following subsections describe in detail the process to trace the I-V curve of each configuration, illustrating it with an application example.

3.2. Evaluation of Possible Array Configurations

To select the array configuration that produces the highest power, all of the possible options must be tested. Since, in general, every panel must be able to be connected to any string, the number of possible configurations (NCF) is given by the combination of all of the PV panels ($M \times N$) and the number of panels in each string (N), as shown in Equation (6).

$$NCF = \frac{(M \times N)!}{N!(M \times N - N)!} \tag{6}$$

Such a number grows exponentially for increments in the number of panels. For example, a PV array formed by two strings of four panels ($N = 4$ and $M = 2$) has 35 possible configurations, while a PV array formed by two strings of twelve panels ($N = 12$ and $M = 2$) has 1,352,078 possible configurations. Since the common approach in the literature is to use models to evaluate the effectiveness of a given configuration, e.g., [37–40], the evaluation of all of the possible configurations is a slow process. Hence, some algorithms have been proposed in the literature to search the best configuration without evaluating all of the possibilities, e.g., [41,42]; however, it is not easy to find a solution that ensures, by means of an analytical proof, that it reaches the optimal configuration. Therefore, the solution proposed in this paper is focused on a different approach: reduce, as much as possible, the time required to evaluate each configuration, which eventually reduces the time required to test all of the possible options. It is clear that testing all of the possibilities ensures the detection of the best configuration.

To speed-up the evaluation of each configuration, complex optimization algorithms and non-linear models must be avoided. Therefore, the proposed solution constructs the array I-V curve from the interpolation of the experimental data measured from the panels using few multiplications and additions, as illustrated in Figure 6: the I-V curve of each string is obtained by adding the voltage generated by the corresponding panels for the same current. This process requires the interpolation of the panel voltage if the requested current does not correspond to a current value measured by the sweep device.

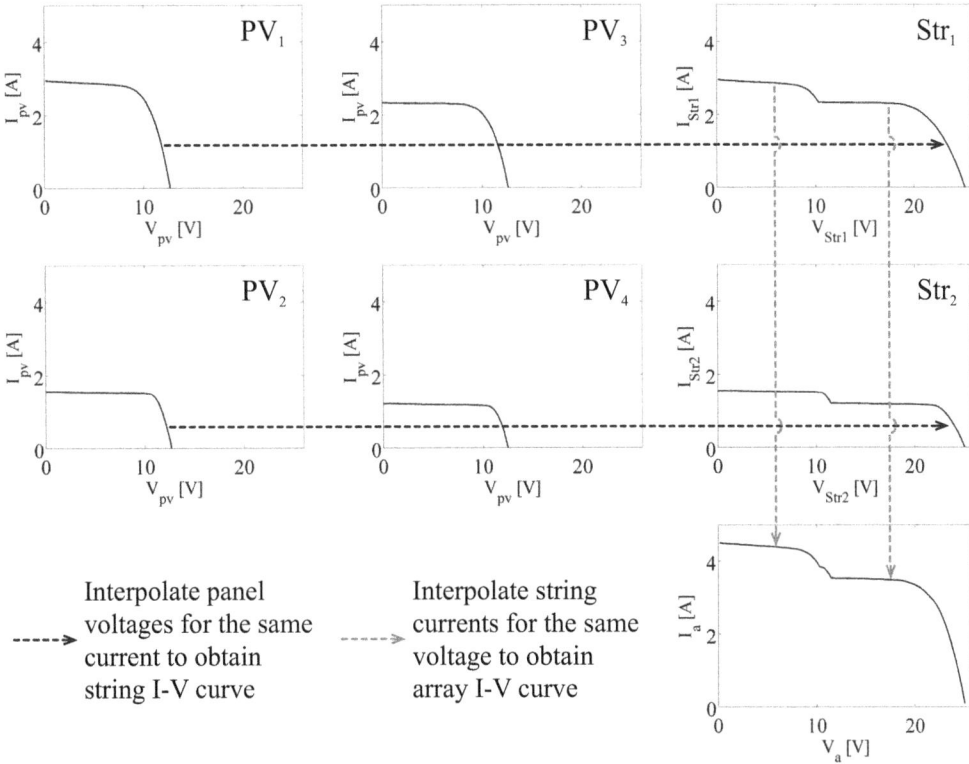

Figure 6. Example of the interpolation procedures to construct the I-V curve of one configuration for an SP array with two strings of two modules each.

Similarly, the I-V curve of the array is obtained by adding the current of each string generated at the same voltage. In this case, the interpolation of the string current is required if the requested voltage does not correspond to a voltage value available in the string data. Since commercial PV inverters have limitations in terms of input voltage, *i.e.*, minimum and maximum array voltage, the array I-V curve is constructed only within that range, which reduces the amount of data required to evaluate.

It is worth noting that the interpolation process is not required to evaluate all of the points of the string I-V curve, since some panels could exhibit the same current values for part of the curve. Similarly, if the voltages of the strings I-V curve match, no interpolation is required to construct the array I-V curve.

The interpolation of a single value requires four additions, one division and one multiplication Equation (7). The division instruction is included in the instructions set architecture (ISA) of some high-level processors; however, other simpler processors require implementing the division using multiplication instructions. Hence, the six floating-point instructions described in Equation (7) must be mapped to fixed-point and float-point instructions in the assembler, including data movement between registers and memory, additions, subtractions, data shift, logic operations, among others [43].

$$m = \frac{y_3 - y_1}{x_3 - x_1}$$

$$y_2 = m \times (x_2 - x_1) + x_1$$

(7)

To estimate the processing time of the reconfiguration algorithm, it is necessary to take into account all of the data points in each I-V curve, the number of panels in each string, the number of strings in the array and the number of possible configurations of the array. Since the reconfiguration algorithm includes both fixed and float point instructions, the performance of the algorithm cannot be measured in instructions; instead, it must be measured in clock cycles. This is because the number of clock cycles required by one fixed-point instruction is, in general, different from the number of clock cycles required by one floating-point instruction.

Nevertheless, to accurately estimate the processing time of the reconfiguration algorithm, it would be necessary to take into account the number of clock cycles required by other sub-processes in the embedded system: function invocation, handle vectors, decision instructions, control flow, communications, among others. However, the number of clock cycles of those sub-processes depends on the technology used to implemented the algorithm, which makes the construction of a mathematical expression difficult to calculate the processing time exactly. Therefore, an accurate evaluation of the processing time required by the reconfiguration algorithm, in therms of clock cycles, should be performed experimentally. This procedure is illustrated in the example presented in Section 5.

3.3. Application Example Based on Commercial Devices

To illustrate the proposed reconfiguration solution, a PV installation based on the commercial PV inverter Sunny Boy 1200 [5] is assumed. Such an inverter has a nominal DC voltage $V_{DC\_nom} = 120$ V, maximum DC voltage and current values equal to 400 V and 12.6 A, respectively, and minimum DC voltage equal to 100 V. The installation uses the commercial PV panels BP-3155 from British Petroleum [44], which are formed by three modules. Each PV panel has nominal MPP voltage (V_{mpp}) and MPP current (I_{mpp}) equal to 34.9 V and 4.5 A, respectively. Hence, to match the nominal requirements of the PV inverter, the PV array must be constructed using two strings of four panels each (eight panels, 24 modules), resulting in an operation voltage and current equal to 139.6 V and 9 A, respectively.

Figure 7 shows, at the top left, the I-V curves of the eight PV panels, which are based on the experimental I-V curves of BP 2150 PV panels [28] subjected to a solar irradiance equal to 625 W/m^2. The PV$_1$ panel is non-shaded; hence, its three modules exhibit the same electrical characteristics. The panel PV$_8$ is uniformly shaded; hence, its three modules also exhibit the same electrical characteristics. Instead, the modules PV$_2$–PV$_7$ are non-uniformly shaded; therefore, the I-V curves of such panels describe irregular electrical characteristics due to the activation of the bypass diodes of some modules.

Using samples of such I-V curves, the algorithm tests all possible configurations following the control algorithm given in Figure 5b. Such a procedure detects the following configuration as the best (CFB): first string as $[PV_1\ PV_2\ PV_3\ PV_6]$ and second string $[PV_4\ PV_5\ PV_7\ PV_8]$. That configuration provides 370.1 W at an array voltage $V_a = 110$ V, which is within the operation range of the PV inverter. To illustrate the power increment achieved by the proposed solution, the worst configuration (CFW) for this PV installation is given by the strings $[PV_1\ PV_2\ PV_3\ PV_7]$ and $[PV_4\ PV_5\ PV_6\ PV_8]$, providing 313.2 W at an array voltage $V_a = 135$ V. It is worth noting that the difference between CFB and CFW is only in the permutation of PV$_6$ and PV$_7$, but such a modification generates a difference of $\Delta\eta = 18.14\%$ in the power production. Similarly, if the default configuration of the array is different from CFB, the proposed reconfiguration system will optimize the panels' connections to improve the power delivered to the inverter, which eventually reduces the return-of-the-investment time by increasing the energy produced.

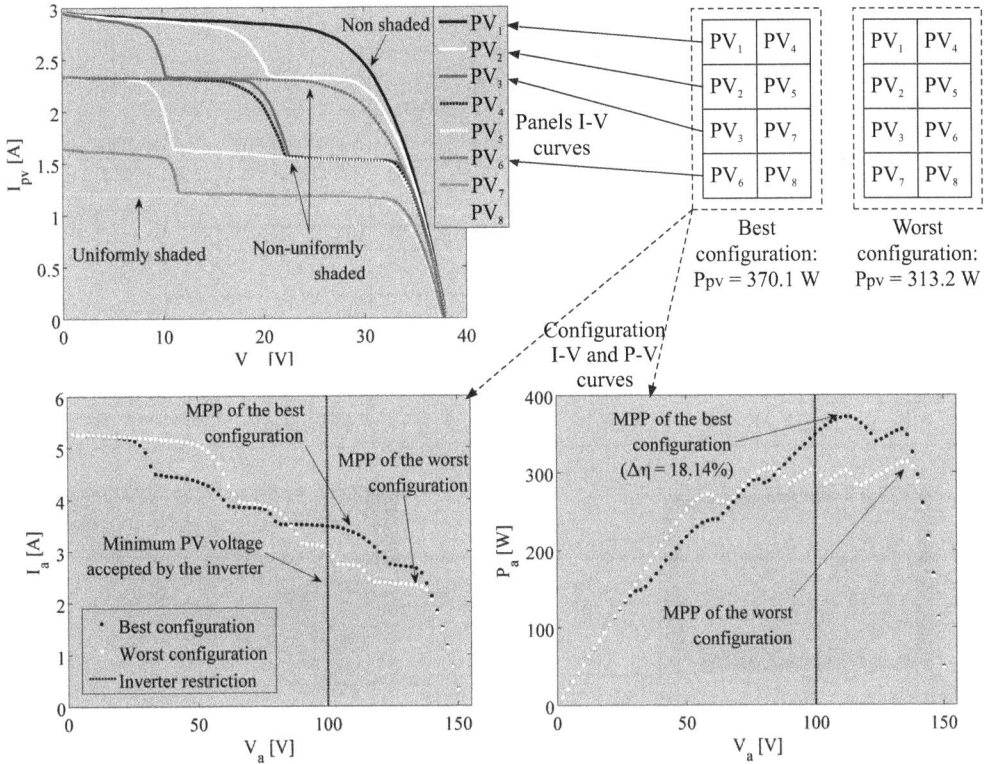

Figure 7. Best and worst configurations for the application example.

4. Simulation Platform Implementation Using Commercial Devices

With the aim of illustrating the performance of the PV array reconfiguration system, a hardware-in-the-loop (HIL) simulator was used to test the proposed solution implemented on a practical embedded device. HIL simulation was used because it ensures the same irradiance and shading conditions for multiple experiments [45], which is not possible using PV panels due to the variability of the solar irradiance. Moreover, the HIL simulator enables one to test the behavior of the proposed reconfiguration solution implemented on a real embedded device to reproduce the conditions of a real commercial deployment.

The proposed simulation platform is illustrated in Figure 8. This platform is composed of two main blocks: the HIL simulator and the embedded device processing the reconfiguration solution. The first block (HIL) emulates the PV system, and it is implemented in Simulink Desktop Real-Time (SDRT) [46]. This HIL simulator processes the models of the PV array (composed of two strings of four PV panels each, *i.e.*, 4 × 2), the switches' matrix, a DC/DC boost converter to perform the maximum power point tracking (MPPT) and the load. The second block is the embedded controller implemented on a DSP TMS320F28335 from Texas Instruments, which processes both the proposed reconfiguration algorithm and the MPPT technique.

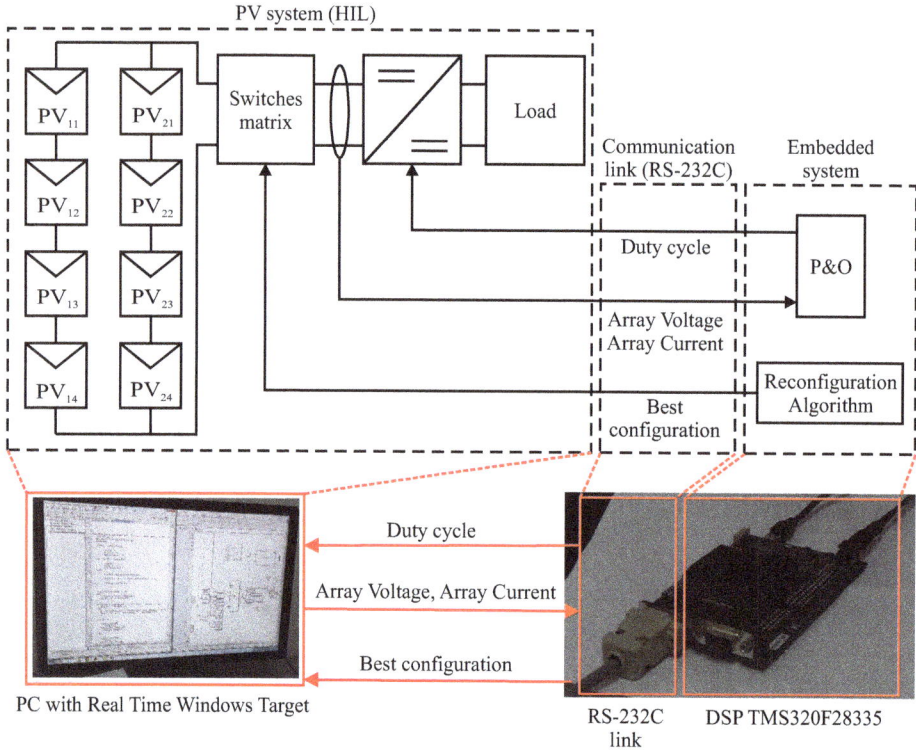

Figure 8. Block diagram and hardware implementation of the proposed simulation platform.

The PV array is implemented in the HIL using a MATLAB function based on a database of the I-V curves of the different configurations for each shading condition. The PV array function receives two parameters: the configuration defined by the switches' matrix and the operating voltage imposed by the power converter; moreover, it provides the array current according to the array I-V curve. Similarly, the switches' matrix is implemented using a MATLAB function that receives a vector with the state of all of the switches (S_{SW}) from the embedded controller, and it provides the configuration to the PV array function. The power converter is implemented using the ideal state-space average model on an S-function. Its input current is provided by the PV array function, while the duty cycle is provided by the embedded controller. Finally, the power converter imposes the voltage to the PV array, and it delivers the generated power to a resistive load.

The PV system emulator and the embedded controller exchange information using an RS-232C channel instead of discrete analog and digital signals. This structure enable one to reduce the wiring between the two devices without impacting the validity of the experiments, since the same information is transferred: the reconfiguration algorithm, on the embedded device, sends the optimal configuration of the array to the switches' matrix of the PV system; while the PV system sends the current and voltage of the PV array to the embedded system, which are used by the MPPT technique (implemented in the controller) to update the duty cycle of the Boostconverter implemented in the HIL simulator. The adopted MPPT algorithm is the perturb and observe (P&O) [47]. It is worth noting that the P&O may be trapped in a local MPP or in the global MPP depending on the starting point (duty cycle) after a new configuration is set by the switches' matrix.

The HIL system is processed in a Dell precision T7600 workstation with the following characteristics: processor Intel Xeon E5-2667 of 2.9 GHz, 32 GB of RAM memory, one COM Intel C600/X79 and a real-time operation imposed by the SDRT. The embedded device, TMS320F28335 DSP [48], used to implement both reconfiguration and MPPT algorithms has the following characteristics: CPU of 32 bits of 150 MHz, IEEE-754 single-precision floating-point unit (FPU) with 150 MIPS, 68 kB RAM memory, 512 kB Flash memory, 18 PWM channels with high resolution, 12-bit 12.5 MSPSADC and serial connectivity (three UART, one SPI, one I2C, one CANand one McBSP).

The sequence diagram in Figure 9 describes the run-time interaction between the PV system emulator and the embedded controller. The controller has four main software objects: Timer, Reconfiguration, Linear Interpolation and P&O; while the emulator has four main objects: PV array, Switches Matrix and DC/DC Converter.

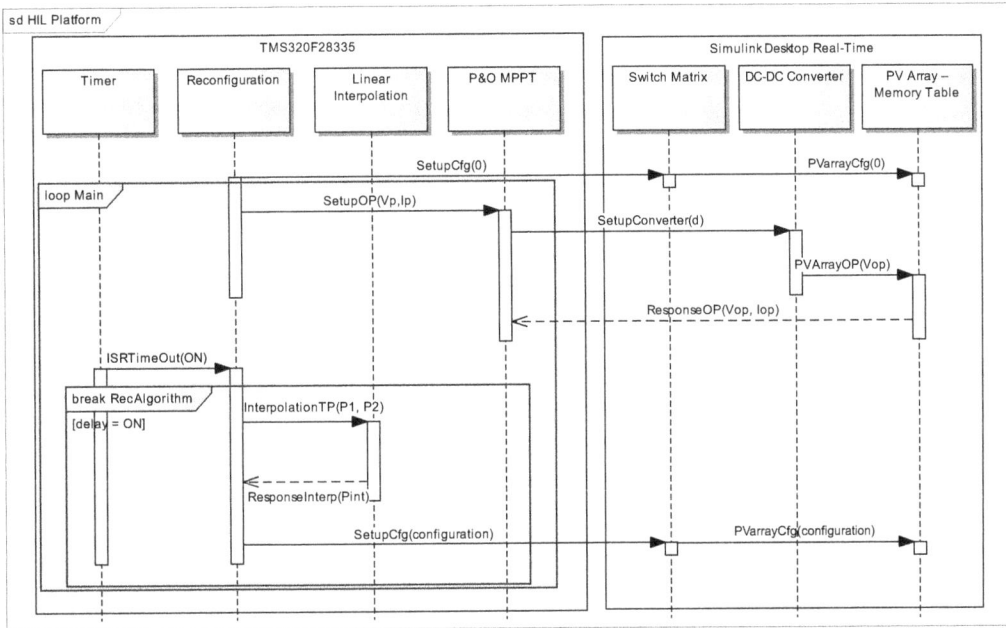

Figure 9. Sequence diagram of the simulation platform.

The simulation starts with the Reconfiguration object sending the state of all the switches to the Switches Matrix object with the message SetupCfg(0) to set the initial configuration of the PV array. From such information, the Switches Matrix determines the array configuration to be set in the PV array by the message PVarrayCfg(0). Then, an infinite loop (loop Main) starts on the controller, where Reconfiguration calculates the optimal voltage (Vp) and current (Ip) for the actual configuration and operating condition. Later, Reconfiguration sends a message (SetupOP(Vp,Ip)) to P&O to set the starting point of the P&O algorithm close to the global MPP. This is why the proposed structure enables the classical P&O algorithm to overcome the multi-maximum problem present in mismatched PV arrays. This is a major advantage over classical PV systems with no updated information about the approximated position of the global MPP.

Once Vp is received by P&O, it calculates the duty cycle of the DC/DC Converter (d), required to set the PV array voltage (Vop). Then, that duty cycle is set to DC/DC Converter using message SetupConverter(d). The DC/DC Converter sets the operating voltage (Vop) to the PV Array using PVArrayOP(Vop) message. The PV array function returns the operating voltage and current (Vop,Iop) to the P&O obtained with d using message ResponseOP(Vop,Iop). Again, the P&O uses Vop and Iop to calculate a new duty cycle in order to track the closest MPP.

The Timer is configured to set the variable *delay* to ON every 90 s to start the combined fragment break (RecAlgorithm), which executes the reconfiguration algorithm. When an interpolation is required within the reconfiguration algorithm, Reconfiguration sends the two points (*P1* and *P2*) to the Linear Interpolation using InterpolationTP(*P1,P2*), which replies with the interpolated point (*Pint*) using ResponseInterp(*Pint*). Once the optimal configuration (*configuration*) has been determined, the reconfiguration algorithm ends by sending the state of all of the switches to the Switches Matrix using the SetupCfg(*configuration*) message. Finally, the Switches Matrix sets the best configuration to the PV array (PVarrayCfg(*configuration*)). It is worth noting that while the reconfiguration algorithm is running, the P&O algorithm does not run; therefore, the converter duty cycle remains constant.

It must be highlighted that both reconfiguration and MPPT algorithms are processed by the TMS320F28335 DSP; hence, that embedded device could be used without modifications in a real application (commercial deployment). In that case, the HIL simulator will be replaced by the real PV panels, the switches' matrix and the power converter.

5. Simulation Results Based on Experimental Data

This section presents a practical example to illustrate the real-time operation of the proposed reconfiguration system running in the deployment-ready embedded device TMS320F28335 DSP.

The PV array considered for the simulations is composed of two strings of four panels each ($N = 4$ and $M = 2$), which has 35 possible configurations that should be evaluated to find the best one for a given shading condition. The time delay between two consecutive executions of the reconfiguration algorithm ($t_{reconfig}$) was set to 90 s, because one run of the reconfiguration algorithm takes around 36 s; moreover, in some cases, the PV power variability can be up to 60% in 60 s, as reported in [49]. However, it is worth noting that the PV power variability also depends on the area occupied by the PV array, the geometry of that area, the velocity of the clouds, fast objects' movement around the array (animals, bird droppings), *etc.*

5.1. Simulated Shading Conditions

Four different shading conditions were simulated to illustrate the performance of the proposed reconfiguration system. Each shading condition is generated from different I-V curves of the eight PV panels that form the array; then, the I-V curve of the PV array is generated by using the interpolation process described in Section 3.2.

The description of each shading condition is given bellow, and the P-V curves of the eight PV panels for each condition are illustrated in Figure 10.

- Condition No. 1 (SC$_1$ (SC, shading condition): PV$_1$, PV$_6$ and PV$_8$ are uniformly shaded; hence, they have a single MPP. PV$_2$, PV$_4$, PV$_5$ and PV$_7$ have one shaded module (*i.e.*, two MPPs), while PV$_3$ has two partially-shaded modules (*i.e.*, three MPPs).
- Condition No. 2 (SC$_1$): PV$_6$ and PV$_7$ have, each, two shaded modules, while the others panels have one shaded module.
- Condition No. 3 (SC$_3$): PV$_5$ and PV$_8$ are uniformly shaded; moreover, PV$_1$, PV$_2$, PV$_3$ and PV$_7$ have one shaded module, while PV$_4$ and PV$_6$ have two shaded modules.
- Condition No. 4 (SC$_4$): PV$_1$, PV$_5$ and PV$_7$ are uniformly shaded. PV$_2$, PV$_3$, PV$_4$, PV$_6$ and PV$_8$ have one shaded module.

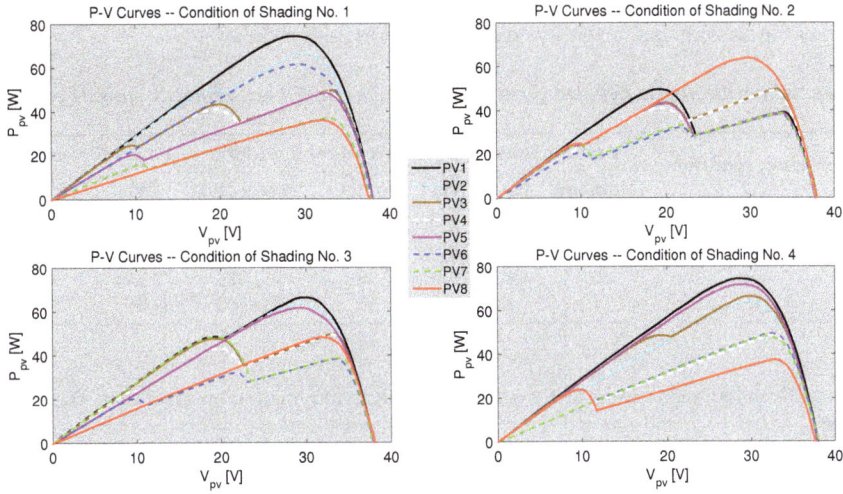

Figure 10. P-V curves of the PV panels for the shading conditions.

The array P-V curves for the best and worst configuration for each shading condition are shown in Figure 11. Note that each P-V curve has multiple local MPPs and one global MPP; however, the global MPP of the best configuration is significantly higher than the global MPP of the worst configuration.

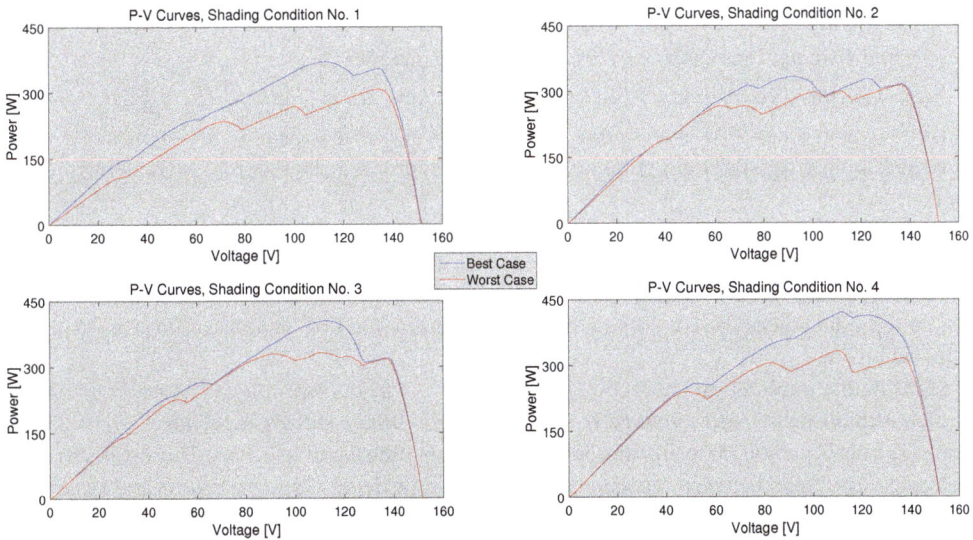

Figure 11. Array P-V curves for best and worst configurations for each shading condition.

The numerical values of the global MPPs of both the best and worst configurations for each shading condition are shown in Table 1. That table also reports the increment in the power production with respect to the worst configuration. Finally, the electrical connections between the PV panels for the best and worst configurations are shown in Table 2.

Table 2 also reports two additional configurations: "Initial with RA" (RA, reconfiguration algorithm) (CF_0) and "Without RA" (CF_{18}). *Initial with RA* corresponds to the initial configuration set in the emulation system for the dynamic simulation of the reconfiguration system presented in

Section 5.3. *Without RA* corresponds to the configuration set in the emulation system for the dynamic simulation of the classical (static) PV system presented in Section 5.5.

Table 1. Information of the best and worst configuration of the PV array for each shading condition.

Shading condition	Best configuration		Worst configuration		Power increment	
	Number	Power (W)	Number	Power (W)	(W)	(%)
1	8	371.426	13	304.982	66.444	21.8
2	14	334.356	31	313.522	20.834	6.6
3	22	405.929	28	332.975	72.953	21.9
4	4	421.665	13	333.620	88.045	26.4

Table 2. Configurations used in the simulations: RA, reconfiguration algorithm; SC, shading condition.

Configuration	Order of PV panels	Description
CF_0	$[PV_1\ PV_2\ PV_3\ PV_4; PV_5\ PV_6\ PV_7\ PV_8]$	Initial with RA
CF_4	$[PV_1\ PV_2\ PV_3\ PV_5; PV_4\ PV_6\ PV_7\ PV_8]$	Best for SC_4
CF_8	$[PV_1\ PV_2\ PV_3\ PV_6; PV_5\ PV_4\ PV_7\ PV_8]$	Best for SC_1
CF_{13}	$[PV_8\ PV_2\ PV_3\ PV_4; PV_5\ PV_6\ PV_7\ PV_1]$	Worst for SC_1 and SC_4
CF_{14}	$[PV_1\ PV_8\ PV_3\ PV_4; PV_5\ PV_6\ PV_7\ PV_2]$	Best for SC_2
CF_{18}	$[PV_1\ PV_5\ PV_6\ PV_4; PV_2\ PV_3\ PV_7\ PV_8]$	Without RA
CF_{22}	$[PV_1\ PV_2\ PV_5\ PV_7; PV_3\ PV_6\ PV_4\ PV_8]$	Best for SC_3
CF_{29}	$[PV_6\ PV_8\ PV_3\ PV_4; PV_5\ PV_1\ PV_7\ PV_2]$	Worst for SC_3
CF_{31}	$[PV_1\ PV_2\ PV_6\ PV_8; PV_5\ PV_3\ PV_7\ PV_4]$	Worst for SC_2

5.2. Sampling Time and Time Delay of the Sweep Device and Switches

The TMS320F28335 DSP has a fast 12-bit ADC capable of sampling analog signal at 12.5 MHz (80-ns conversion time). However, since a single ADC is available to acquire two signals (PV voltage and current), those signals could be sampled at a maximum frequency of 6.25 MHz, which remains very fast.

To define the sampling frequency for tracing the I-V curves, the following conditions must be taken into account: first, the time required to trace the I-V curve depends on the sweep device capacitor C_{SD}, as given in Equation (3); second, the number of samples per curve must enable an accurate reproduction of the I-V curve without overloading the DSP memory. Those conditions are addressed below.

Considering a minimum short-circuit current of 1 A for this example (as depicted in Figure 7) and a maximum open-circuit voltage equal to 38 V, the sweep device capacitor $C_{SD} = 39\ \mu F$ is calculated from Equation (3) to ensure a maximum sweep time equal to 1.5 ms. Those values provide an acceptable trade-off between capacitor size, cost and time delay. Other capacitors and sweep times can be adopted taking into account the following conditions: larger capacitors will require larger time delays, which in turn increases the reconfiguration time and capacitor cost; while smaller capacitors enable one to reduce the reconfiguration time; however, higher sampling frequencies and shorter processing times are required to acquire the I-V curve data.

The commercially available capacitor MKT1820639015 fulfills the requirements of capacitance and maximum voltage for this application (39 μF, 63 V). Another option is to implement C_{SD} using multiple capacitors in parallel to improve the system reliability. In that case, three capacitors MKT1820615015 (15 μF, 63 V) can be used in parallel to provide a much higher maximum current derivative, so that the capacitors stress is reduced. However, this last solution increases the cost of the sweep device. To discharge the capacitor C_{SD} without a significant stress, the additional resistor $R_{SD} = 8.25\ \Omega$ is calculated from Equation (4) to ensure a maximum discharge time $T_{dis} = 1.5$ ms.

Resistor R_{SD} must support a maximum power $R_{SD} \times (I_{SC.max})^2 = 3$ W, where $I_{SC.max} = 5$ A is the maximum short-circuit current possible in the adopted PV panels. The commercially-available resistor 83F8R25 fulfills those conditions.

Taking into account that a PV curve must be acquired for each PV panel and a PV curve must be constructed for each string and the array, those PV curves are defined to have 160 samples each, so that the DSP memory is not overloaded. Since the PV curves are acquired in 1.5 ms, the sampling frequency of each channel of the ADC is set to 110 kHz.

Another time delay present in the hardware of the reconfiguration system concerns the switches' operation. To provide an inexpensive solution, general purpose electro-mechanical relays, such as the G5LE-1A4-DC24, can be adopted. Those types of relays have an average operation time of 15 ms. Based on the previous time delays of the sweep device and switches, the total time required to acquire all the PV panels I-V curves is calculated from Equation (5) as $T_{sweep} = 264$ ms. Moreover, each reconfiguration cycle requires opening and closing, simultaneously, some relays in the switches' matrix to set the new configuration; therefore, an additional 15-ms delay is imposed by the switching matrix. Finally, the time required by the hardware of the reconfiguration system is 279 ms for each reconfiguration cycle.

5.3. Dynamic Simulation of the Reconfiguration System

A dynamic simulation of the proposed reconfiguration system was carried out considering the PV array operating under the four different shading conditions described in Section 5.1. The simulation results are presented in Figure 12a: the plots at the top and center correspond to the power and voltage of the PV array, respectively, while the plot at the bottom reports the shading condition (SC) and the time intervals in which the reconfiguration algorithm (RA) is running (Run) and stopped (Stop). Moreover, the black dashed lines illustrate the instants in which a run of the reconfiguration algorithm finishes, while the purple dashed lines report the instants in which the shadowing conditions change.

The simulation starts with the PV array in Configuration 0 (CF_0 in Table 2) under the shading condition No. 1 (SC_1), where the P&O algorithm tracks an MPP around 40 V obtaining a power of 150 W. At 29.3 s, the RA starts, and it takes 33.4 s (i.e., $t_{reconfig} = 33.4$ s) to find the best configuration, which, in this case, is CF_8. Such a configuration is set by the switches' matrix at 67.88 s (including the 279-ms delay of the hardware) to produce a significant increment of 147% in the PV array power (from 150 W–371 W).

It is observed that each time the RA ends (black dashed lines), the PV power is increased or remains constant, depending on the existence of a change of the shading conditions. Moreover, each time the shading condition changes, there is a reduction in the PV array power, which means that the array configuration must be optimized again.

As explained at the beginning of Section 5, the time period selected to run the RA consecutively is 90.0 s (i.e., $t_{ck\_reconfig} = 90.0$ s); therefore, at 119.3 s, the RA starts again with $t_{reconfig} = 37.2$ s. At that time, the shading conditions have not changed; hence, there is no change in the configuration of the PV array and, as consequence, no change in the PV array power and voltage.

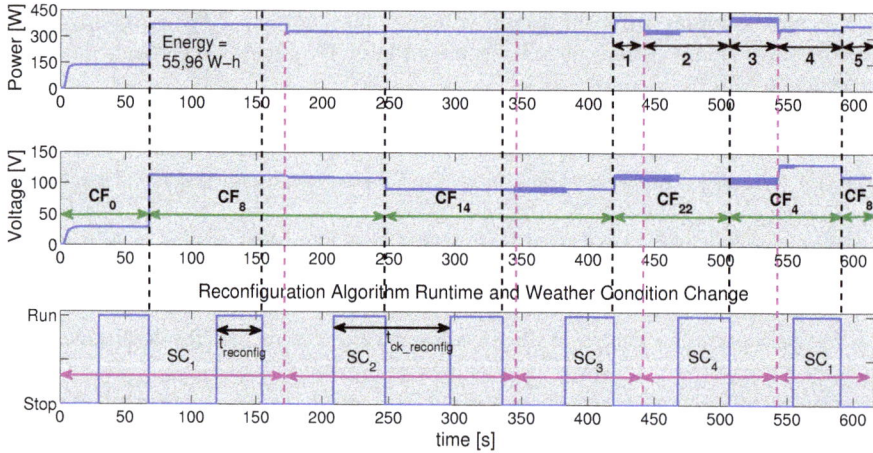

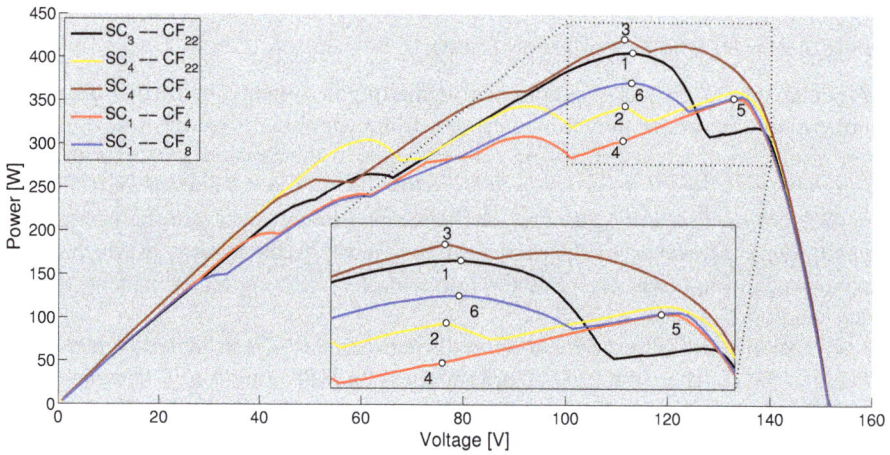

Figure 12. Behavior of the PV array with the proposed reconfiguration algorithm. (**a**) Dynamic behavior of PV array power, voltage and reconfiguration algorithm runtime; (**b**) Operating points in the P-V curves from 430 s.

The first change in the shading conditions, from SC_1 to SC_2, is produced at 172.3 s, which leads to a reduction in the PV array power and voltage. Then, the RA starts at 208.5 s, with $t_{reconfig} = 36.2$ s, obtaining that the best configuration for SC_2 is CF_{14} (see Table 2). CF_{14} is set at 247.56 s, producing a small increase in the PV power (1%) and a reduction in the PV voltage. Such a reduction shows that the voltage of the global MPP for CF_{14} under SC_2 is lower than the global MPP voltage for CF_8 under the same shading profile. The RA runs again between 296.6 s and 335.7 s ($t_{reconfig} = 38.4$ s); however, the configuration does not change since the shading condition is the same (SC_2).

The shading condition changes again at 344.3 s, this time from SC_2 to SC_3. Therefore, after the subsequent run of the RA, the new optimal configuration CF_{22} is set at 419.38 s ($t_{reconfig} = 37.6$ s + hardware delay of 279 ms). Once again, the change in the configuration produces an increment in the PV array power (21%).

At 441.3 s, there is a new change in the shading profile from SC_3 to SC_4. The RA set the new optimal configuration is CF_4 at 506.78 s ($t_{reconfig} = 36.6$ s + 279 ms). The configuration change produces a PV power increment of 22%. The last change in the shading conditions, from SC_4 to

SC_1, is produced at 541.9 s. As expected, the RA determines the new optimal configuration CF_8, and the PV power and voltage are the same obtained after the first run of the RA. This time CF_8 is set at 588.08 s ($t_{reconfig}$ = 33.6 s + 279 ms).

To illustrate the behavior of both the reconfiguration system and P&O algorithm, Figure 12b shows the trajectory of the operating point, on the array P-V curve, for the dynamic conditions between 430 s and 630 s of the previous simulation. The operating points are also illustrated in Figure 12a in the plot at the top of the Figure. The sequence starts with the PV array at Point 1 (112.9 V, 405.9 W) that correspond to global MPP of CF_{22} under SC_3 (black line). The change from SC_3 to SC_4, at 441.3 s, changes the P-V curve to the yellow trace; therefore, the P&O tracks the local MPP closest to the last global MPP tracked (112.9 V), which is Point 2 (111.4 V, 344.7 W).

After the run of the RA (506.78 s), the P-V curve changes to the brown trace, which corresponds to CF_4, and the P&O algorithm tracks the new operating Point 3 (111.3 V, 421.7 W). With the change in the shading conditions from SC_4 to SC_1, at 541.9 s, the new operating point is 4 (111.0 V, 304.9 W); then, the P&O algorithm tracks the closest MPP: Point 5 (132.7 V, 353.6 W). Finally, the RA runs and determines that CF_8 is the best configuration for SC_1, and it sets the operating point to 6 (112.7 V, 371.4 W), which is the global MPP.

5.4. Calculation Burden of the Reconfiguration Algorithm

The software objects implemented on the embedded controller (ES in Figure 9) were programmed in the C language and compiled by using Code Composer Studio [43], which is an optimized compiler for the TMS320F28335 DSP. That compilation transforms each operation written in the C language (e.g., interpolation, communication with SDRT, etc.) into multiple fixed-point and floating-point instructions, including also control flow operations, which require different numbers of clock cycles. Moreover, the number of operations performed depends on the shading conditions; therefore, the performance of the RA is analyzed by using the number of clock cycles (ck_{cycles}) and the time required to find the best configuration $t_{reconfig}$.

For example, in the results presented in Section 5.3, the first execution of RA requires 5,050,855,766 clock cycles to evaluate the 35 possible configurations and to determine the best configuration (SC_1). In the TMS320F28335, such a number of clock cycles is translated into 33.672 s, because each cycle takes 6.67 ns (i.e., clock of 150 MHz).

The calculation burden of the proposed RA is evaluated by using $t_{reconfig_i}$ and ck_{cycles_i} for each run of the RA in the simulation results presented in Figure 12a. The sub-index i represents the number of the RA executions. The values of $t_{reconfig_i}$ and ck_{cycles_i} are shown in Table 3, which also reports the deviation with respect to the average values (36.128 s and 5,416,534,590 clock cycles, respectively).

Table 3 shows that $t_{reconfig_i}$ and ck_{cycles_i} change depending on the shading conditions. However, those measurements also depend on other factors, like the communication delays, since even when the shading condition is the same, the simulation time is slightly different.

Table 3. Reconfiguration algorithm performance.

Shading condition	i	$t_{reconfig_i}$ (s)	ck_{cycles_i}	Deviation from average
SC_1	1	33.409	5,008,845,577	−7.5%
SC_1	2	37.180	5,574,212,894	2.9%
SC_2	3	36.183	5,424,737,631	0.2%
SC_2	4	38.388	5,755,322,339	6.3%
SC_3	5	37.591	5,635,832,084	4.0%
SC_4	6	36.589	5,485,607,196	1.3%
SC_1	7	33.558	5,031,184,408	−7.1%

Finally, the longest processing time $t_{reconfig,MAX}$ must be shorter than the time period selected to run the RA consecutively. In this example, $t_{reconfig,MAX}$ = 38.388 s < 90 s. It must be noted that the processing time will increase for a higher number of panels; therefore, an evaluation of the calculation burden must be performed for each case to ensure a correct selection of the embedded device and RA time period.

5.5. Tests without Accounting for Array Reconfiguration

With the aim of illustrating the advantage of the proposed reconfiguration system over the classical static PV array, the simulation performed in Section 5.3 is repeated considering a PV array with a fixed configuration (CF_{18}). CF_{18} is selected because it is not the best or the worst in terms of power production. The test conditions are the same ones used to evaluate the PV system with reconfiguration, namely a duration close to 613 s and changes on the shading conditions at the same instants of time. The simulation results for this classical PV system are shown in Figure 13a.

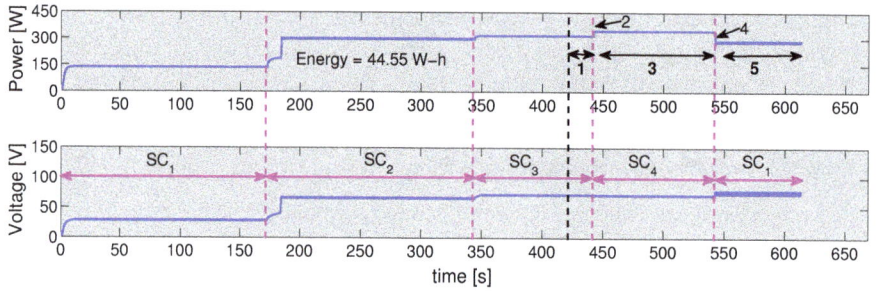

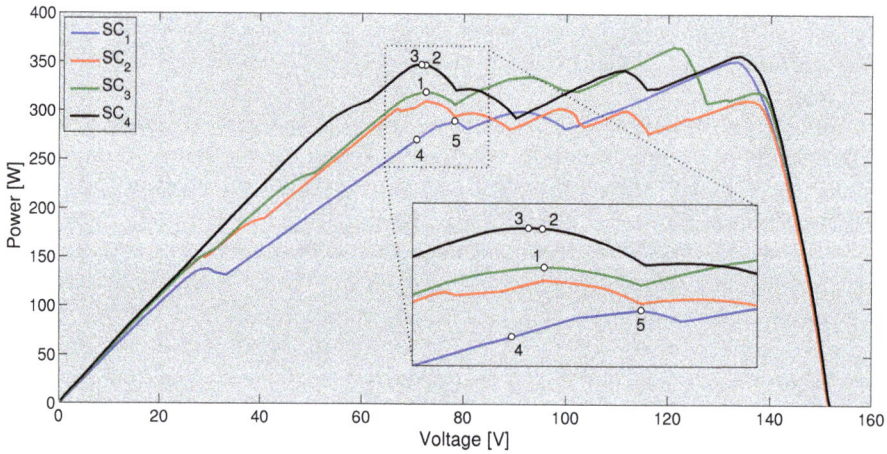

Figure 13. Behavior of the PV array with fixed configuration. (**a**) Dynamic behavior of PV array power and voltage; (**b**) Operating points in the P-V curves.

At the simulation start, the output power of the PV systems with and without reconfiguration is similar prior to the first run of the RA (29.91 s). Nonetheless, the output power of the PV system without reconfiguration is lower in comparison to the proposed solution after the first run of the RA. In the simulation, the shading condition sequence is: SC_1, SC_2, SC_3, SC_4 and SC_1. For such a sequence, the static PV system produces 230 W, 34 W, 85 W, 70 W and 80 W less power in comparison to the reconfiguration system. It is worth noting that the difference in the power wasted for SC_1 at the beginning and at the end of the simulation indicates that the MPPT tracks different local MPPs in both cases. This is due to the MPP detected by the P&O depends on the last operating point prior to the change in the shading profile as reported in Figure 13b, which shows the evolution of the operating points in the simulation.

The reduced power production is translated into a significant reduction in the energy delivered to the load. The system with reconfiguration produced 55.96 Wh during the simulation, while the system without reconfiguration only produced 44.55 Wh. Therefore, the increment in the energy production with the proposed RA is 25.61%.

The operating points of the static PV array during the simulation, starting from 430 s, have the following trajectory, as reported in Figure 13b; moreover, those operating points are also illustrated in Figure 13a. The sequence starts with the PV array at Point 1, which corresponds to a local maximum for CF_{18} under SC_3. After the change on the shading conditions, the operating point moves to 2 on the P-V curve of SC_4. Then, the MPPT moves the operating point to 3, which is the closest local MPP. For the next change of the shading profile, the operating point moves to 4 and then to 5. It should be noted that the operating point never reaches the global MPP, which explains why the power produced by this static solution is significantly lower in comparison with the PV reconfiguration system.

With the aim of illustrating the cost-benefit of the reconfiguration solution over the static PV solution, the costs of both solutions are analyzed. The cost of the PV inverter is U.S. $680, and each PV panel has a cost of U.S. $85. The TMS320F28335 DSP has a cost of U.S. $35. The sweep device requires a capacitor with a costs equal to U.S. $10, a resistor with a cost equal to U.S. $2 and a relay to discharge the capacitor. Moreover, the switches' matrix must support four panels per strings and three strings (one string is used for the sweep device), which requires 82 relays according to [15]. In addition, the DSP, switches' matrix and sweep device must be placed in a PCB with a costs near U.S. $10. Taking into account that each relay has a cost of U.S. $1.3, the reconfiguration system (DSP, switches' matrix and sweep device) has a costs close to U.S. $165. Therefore, the complete PV installation with reconfiguration capabilities (including the inverter and eight panels) has a total cost of U.S. $1525. Taking into account that, in this example, the reconfiguration solution produces 25.61% more energy, the static PV system will need, approximately, three or four additional PV panels (depending on the strings' configuration) to produce the same energy provided by the reconfigurable PV system. This means that an equivalent static PV system (with three and four additional PV panels) will costs between U.S. $1615 and U.S. $1700, which is between 6% and 11.5% more expensive. In this case, the reconfiguration solution is the option with the lower cost per kWh.

Finally, the example presented in this section is a particular case of study to demonstrate the operation and benefits of the proposed reconfiguration system. However, depending on the shading profiles present in a particular PV installation, the dynamic behavior of the reconfiguration system (Section 5.3) could produce a negligible, or even null, power increment with respect to the static PV array. Therefore, it is required to analyze the economic viability of the reconfiguration system for the particular geographic location and shading patterns exhibited in a given application. This procedure can be done using simulations of the annual energy production and costs analyses for the PV installation with and without the reconfiguration system, so that the effective cost-benefit of this solution can be evaluated.

6. Conclusions

A reconfiguration algorithm for PV arrays in a series-parallel structure, based on the I-V curves of the PV panels, was proposed. This algorithm starts measuring the I-V curves of the PV panels using a simple and inexpensive sweep device. Then, the P-V curves of all possible configurations are constructed from the I-V curves of the panels to determine the configuration that provides the highest global MPP. The proposed algorithm uses interpolation to minimize the arithmetic operations executed in the embedded system in order to reduce, as much as possible, the time required to evaluate each configuration. The reduction in the calculation time enables the evaluation of all of the possible configurations in a reasonable time using inexpensive embedded devices.

The proposed algorithm was implemented on the embedded system TMS320F28335 DSP. The PV system, *i.e.*, PV array, switches' matrix, DC/DC converter and load, was emulated using a the Simulation Desktop Real-Time toolbox from MATLAB. This approach enables one to evaluate the performance of the proposed solution in a hardware-in-the-loop environment, to guarantee the repetitiveness of the operating conditions to provide a fair comparison of different solutions.

The real-time operation and benefits of the reconfiguration solution were evaluated using a practical example based on commercial devices. Those HIL tests were performed considering a PV array formed by two strings of four panels each (4 × 2). The experiments consider four changes in the shading conditions to evaluate the reconfiguration speed, where the average time required to find the best configuration was 36.128 s plus 279 ms required to obtain the experimental I-V curves of all of the PV panels. Those results were compared with a classical static PV system with a typical MPPT technique. In this particular case, the reconfigurable PV system produces an increment of 25.61% in the energy generated. To obtain the same energy with a fixed PV system, it would be necessary to add three or four panels to the array depending on the strings' configuration, which would be more expensive than the hardware required to perform the reconfiguration. However, it is important to note that the economical viability of the proposed reconfiguration system has to be evaluated for each particular application.

It is worth noting that the reconfiguration solution can be applied to small PV arrays (e.g., residential applications) in order to obtain a reasonable number of possible configurations and switches. Moreover, this solution can be improved in the following ways: the time between two consecutive runs of the algorithm could be optimized by introducing a circuit to detect a change in the shading conditions, so that the reconfiguration only runs when it is required. Other possible improvement would be to reduce, even more, the calculation time by introducing a combinational optimization algorithm. In this case, the reconfiguration solution could be suitable for larger PV arrays.

Acknowledgments: This work was supported by the Automatic, Electronic and Computer Science research group of the Instituto Tecnológico Metropolitano (ITM), the GAUNAL research group of the Universidad Nacional de Colombia and the CEMOS research group of the Universidad Industrial de Santander under Projects P14215, P14220, UNAL-ITM-26281 and UNAL-ITM-26283.

Author Contributions: All the authors conceived the paper contributions, designed and perform the simulations and experiments. Moreover, All authors contributed to the writing of the paper.

Conflicts of Interest: The authors declare no conflict of interest.

References

1. La Manna, D.; li Vigni, V.; Riva Sanseverino, E.; di Dio, V.; Romano, P. Reconfigurable electrical interconnection strategies for photovoltaic arrays: A review. *Renew. Sustain. Energy Rev.* **2014**, *33*, 412–426.
2. Parlak, K.S. PV array reconfiguration method under partial shading conditions. *Int. J. Electr. Power Energy Syst.* **2014**, *63*, 713–721.
3. Fathabadi, H. Lambert W function-based technique for tracking the maximum power point of PV modules connected in various configurations. *Renew. Energy* **2015**, *74*, 214–226.
4. Ramaprabha, R.; Mathur, B.L. A comprehensive review and analysis of solar photovoltaic array configurations under partial shaded conditions. *Int. J. Photoenergy* **2012**, *2012*, 16, doi:10.1155/2012/120214.
5. SMA Solar Technology AG. Sunny Boy 1200/1700/2500/3000. In *Datasheet SB1200_3000-DEN110712*; SMA Solar Technology AG: Niestetal, Germany, 2004. Available online: http://www.el-tec.nl/include/nl/downloads/SB1200_3000-DEN110712W.pdf (accessed on 1 October 2015).
6. SolarEdge Technologies Inc. *SolarEdge Single Phase Inverters SE2200-SE6000*; SolarEdge Technologies Inc.: Fremont, CA, USA, 2014.
7. Balato, M.; Vitelli, M. A new control strategy for the optimization of distributed MPPT in PV applications. *Int. J. Electr. Power Energy Syst.* **2014**, *62*, 763–773.

8. Balato, M.; Vitelli, M. Optimization of distributed maximum power point tracking PV applications: The scan of the power *vs.* voltage input characteristic of the inverter. *Int. J. Electr. Power Energy Syst.* **2014**, *60*, 334–346.

9. Subudhi, B.; Pradhan, R. A comparative study on maximum power point tracking techniques for photovoltaic power systems. *IEEE Trans. Sustain. Energy* **2013**, *4*, 89–98.

10. Gules, R.; de Pellegrin Pacheco, J.; Hey, H.; Imhoff, J. A maximum power point tracking system with parallel connection for PV stand-alone applications. *IEEE Trans. Ind. Electr.* **2008**, *55*, 2674–2683.

11. Esram, T.; Chapman, P. Comparison of photovoltaic array maximum power point tracking techniques. *IEEE Trans. Energy Convers.* **2007**, *22*, 439–449.

12. Qi, J.; Zhang, Y.; Chen, Y. Modeling and maximum power point tracking (MPPT) method for PV array under partial shade conditions. *Renew. Energy* **2014**, *66*, 337–345.

13. Bastidas-Rodriguez, J.; Franco, E.; Petrone, G.; Andrés Ramos-Paja, C.; Spagnuolo, G. Maximum power point tracking architectures for photovoltaic systems in mismatching conditions: A review. *IET Power Electr.* **2014**, *7*, 1396–1413.

14. Vincenzo, M.C.D.; Infield, D. Detailed PV array model for non-uniform irradiance and its validation against experimental data. *Sol. Energy* **2013**, *97*, 314–331.

15. Bastidas-Rodriguez, J.D.; Ramos-Paja, C.A.; Saavedra-Montes, A.J. Reconfiguration analysis of photovoltaic arrays based on parameters estimation. *Simul. Model. Pract. Theory* **2013**, *35*, 50–68.

16. El-Dein, M.; Kazerani, M.; Salama, M. Optimal photovoltaic array reconfiguration to reduce partial shading losses. *IEEE Trans. Sustain. Energy* **2013**, *4*, 145–153.

17. Trina Solar. PDG5: The Dual Glass Module. In *Datasheet TSM_EN_August_2014_A(US)*; Trina Solar Limited: 100 Century Center: San Jose, CA, USA, 2014. Available online:http://www.trinasolar.com/HtmlData/downloads/us/US_Datasheet_PDG5.pdf (accessed on 30 September 2015).

18. KYOCERA Solar, Inc. KD F Series Family. In *Datasheet KD Modules 032114*; KYOCERA Solar, Inc.: Scottsdale, AZ, USA, 2013. Available online: http://www.kyocerasolar.com/assets/001/5133.pdf (accessed on 30 september 2015).

19. Ji, Y.H.; Jung, D.Y.; Kim, J.G.; Kim, J.H.; Lee, T.W.; Won, C.Y. A real maximum power point tracking method for mismatching compensation in PV array under partially shaded conditions. *IEEE Trans. Power Electr.* **2011**, *26*, 1001–1009.

20. Picault, D.; Raison, B.; Bacha, S.; de la Casa, J.; Aguilera, J. Forecasting photovoltaic array power production subject to mismatch losses. *Sol. Energy* **2010**, *84*, 1301–1309.

21. Velasco-Quesada, G.; Guinjoan-Gispert, F.; Pique-Lopez, R.; Roman-Lumbreras, M.; Conesa-Roca, A. Electrical PV array reconfiguration strategy for energy extraction improvement in grid-connected PV systems. *IEEE Trans. Ind. Electr.* **2009**, *56*, 4319–4331.

22. Karatepe, E.; Boztepe, M.; çolak, M. Development of a suitable model for characterizing photovoltaic arrays with shaded solar cells. *Sol. Energy* **2007**, *81*, 977–992.

23. Bishop, J.W. Computer simulation of the effects of electrical mismatches in photovoltaic cell interconnection circuits. *Sol. Cells* **1988**, *25*, 73–89.

24. Petrone, G.; Ramos-Paja, C. Modeling of photovoltaic fields in mismatched conditions for energy yield evaluations. *Electr. Power Syst. Res.* **2011**, *81*, 1003–1013.

25. Bastidas, J.D.; Franco, E.; Petrone, G.; Ramos-Paja, C.A.; Spagnuolo, G. A model of photovoltaic fields in mismatching conditions featuring an improved calculation speed. *Electr. Power Syst. Res.* **2013**, *96*, 81–90.

26. National Instruments. NI 2810/2811/2812/2813/2814 Specifications. In *Datasheet 375527E*; National Instruments Corporation: Austin, TX, USA, 2012.

27. Pickering Interfaces Ltd. 40-550 *Power Matrix Module*; Issue 7.1, May 2014. Available online: http://www.pickeringtest.com/content/private/datasheets/40-550D.pdf (accessed on 9 October 2015).

28. BP Solar. BP 2150S. In *Datasheet 01-3001-2A*; BP group: 501 Westlake Park Boulevard Houston: Houston TX, USA, 2002. Available online: http://www.solarcellsales.com/techinfo/docs/BP2150S-DataSheet.pdf (accessed on 19 August 2015)

29. SolarEdge Technologies Inc. *SolarEdge Power Optimizer Module Add-On P300/P350/P404/P405/P500*; 47505 Seabridge Drive, Fremont, CA, USA, 2015.

30. Ramos-Paja, C.A.; Bastidas-Rodriguez, J.D.; Saavedra-Montes, A.J. Experimental validation of a model for photovoltaic arrays in total cross-tied configuration. *DYNA* **2013**, *80*, 191–199.

31. Ramabadran, R.; Gandhi Salai, R.; Mathur, B. Effect of shading on series and parallel connected solar PV modules. *Mod. Appl. Sci.* **2009**, *3*, 32–41.

32. Aranda, E.; Galan, J.; de Cardona, M.; Marquez, J. Measuring the I-V curve of PV generators. *IEEE Ind. Electr. Mag.* **2009**, *3*, 4–14.

33. Bitron. Optimization, monitoring and safety for photovoltaics; In *Datasheet endana, Intelligence applied to photovoltaics*; Bitron SPA: Grugliasco, TO, Italy. Available online: http://www.smartblue.de/wp-content/uploads/2012/06/ENDANA-Brochure-EN.pdf (accessed on 12 October 2015).

34. Alahmad, M.; Chaaban, M.A.; Lau, S.K.; Shi, J.; Neal, J. An adaptive utility interactive photovoltaic system based on a flexible switch matrix to optimize performance in real-time. *Sol. Energy* **2012**, *86*, 951–963.

35. Nguyen, D.; Lehman, B. An adaptive solar photovoltaic array using model-based reconfiguration algorithm. *IEEE Trans. Ind. Electr.* **2008**, *55*, 2644–2654.

36. Romero-Cadaval, E.; Spagnuolo, G.; Garcia Franquelo, L.; Ramos-Paja, C.; Suntio, T.; Xiao, W. Grid-connected photovoltaic generation plants: Components and operation. *IEEE Ind. Electr. Mag.* **2013**, *7*, 6–20.

37. Lo Brano, V.; Orioli, A.; Ciulla, G. On the experimental validation of an improved five-parameter model for silicon photovoltaic modules. *Sol. Energy Mater. Sol. Cells* **2012**, *105*, 27–39.

38. Salmi, T.; Bouzguenda, M.; Gastli, A.; Masmoudi, A. MATLAB/simulink based modeling of photovoltaic cell. *Int. J. Renew. Energy Res.* **2012**, *2*, 213–218.

39. Villalva, M.G.; Gazoli, J.R.; Filho, E. Comprehensive approach to modeling and simulation of photovoltaic arrays. *IEEE Trans. Power Electr.* **2009**, *24*, 1198–1208.

40. Wang, Y.J.; Hsu, P.C. Analysis of partially shaded PV modules using piecewise linear parallel branches model. *World Acad. Sci. Eng. Technol.* **2009**, *3*, 2360–2366.

41. Obane, H.; Okajima, K.; Oozeki, T.; Ishii, T. PV system with reconnection to improve output under nonuniform illumination. *IEEE J. Photovolt.* **2012**, *2*, 341–347.

42. Villa, L.F.L. and Picault, D.; Raison, B.; Bacha, S.; Labonne, A. Maximizing the power output of partially shaded photovoltaic plants through optimization of the interconnections among its modules. *IEEE J. Photovolt.* **2012**, *2*, 154–163.

43. Texas Instruments Incorporated. Code Composer Studio (CCS) Integrated Development Environment (IDE)-CCSTUDIO-TI Tool Folder. Avaiable online: http://www.ti.com/tool/ccstudio (accessed on 31 July 2015).

44. BP Solat Pty Ltd. BP 3165. In *Datasheet 4039A-2*, Houston TX, USA, 2007. Avaiable online: https://www.solarpanelsaustralia.com.au/downloads/bpsolar_bp3165.pdf (accessed on 5 August 2015).

45. Chris Washington. *Real-Time Testing: Hardware-in-the-Loop and Beyond*; National Instruments Corporation: Austin, TX, USA, 2010.

46. MathWorks, Inc. Simulink Desktop Real-Time, 2015. Avaiable online: http://cn.mathworks.com/products/simulink-desktop-real-time/ (accessed on 19 October 2015).

47. Malla, S. Perturb and Observe (P & O) Algorithm for PV MPPT, 26 December 2012. Avaiable online: http://www.mathworks.com/matlabcentral/fileexchange/39641-perturb-and-observe–p-o–algorithm-for-pv-mppt (accessed on 24 June 2015).

48. Texas Instruments. TMS320F28335, TMS320F28334, TMS320F28332, TMS320F28235, TMS320F28234, TMS320F28232 Digital Signal Controllers (DSCs). In *Data Manual SPRS439M*; Texas Instruments Incorporated: Dallas, TX, USA, 2007.

49. Hoff, T.E.; Perez, R. Quantifying PV power output variability. *Sol. Energy* **2010**, *84*, 1782–1793.

PERMISSIONS

LIST OF CONTRIBUTORS

Jesús Muñoz-Cruzado-Alba, Javier Villegas-Núñez and José Alberto Vite-Frías
R & D Department, GPTech, Av. Camas N26, Bollullos de la Mitacion 41703, Spain

Juan Manuel Carrasco Solís
Electronics Engineering Department, Seville University, Av. de los Descubrimientos S/N, Seville 41092, Spain

Stéphane Guichard
Research Institute in Innovation and Business Sciences (IRISE) Laboratory/Superior Industrial Center Study (CESI)-Reunion Chamber of Commerce and Industry (CCIR)/ Regional Centre for the Innovation and Transfer of Technologies (CRITT), The CESI engineering school, Campus Pro—CCIR 65 rue du Père Lafosse-Boîte n4, Saint-Pierre 97410, France

Frédéric Miranville, Dimitri Bigot, Bruno Malet-Damour,
Teddy Libelle and Harry Boyer
Physics and Mathematical Engineering Laboratory for Energy, Environment and Building (PIMENT), University of Reunion, 117, rue du Général Ailleret Le Tampon 97430, France

Laurent Malys and Marjorie Musy
L'Université Nantes Angers Le Mans, ensa Nantes, UMR CNRS 1563, Centre de REcherche Nantais Architecture Urbanité, 6 quai F. Mitterrand, Nantes 44000, France
Institut de Recherche en Sciences et Techniques de la Ville, FR CNRS 2488, 1 rue de La Noé, Nantes 44000, France

Christian Inard
Institut de Recherche en Sciences et Techniques de la Ville, FR CNRS 2488, 1 rue de La Noé, Nantes 44000, France

Laboratoire des Sciences de l'Ingénieur pour l'Environnement, Université de La Rochelle, UMR CNRS 7356, Avenue M. Crépeau, 17042 La Rochelle Cedex 1, France

Hussein J. Akeiber, Seyed Ehsan Hosseini, Mazlan A. Wahid
High-Speed Reacting Flow Laboratory, Faculty of Mechanical Engineering, Universiti Teknologi Malaysia, 81310 UTM Skudai, Johor 81310, Malaysia

Hasanen M. Hussen
Machine and Mechanical Department, University of Technology, Baghdad 35023, Iraq

Abdulrahman Th. Mohammad
Baqubah Technical Institute, Middle Technical University, Baghdad 06800, Iraq

Jinsang Kim
Blue Economy Strategy Institute, Suite 911, 55 Digital-Ro 34-Gil, Seoul 08378, Korea

Yujin Nam
Department of Architectural Engineering, Pusan National University, Jangjun-2 Dong, Busan 609-735, Korea

Mitsuoki Hishida
Automotive & Industrial Systems Company, Panasonic Corporation, Kadoma, Osaka 571-8506, Japan

Takeyuki Sekimoto
Advanced Research Division, Panasonic Corporation, Seika, Kyoto 619-0237, Japan

Mitsuhiro Matsumoto and Akira Terakawa
Eco Solution Company, Panasonic Corporation, Kaizuka, Osaka 597-0094, Japan

Enrico Telaretti, Mariano Ippolito and Luigi Dusonchet
Department of Energy, Information Engineering and Mathematical Models, University of Palermo, Viale delle Scienze, 90128 Palermo, Italy

Burcin Atilgan and Adisa Azapagic
School of Chemical Engineering and Analytical Science, The University of Manchester, The Mill, Room C16, Sackville Street, Manchester M13 9PL, UK

Honglu Zhu, Xu Li and Jianxi Yao
School of Renewable Energy, North China Electric Power University, Beijing 102206, China

Qiao Sun and Ling Nie
Beijing Guodiantong Network Technology Co., Ltd., Beijing 100070, China

Gang Zhao
School of Electronic Engineering, Xidian University, Xian 710071, China

Gianfranco Chicco, Valeria Cocina, Paolo Di Leo and Filippo Spertino
Energy Department, Politecnico di Torino, corso Duca degli Abruzzi 24, Torino 10129, Italy

Alessandro Massi Pavan
Department of Engineering and Architecture, University of Trieste, Via Valerio 10, Trieste 34127, Italy

Svitlana Karamshuk, Norberto Manfredi, Matteo Salamone, Riccardo Ruffo and Alessandro Abbotto
Department of Materials Science and Milano-Bicocca Solar Energy Research Center — MIB-Solar, University of Milano-Bicocca, INSTM Unit, Via Cozzi 55, 20125 Milano, Italy

Stefano Caramori, Stefano Carli, Carlo A. Bignozzi
Department of Chemistry, University of Ferrara, Via L. Borsari 46, 44121 Ferrara, Italy

Sergio Ignacio Serna-Garcés
Departamento de Electrónica y Telecomunicaciones, Instituto Tecnológico Metropolitano, Cra. 31 No. 54-10, Medellín 050013, Colombia

Juan David Bastidas-Rodríguez
Escuela de Ingenierías Eléctrica, Electrónica y de Telecomunicaciones, Universidad Industrial de Santander, Cra. 27 Calle 9, Bucaramanga 680002, Colombia

Carlos Andrés Ramos-Paja
Departamento de Energía Eléctrica y Automática, Universidad Nacional de Colombia, Cra. 80 No. 65-223, Medellín 050041, Colombia

Index

www.ingramcontent.com/pod-product-compliance
Lightning Source LLC
Chambersburg PA
CBHW061951190326
41458CB00009B/2845